Terrapoetic Urbanism

Xi Zheng

Terrapoetic Urbanism

Cities on Regional Landscape Systems

中国建筑出版传媒有限公司
China Architecture Publishing & Media Co.,Ltd.
中国建筑工业出版社

Springer

Xi Zheng
School of Landscape Architecture
Beijing Forestry University
Beijing, China

ISBN 978-981-95-8675-2 ISBN 978-981-95-8676-9 (eBook)
https://doi.org/10.1007/978-981-95-8676-9

Jointly published with China Architecture Publishing & Media Co., Ltd.
The print edition is not for sale in China mainland. Customers from China mainland please order the print book from: China Architecture Publishing & Media Co., Ltd.

This Springer imprint is published by the registered company Springer Nature Singapore Pte Ltd.
The registered company address is: 152 Beach Road, #21-01/04 Gateway East, Singapore 189721, Singapore

Preface: Inscribing the Poetics of the City onto the Text of the Land

The city is an extremely complex composite system and represents the highest material manifestation of humanity's action upon and transformation of nature. Over the long course of civilizational evolution, Chinese cities, shaped by traditional culture, have developed distinctive characteristics. Under the influence of the philosophy of Tianren Heyi (the unity of humanity and nature) and a profound Shanshui worldview, earlier generations created numerous traditional urban forms based on integrated "mountain–water–city" construction models.

Such practices were not merely acts of landscape beautification, but rather profound engagements with geography and ecology. From a regional perspective, the city and its surrounding geographical unit constitute an inseparable whole. This unit is not a pristine or untamed natural environment, but one that has continuously evolved toward greater habitability through sustained human intervention and inhabited landscape reshaped through the interaction of artificial agency, cultural inscription, and natural systems.

The geographer Ian D. Whyte insightfully observed that "landscapes matter because they are among the most enduring products of the relationships between the physical environment and human society. Landscapes are created through human interaction with the surrounding world… To understand landscapes properly, they must be viewed within their specific natural and cultural histories."

The theory of Terrapoetic Urbanism proposed in this book is grounded in a re-recognition of this enduring relationship. Cities are situated upon regional landscape systems, whose evolution profoundly shapes urban form, structure, and development. Here, terrapoetic refers to a bidirectional process of formation: on the one hand, the land (terra) constrains and nourishes the city as its physical foundation; on the other hand, urban construction and cultural inscription understood as a form of poetics continuously refine and enhance the regional landscape system, enabling a dynamic equilibrium.

Through this reciprocal interaction and mutual adaptation, a human settlement community emerges in which nature and culture are intertwined, and landscape and meaning coexist—namely, the Shanshui city. This book defines the urbanization process that shapes land and enables poetic dwelling as Shanshui urbanization, or

Terrapoetic Urbanism. It emphasizes the regional landscape system as both the foundation and the medium of human settlement formation, describing how moderate and sustained artificial intervention transforms natural environments into regional landscape systems that support urban development.

The temporal and spatial trajectories of human settlement construction in China are extraordinarily rich and complex, giving rise to a remarkable diversity of cities and landscapes. In the contemporary context, by synthesizing this traditional wisdom, the book seeks to inspire planners and designers to rediscover ancient, land-responsive strategies, offering foundational insights for addressing today's global challenges of ecological crisis and cultural discontinuity. How to inherit and sustain human settlement practices grounded in site-specific adaptation and learning from nature, how to safeguard regional landscape systems, and how to foster cities with stronger identity and resilience remain questions requiring continued and in-depth inquiry.

Focusing on the evolution of regional landscape systems and the urban development they support, this book synthesizes Shanshui thought and analyzes the processes and pathways through which humans intervene in nature—including the construction of hydraulic infrastructure, urban site selection and spatial organization, and the making of landscapes and gardens. Through this approach, it establishes a research framework for Terrapoetic Urbanism, one that examines cities through the lens of regional landscape systems and reads the development of human settlements as an evolving process. The aim is to encourage contemporary scholars and practitioners to rediscover traditional wisdom and cultural values and to apply them more effectively to present and future human settlement development.

Contemporary urban development increasingly emphasizes harmonious coexistence between humanity and nature, placing the coupling of ecological and social systems at the center of human settlement planning. What is required today is to draw upon the strategies embedded in traditional urban construction forms of wisdom that align human activity with natural rhythms and to ensure their continuation and transformation. At the same time, based on careful historical understanding, the question of how to pursue adaptive inheritance in response to global and regional challenges and opportunities facing cities and their surrounding landscapes remains a critical subject for further exploration.

Beijing, China Xi Zheng

Introduction: Conceptual Reconstruction and Theoretical Framework

1. Theoretical Origins: The Dialectic of Civilization and Dwelling

From the very beginning of human existence, people have engaged in the selection of settlement sites. This process involved the use of tools, moderate intervention in nature, and the perception, transformation, and utilization of surrounding environments marking the emergence of civilization itself. Throughout long historical trajectories, humanity has continually explored and expanded suitable living environments, gradually forming enduring modes of habitation and urban construction traditions that have been transmitted for millennia.

China is a predominantly mountainous country, with approximately 70% of its territory consisting of mountains and hills. Where there are mountains, there is water. This distinctive geographical condition fostered an early and sustained awareness of Shanshui (mountain–water) among Chinese ancestors. Over time, this awareness became deeply embedded in cultural consciousness, forming an aesthetic and ethical tradition expressed in the belief that "mountains and waters possess great virtue without speech." From the appreciation of national territory shaped by dramatic landforms, to the selection of city sites, the construction of scenic places, and the making of gardens, Chinese spatial practices have been inseparable from the Shanshui environment, serving as a concrete manifestation of the philosophy of Tianren Heyi.

2. Core Definition: Terrapoetic Urbanism

The "mountain–water–city" spatial configuration represents a distinctive Chinese model of human settlement construction. In order to integrate this Eastern wisdom into contemporary urban theory, this book examines cities and their geographical settings from a regional perspective, interpreting the formation of human settlement environments as a dynamic ecological process. This process is defined here as Terrapoetic Urbanism, or Shanshui urbanization.

This theoretical framework is structured around three interrelated dimensions that form a coherent logical system:

Substrate: The regional landscape system. The regional landscape system referred to here as Shanshui in the terrapoetic sense represents the combined outcome of natural environments and sustained human intervention, particularly within low-mountain and plain water network regions. It is a composite concept that encompasses both physical morphology and cultural meaning. In this book, it is defined as the regional landscape system, which serves not merely as the background of urban development but as the material foundation upon which cities depend. Understanding and analyzing the regional landscape system constitutes the fundamental basis of this research. It refers to the integrated assemblage of natural Shanshui environments and artificial water networks within a defined settlement region, composed of mountain ranges (primarily low hills and gentle mountains), river systems, vegetation, and a series of hydraulic infrastructures including ponds, reservoirs, canals, lakes, and navigational waterways forming a comprehensive Shanshui resource system.

Process: Urbanization as constructive practice. Urbanization is understood here as a constructive process aimed at realizing the ideal of poetic dwelling. The term "urbanism" refers broadly to human settlement forms, encompassing both cities and their surrounding environments. In this book, "the urban" is distinguished from "the city," transcending the defensive boundary of city walls and instead conceptualizing the city and its hinterland as a unified geographical unit. Urbanization describes the mechanisms through which regional landscape systems shape the evolution of cities. When Shanshui and urbanization are considered together, they articulate a dynamic process in which cities grow upon regional landscape systems, jointly forming human settlement ecologies. This process constitutes a resilient socio-ecological system, characterized by continuous negotiation and mutual adaptation between artificial intervention and natural forces in response to regional development challenges.

Goal: The Shanshui city (symbiosis). A Shanshui city refers to a human settlement environment informed by traditional Chinese ecological thought. It represents the ultimate outcome of the terrapoetic urbanization process and the normative goal of urban construction. In essence, the Shanshui city embodies a symbiotic relationship between nature and culture.

In summary, the Shanshui city is the objective, terrapoetic urbanization is the process, and the regional landscape system is the substrate. Together, these three dimensions form the theoretical closed loop of Terrapoetic Urbanism (Fig. 1).

3. Research Structure of the Book

The main body of this book is organized into four parts: an analytical overview of terrapoetic urbanization, its historical traditions, its constituent elements and system principles, and case-based empirical analyses.

Part I Theoretical Overview: This section explicates the core concepts of terrapoetic urbanization, including Shanshui city, terrapoetic urbanization, and city development on regional landscape systems. It also introduces the research methodologies employed in the book through five thematic perspectives.

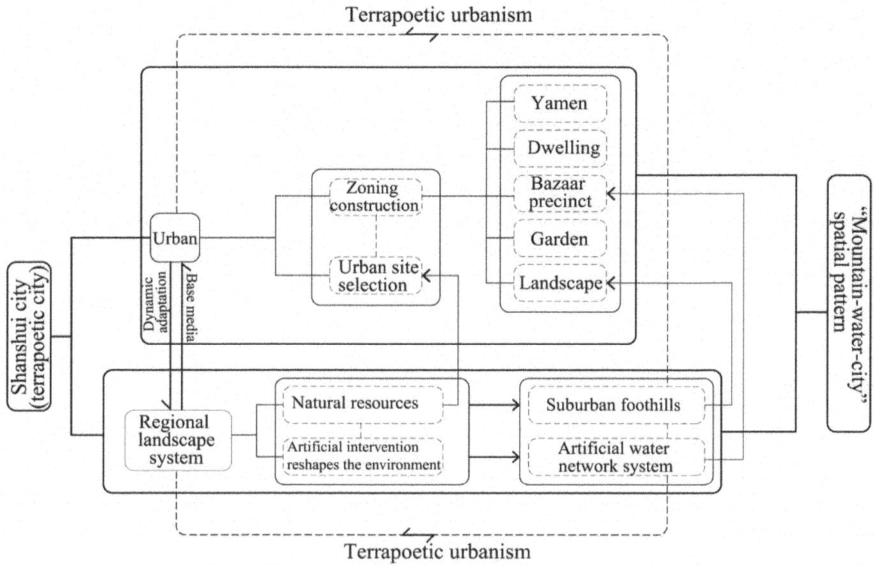

Fig. 1 Conceptual diagram of the connotation and process of Terrapoetic Urbanism

Part II Rediscovering Tradition: This section traces the historical tradition of terrapoetic urbanization. Many ancient cities were planned with landscape as a guiding principle; consequently, the history of urban construction is simultaneously a history of Shanshui culture that is, the development of cities and settlements grounded in landscape construction.

By examining cities and their surrounding Shanshui environments as integrated wholes, this book analyzes the urbanization process across historical periods. Early cartographic practices already demonstrate the high value placed on landscape in city construction. The legendary flood control efforts of Yu the Great illustrate settlement formation following the transformation and utilization of nature. Pre-Qin texts such as the Guanzi emphasize the integration of urban construction with Shanshui environments. During the Northern and Southern Song dynasties, the dissolution of the ward-market system allowed closer integration between urban functional zones and scenic systems, leading to increased city–suburb continuity.

Historical development reveals a trajectory in which human understanding of settlement environments evolved from reverence for nature to moderate intervention and harmonious coexistence. Under the combined influence of spiritual and aesthetic consciousness, Shanshui gradually became an ecological intuition embedded in traditional Chinese culture, forming an ethical foundation for human–nature relations.

Part III Constituent Elements and System Principles: This part analyzes the constituent elements of terrapoetic urbanization. Both the regional landscape system and the cities developed upon it serve as substrates of the urbanization process, dynamically influencing and adapting to one another. Unlike conventional garden-based green-space analysis, this approach examines cities from the perspective of their hinterland the geographical unit in which they are situated. Cities are constructed in accordance with natural topography and hydrological conditions, guiding water flows and shaping habitable Shanshui configurations before building urban forms upon them. Urban functional zones such as administrative precincts, markets, wharfs, residences, public squares, temples, and pagodas were closely integrated with shallow mountain–water networks. Urban construction also stimulated the development of inner city gardens and outer scenic landscapes, forming interconnected systems of water networks, gardens, scenic hills, and landmark pagodas that spanned city interiors and exteriors. Through long-term cultivation and management, these systems enriched the cultural meanings of Shanshui configurations, ultimately giving rise to the Shanshui city.

Part IV Case Analysis and Empirical Evidence: This part consists of the analysis of specific cases. On the basis of the preceding three parts, it explains the role of Terrapoetic Urbanism in the process of ancient Chinese city construction by presenting cases of different scales and types, including the regional scale, urban scale, district scale, and watershed scale. The formation of a city is neither achieved overnight nor driven by a single factor; rather, it represents a comprehensive solution to complex issues arising from economic, social, and ecological changes across different historical periods. This process includes site selection based on the configuration of mountains and waters, as well as spatial arrangements derived from ritual systems, such as the principle of placing the ancestral temple on the left and the altar of soil and grain on the right. As the underlying mountain–water environment that supports urban development, the landscape exerts a deeper and more lasting influence on urban form. Through these cases, it is illustrated that, in the course of history, regional landscape systems formed through sustained human intervention constitute the cultural foundation and ecological basis for urban development. The regional landscape system in which a city is embedded and the process of urbanization maintain a relationship of dynamic adaptation (Fig. 2).

The left panel illustrates a city in which urban functional zoning (including garden and scenic systems) develops upon the regional landscape system as the foundational substrate. The right panel shows a city shaped by modernist planning, where residential, transportation, recreation, and work functions are strictly segregated, resulting in spatial and functional separation.

Suzhou Ancient City is characterized by a nearly square city wall whose four corners are designed with curved forms. One of the main reasons for this design was consideration of the waterways surrounding the city, which helped prevent fast-flowing water from eroding the city walls while also facilitating smooth water circulation and navigability. This was not merely a matter of formal aesthetics, but a precise response to hydrological and geological conditions.

Shaoxing Ancient City exhibits similar characteristics. The irregular boundary of the entire ancient city was largely constrained by the water network environment of the Shanshui Plain in which the city is located.

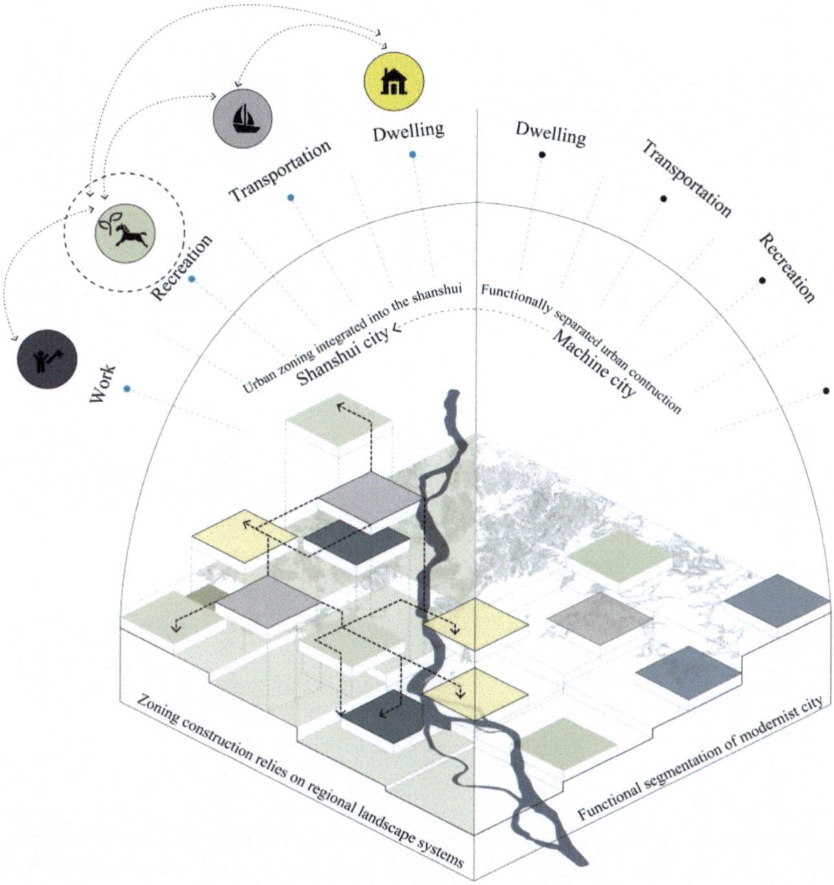

Fig. 2 Comparative diagram of the Shanshui city mediated by the regional landscape system and the functional city under modernist influence

The Six Wells of Li Mi in Hangzhou City, more than a thousand years ago, groundwater in Hangzhou was saline and unsuitable for drinking. Li Mi, then the prefectural governor, excavated six wells within the city, channeling freshwater stored in West Lake into the urban area as a source of drinking water for residents. West Lake thus became the infrastructure that genuinely supported everyday urban life and, through sustained landscape construction, evolved into a renowned scenic and cultural site.

These cases demonstrate that planners two thousand years ago already embodied a high level of ecological awareness, holistic thinking, and context-sensitive, dynamically adaptive approaches, employing Shanshui thought and the ecological wisdom of Tianren Heyi (the unity of humanity and nature) to address urban issues in different periods. Li Mi, when Hangzhou faced developmental challenges, creatively proposed solutions that enabled West Lake to function as an environmental support system in the city's development process.

As historical conditions evolved, modes of urban development also changed and adjusted accordingly, with each period dynamically identifying strategies suited to its specific context. Studying history enables the exploration of underlying patterns, the wisdom of earlier generations, and the measures they adopted aligned with natural principles to address the development challenges of their own times.

Contents

Part I
A Summary Analysis of "Terrapoetic Urbanism"

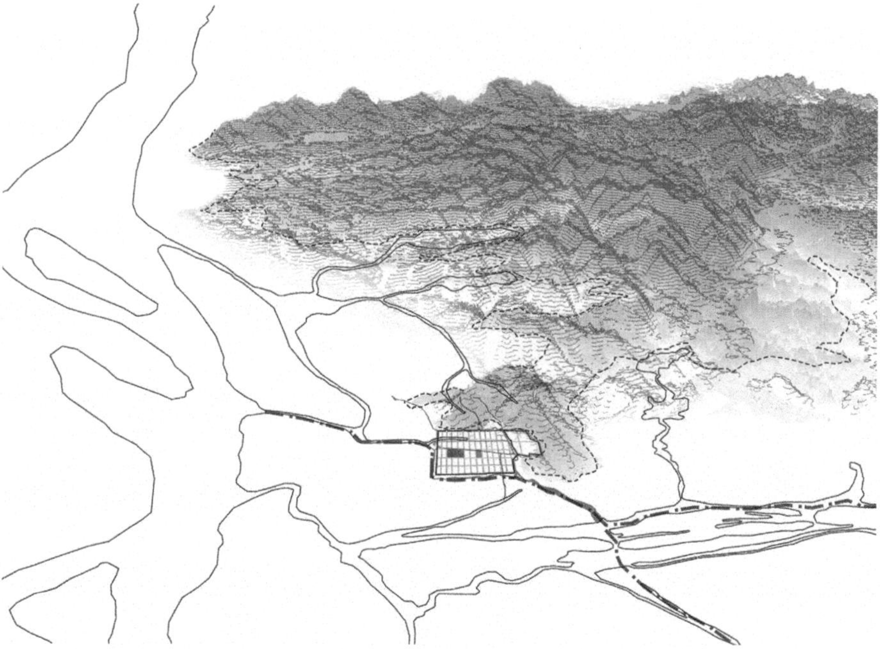

Chapter 1
Explanation

Abstract This chapter introduces a theoretical framework for understanding Chinese urban development through the concepts of the Shanshui city and Terrapoetic Urbanism. Drawing on traditional Chinese views of nature and the philosophy of the unity of humans and nature (Tianren Heyi), it conceptualizes cities as integrated systems of mountains, waters, and human settlements rather than isolated built forms. The chapter defines the regional landscape system as a composite of natural landforms, water networks, vegetation, and human-made hydraulic infrastructures that extends beyond the urban core. It argues that Chinese cities historically emerged and evolved through long-term human intervention and adaptive interaction with these regional systems. By emphasizing regionality, process, and landscape mediation, the chapter provides a lens for interpreting the spatial evolution of Chinese cities and offers theoretical insights relevant to contemporary landscape-based and ecological urbanism.

1 Shanshui City

The concept of the shanshui city was first proposed by Qian Xuesen[1] in a 1990 letter to Wu Liangyong.[2] Drawing on China's traditional Shanshui thought and the philosophical idea of tianren heyi, Qian envisioned a future city that integrates urban space, gardens, and culture to create a Shanshui-style settlement space. Wu Liangyong later stated that the ultimate aim of the Shanshui city is to establish a human settlement environment in which the artificial environment is harmoniously integrated with the natural environment represented by Shanshui [1]. The Shanshui city features the integration of natural ecological environment and cultural landscape, with a planning intention that respects nature and pursues a mountain-encircled and water-surrounded spatial ideal, inheriting the traditions of Chinese urban development. Mountain and water possess great virtue without words. In the Chinese context, Shanshui embodies the intertwining of nature and culture. From urban site selection to construction and transformation of the surrounding natural environment, the city has experienced

X. Zheng, *Terrapoetic Urbanism*, https://doi.org/10.1007/978-981-95-8676-9_1

a long process of integrated development, undergoing both Shanshui-ization and urbanism.

In recent years, with the continuous deepening and refinement of research on the theory of Shanshui city, the contemporary relevance of the Shanshui city has gone beyond mere conception, attracting increasing attention from scholars both in China and abroad. Exploring and summarizing the formation and development of the ancient prototypes of the shanshui city from a historical perspective provides important guidance for the future construction of the Shanshui city.

2 Terrapoetic Urbanism

The core of Terrapoetic Urbanism is the construction of the Shanshui city. The Shanshui city explicitly highlights the urban development objectives and outcomes based on the regional landscape system. This book adopts the term Terrapoetic Urbanism because urbanism embodies the process-oriented connotation of urban development and construction, emphasizing the laws of development, mechanisms, and the evolution of elements.

The process of urban development extends beyond the scope of the built-up area and belongs to the regional category, capable of driving the development of the entire geographical unit of the city site. This is the concept of urbanism. For example, in history, the development of Beijing was closely related to the western suburban low mountains, the water system, and the Grand Canal in the eastern suburbs (Fig. 1).

The construction site of a city is not a completely natural environment; it necessarily exists within a system that develops in coordination with the surrounding geographical environment, and the city represents the concentrated manifestation of highly developed civilization in this region.

China is a country with abundant mountains and waters, and the Shanshui structure represents the typical features of the national landscape. This book adopts the term Shanshui instead of natural system or natural environment, because Shanshui carries the connotation of culture and of artificial intervention in natural processes; picturesque mountain-and-water landscape is also a characteristic of China's traditional geography and cultural heritage. The choice of the term urbanism reflects the fact that in the historical context of China, the meaning of city was quite clear, referring to an area within city walls with a distinct defensive character. The existence and development of the city depended on the growth of an integrated urban–rural integration regional landscape system, that is, the scope of the urban. In other words, the development of the entire Shanshui environment propelled and supported the development and continuity of the city located within it. Shanshui refers to large-scale landforms where, on the basis of organization and intervention, settlements are built and urban develop, thereby promoting regional development. Urbanism represents the regional development process of urban–rural integration within the framework of the regional landscape system.

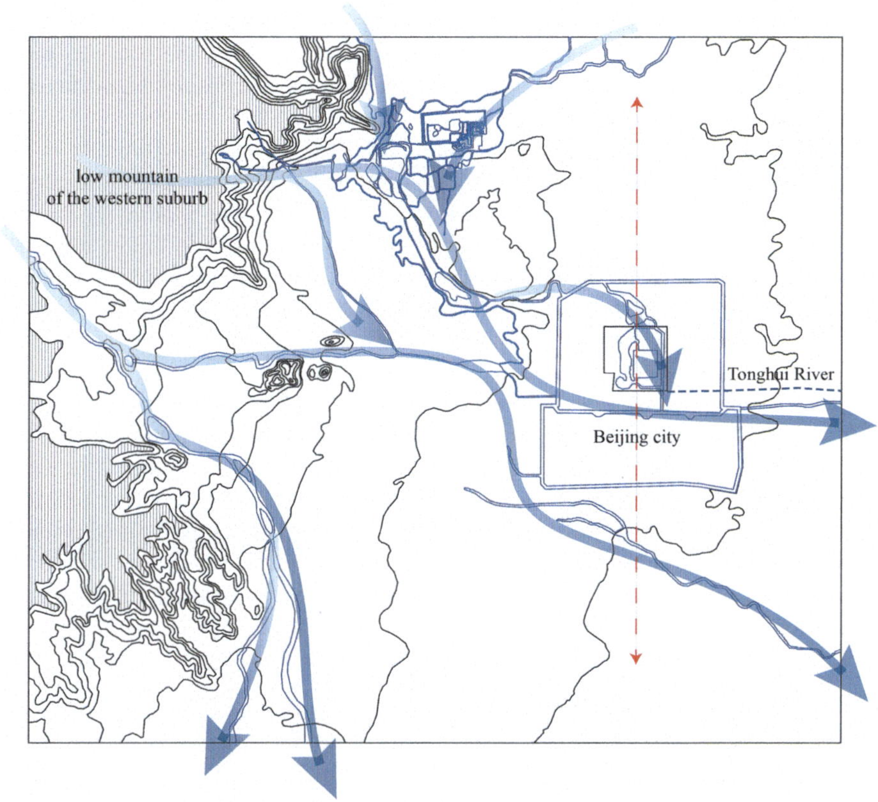

Fig. 1 Schematic diagram of the concept and process of Terrapoetic Urbanism

3 Terrapoetic Urbanism—Cities Based on the Regional Landscape System

3.1 Relationship Between the Regional Landscape System and Urbanism

In the Shanshui city, the term Shanshui refers to a comprehensive concept of mountain-and-water landscape elements that extend beyond the built-up urban area. This book defines it as the regional landscape system, which denotes the totality of the natural Shanshui environment and the artificial water network environment within a given settlement area. It is an integrated complex composed of mountain ranges, river systems, vegetation, and a series of hydraulic facilities, including ponds, reservoirs, rivers, and canals.

Terrapoetic Urbanism is the expression of the interactive process between the city and the regional landscape system. In this process, the regional landscape system serves as the medium for urban development, bearing urban functions and guiding urban growth. That is, Terrapoetic Urbanism describes the process, while the regional landscape system provides the foundation and carries for the occurrence and development of this process.

The city relies on its natural environment, yet the natural environment is not entirely suited to human habitation. Through sustained artificial intervention, people have created an ever-evolving and adaptive regional landscape character, thereby constructing the regional landscape system. The city exists within this system and continually adjusts and grows upon it. For example, the regional water system has been organized with primary functions such as flood discharge and diversion, rainwater storage, irrigation agriculture, and Grand Canal transport system, reshaping the regional natural environment, expressing humanistic beauty, and forming the regional landscape system that provides the foundation for site selection and urban construction. The city is thus constructed within a Shanshui environment that has been modified by human agency and continues to grow and develop. The urban zoning construction, such as government office areas, commercial bazaar precinct, wharf districts, and neighborhood and district system residential wards, extends across the urban–rural boundary. At the same time, gardens are created within the city and scenic landscapes are developed outside the city, forming an integrated mountain–water–city human settlement environment.

3.2 Regional and Dynamic Perspectives of Terrapoetic Urbanism

Terrapoetic Urbanism emphasizes both a regional perspective of development and a dynamic perspective of adaptation. On the one hand, the city exists by relying on nature, which provides fertile ground for human survival. Here, nature includes the regional overall environment formed by grand mountains and waters as well as small-scale mountains and waters inside and outside the city, and regionality is an inherent attribute of the city. On the other hand, the relationship between the regional landscape system and the city is not static but is in a process of dynamic adaptation. Through sustained artificial intervention and limited engineering measures, humans intervene in and reshape nature, constructing an adaptive regional landscape system on which the city continually adjusts and grows.

(1) Examining urban development from a regional perspective

From a broad geographical scope, every city exists within nature, and the influence of the external environment penetrates comprehensively and profoundly into political, economic, military, and social spheres. The city and its surrounding environment are in fact combined as an integrated urban–rural integration regional space for joint development [2].

China is a country with an extremely rich geographical environment. Across its vast territory, different Shanshui environments have created different urban forms and urban cultures, such as the mountain city of Chongqing and the water city of Suzhou. Considering urban development together with its natural environment from a regional perspective, each region displays a unique Shanshui system, constituent elements, and its own laws of evolution. Urban construction must therefore comprehensively address the influences and interactions among these diverse elements. From urban site selection to zoned construction, from garden creation within the city to scenic construction outside the city, the Shanshui environment, namely the regional landscape system, has always served as a medium. This demonstrates that the city is a comprehensive entity that develops and evolves within a specific regional environment. Regionality is therefore an objective attribute that cannot be ignored. Examining urban development from a regional perspective allows tracing the internal laws governing the interaction between the regional environment and urban development.

(2) Exploring the interactive relationship between the city and nature in the process of adaptation

The natural environment is where human beings live and thrive, providing abundant benefits from mountains, marshes, and aquatic resources as well as the convenience of navigation and irrigation. Historically, the earliest human settlement in various regions were often established by making use of low mountain terrain and selecting sites near water, south-facing, and wind shelter, consistently demonstrating humanity's dependence on and closeness to Shanshui [3]. Human selection and utilization of nature have in fact been a continuous process of transforming nature and adapting to nature. The city, as the highest material embodiment of human transformation of nature, has never been in a static state. In the course of utilizing and transforming nature, human needs, wisdom, and capabilities have been integrated into Shanshui, thereby constructing a suitable human settlement environment. Meanwhile, through ongoing interaction with nature, human spiritual life has been continuously enriched, giving rise to distinctive regional cultures and deepening the connotations of the Shanshui environment. The two have always maintained a state of relative stability and dynamic adaptation. Understanding the interactive relationship between the city and nature enables a clearer grasp of how regional culture is continually deposited, accumulated, and transmitted in this process. One of the intrinsic driving forces that allows a city to remain prosperous for millennia is the harmonious mutual dependence between humans and nature, a process that requires constant exploration and interpretation.

(3) The regional landscape system as a medium for city development

This book analyzes the process of mutual intervention and interaction between the evolution of urban spatial form and the regional landscape system in which it is located, explaining that Terrapoetic Urbanism in which an artificially intervened regional landscape system serves as the medium to promote and define the process of urban spatial evolution.

The city relies on the regional landscape system for its existence. Human activities and interventions influence the succession of the natural environment, guiding it as much as possible toward a state suitable for urbanization. To meet the requirements

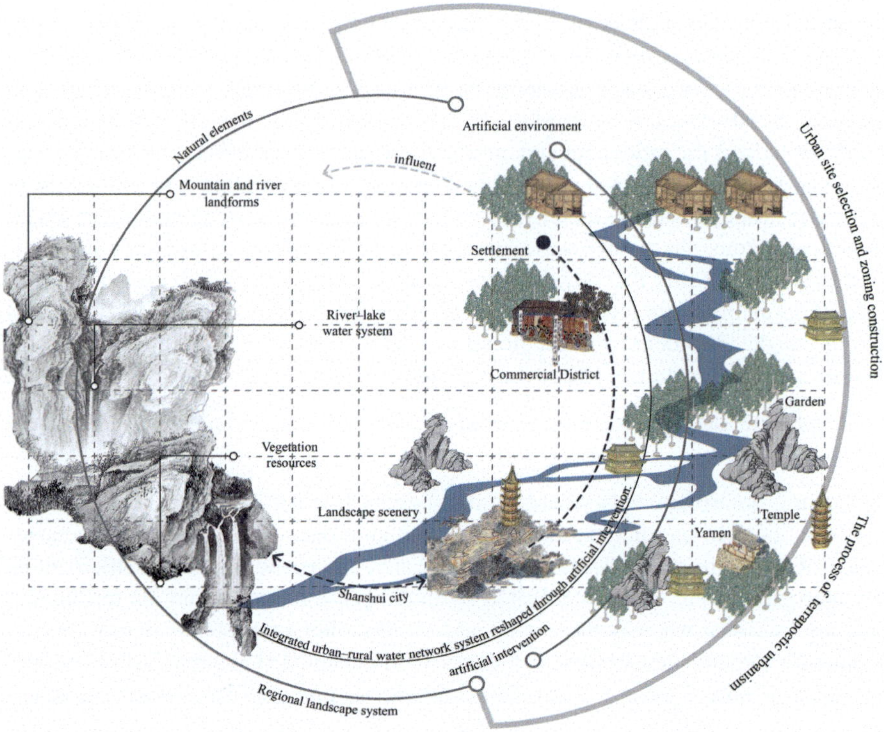

Fig. 2 Diagram illustrating the interrelationship among Terrapoetic Urbanism, regional landscape systems, and Shanshui cities. Image source: Qiu Ying[3] ((Ming Dynasty, 1368–1644)), *Prosperous Scenes of the Southern Capital*

of the urban human settlement environment, people employ a series of landscape intervention measures to effectively alleviate urban environmental and development problems, while the urban pattern dependent on the regional landscape system continuously evolves in this process. Through continuous human intervention in the natural environment, new landscape patterns and features are reconstructed, forming a regional landscape system that serves as the medium for urban growth and helps shape the optimal urban form of mutual adaptation between humans and nature. Thus, using landscape intervention to reconcile the artificial and the natural is a process of engineering-based environmental reshaping and re-adaptation of nature. The construction of urban form should transcend the boundaries of the built-up area and, from a regional perspective, guide the continuous development of cities growing within natural regions through the medium of landscape (Fig. 2).

Notes

1. Qian Xuesen (1911–2009) was a Chinese scientist and systems thinker best known as the founder of China's aerospace and missile programs. Trained at MIT and Caltech in the United States, he made major contributions to aerodynamics, systems engineering, and cybernetics before returning to China in the 1950s. Beyond engineering, Qian played a pivotal role in introducing systems science and holistic thinking into Chinese academic and planning discourse.
2. Wu Liangyong (1922–2021) was a leading Chinese architect, urban planner, and educator, recognized as one of the pioneers of modern urban planning and architectural education in China. Wu contributed extensively to urban planning theory, residential community design, and the integration of traditional Chinese spatial concepts with modern planning methodologies. His work emphasized harmony between natural landscapes, urban form, and human activity, advocating regional and ecological perspectives in planning.
3. Qiu Ying (Ming Dynasty, 1368–1644): Painter whose detailed cityscapes, like Prosperous Scenes of the Southern Capital, visually document Ming urban layout, waterways, and gardens.

References

1. Bao, S., & Gu, M. (1996). *Urban studies and landscape cities: Qian Xuesen's theory*. China Architecture & Building Press.
2. Wang, D. (2002). *Chinese landscape culture and urban planning*. Southeast University Press.
3. Fu, L. (2004). *Studies on landscape cities*. Hubei Science and Technology Press.

Chapter 2
Methodology

Abstract Terrapoetic Urbanism offers a regional and process-oriented framework for understanding urban development through the interaction between cities and regional landscape systems. Rather than treating the city as an isolated entity, this book conceptualizes urbanization as an evolving relationship embedded within Shanshui environments shaped jointly by natural conditions and long-term human intervention. By integrating regional and dynamic analyses with theoretical inquiry and empirical case studies, the book establishes a systematic approach to interpreting urban form, spatial organization, and landscape transformation. It examines how mountains, water systems, and human activities interact across scales to guide urban site selection, zoning, and environmental adaptation. Through representative cases spanning different historical periods and geographic contexts, Terrapoetic Urbanism reveals the "mountain–water–city" model as a key source of urban resilience and cultural continuity, offering critical insights for contemporary ecological civilization and resilient human settlement strategies.

Terrapoetic Urbanism constitutes a cognitive and expressive framework grounded in the interplay between regional landscape systems and processes of urbanization. Adopting an approach that integrates regional and dynamic analyses and combines theoretical inquiry with exemplary empirical studies, this book advances a framework for interpreting urban development processes through regional landscape system analysis.

First, the Shanshui geographical and environmental conditions of the proposed urban site are systematically reviewed, thereby establishing—at the regional scale—the foundational basis for constructing the human settlement environment. Second, correlations are established between the city and the elements of the regional landscape system, including the influence of landscape elements on site selection and spatial subdivision for urban construction, as well as the ways in which human activities drive the evolution of the regional landscape system. This constitutes an analytical elaboration of the processes guided by the concept of Terrapoetic Urbanism.

X. Zheng, *Terrapoetic Urbanism*, https://doi.org/10.1007/978-981-95-8676-9_2

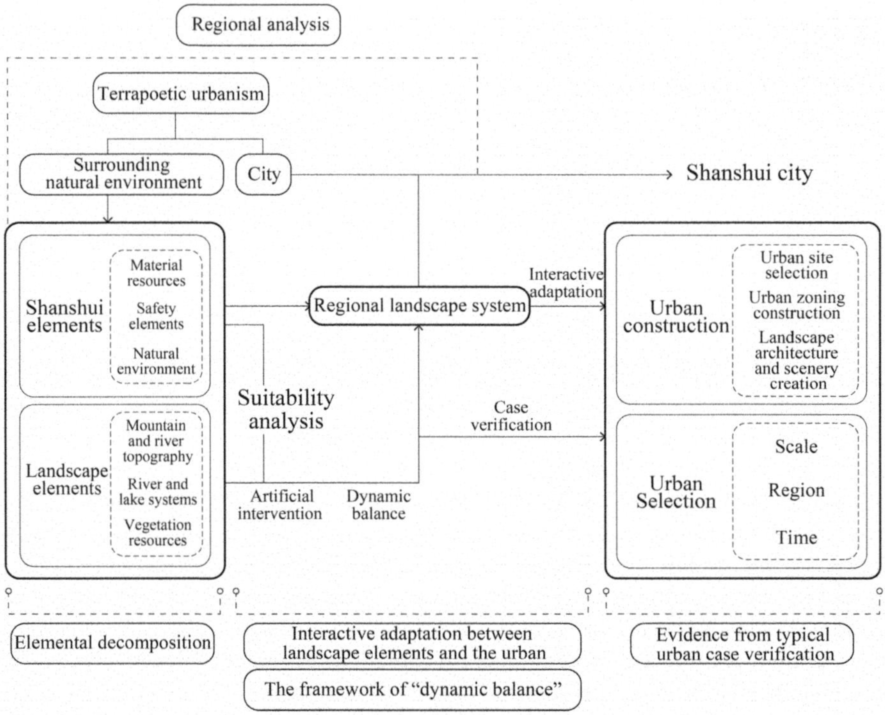

Fig. 1 Schematic framework for analyzing city development processes based on regional landscape systems under the concept of "Terrapoetic Urbanism"

Finally, representative cases are subjected to in-depth empirical analysis and validation, in order to elucidate Terrapoetic Urbanism. The detailed analytical framework is illustrated in Fig. 1.

1 Regional Analysis: Systematizing Foundational Substrate for Urban Construction

The city is not an isolated entity; it is grounded in the natural environment, and regionality constitutes an inherent attribute of the urban condition. Because the natural environments in which cities are situated differ, they cannot be subsumed under a single, uniform generalization. Accordingly, "region" is here defined as a regional landscape system composed of the city and its environs, encompassing the natural environment, Shanshui elements, and related constituents.

Beginning with the natural environment, the analysis traces the process by which, under artificial intervention, it evolves into a Shanshui environment together with urban construction. The natural environment is the prerequisite for urban site selection and construction: it must provide the material resources necessary for human survival (including water and food) and be sufficiently safe, avoiding factors that threaten security. Yet natural environments are often not fully suited to human habitation, thereby prompting intervention—undertaken primarily through the systematization of the water system to discharge and divert floods, excavate ponds to store rainwater, channel water for agricultural irrigation, and open rivers for navigation. In this way, the regional water network environment is reshaped, the beauty of artifice is expressed, and sites are selected for building the city, such that urban growth occurs within a human-modified natural environment. Through a genealogical inquiry into the development of the shanshui system in the urban periphery and the urban–suburban regional landscape system, it becomes evident that anthropogenic adaptation and transformation—from grand mountains and waters to small-scale mountains and waters within the region—constitute the very formation of the medium that guides urban development.

2 Elemental Decomposition: Explicating the Composition of the Regional Landscape System

Research on the regional landscape system proceeds by analyzing its constitutive elements; employing an elemental decomposition method facilitates the establishment of relationships between each element and the city, enables the development of generalizable descriptions, and helps avoid the limitations of a singular line of argumentation.

The regional landscape system constitutes a comprehensive landscape entity formed through the interaction of the natural environment and the water network system shaped by artificial intervention. The natural environment primarily encompasses mountain and river landforms, river–lake water systems, and vegetation resources. Furthermore, conducting a suitability analysis of the natural environment—that is, assessing the environmental conditions of a specific region that are favorable for human habitation as well as potential threats—serves as the fundamental prerequisite for urban site selection and city building. This analysis also provides the underlying rationale for promoting artificial intervention in nature.

The water network system after artificial intervention primarily examines the conditions of the water network environment within the region across different historical periods, as well as the processes of evolution of water network landscape elements during those periods—that is, the measures taken by humans to modify the water network. While elaborating on each constituent element individually, equal emphasis is placed on the integrity of interconnections among elements. For instance, the pond and canal system are not independent entities; rather, they function in concert.

3 Interactive Adaptation: Regional Landscape and Urban Zoning Development

The process of urban construction encompasses the determination of the urban form, the implementation of zoning construction, and the creation of gardens and scenic landscapes. Analytical procedures involve integrating the constituent elements of the aforementioned regional landscape system into the various stages of urban construction; examining the interactive processes between each constituent element and the city; providing a detailed description of the morphological characteristics of urban construction on the material level—including the spatial relationships among different functional zones within the city as well as the distinctive appearances manifested by each functional zone; and elucidating the processes by which landscape elements and urban zoning construction promote each other's development in a mutual manner.

4 Dynamic Balance: Synergistic Interaction Between the Regional Landscape and Urban Construction

Through the analysis of the interaction between the regional landscape system and urban development, this book clarifies that the regional landscape system and urban development exist in a state of continuous adaptation and mutual adjustment, namely a state of dynamic balance. The relationship between the regional landscape system and the city evolves continuously within the frameworks of the temporal perspective, the spatial-regional perspective, and the process-oriented perspective. The first reason is that each stage of development faces distinct problems and requires corresponding measures to address them; moreover, under different regional conditions, the challenges encountered in development also vary. The second reason is that human intervention in nature is a gradual process, propelled by the synergistic process of mutual adaptation and continuous adjustment between the regional landscape system and urban development.

5 Exemplary Cases: Empirical Analysis of Terrapoetic Urbanism

The analysis of typical cases serves to verify the research framework through the examination of cities. This book selects empirical cases encompassing cities, districts, and river basins of different scales, historical periods, and geographical contexts.

The discussion in this book unfolds from the foundation of human settlement environment construction, proceeds to examine the interactive process between the regional landscape system and city development, and ultimately culminates in the creation of a pleasant human settlement environment, which is then substantiated through empirical cases. The synthesis of this process constitutes the concept of Terrapoetic Urbanism. Terrapoetic Urbanism encapsulates the patterns of mutual adjustment and dynamic evolution between the regional landscape system and urban construction, offering valuable insights for contemporary discourse on ecological civilization construction (ECC) and strategies for crafting resilient human settlement environments. One of the key factors underpinning the millennia-long continuity of cities is the developmental resilience derived from their surrounding Shanshui environment. The mountain–water–city model demonstrates the capacity to withstand regional environmental pressures and developmental risks, while simultaneously shaping and perpetuating distinctive local cultural characteristics, thereby becoming a paradigm of livability (Fig. 2).

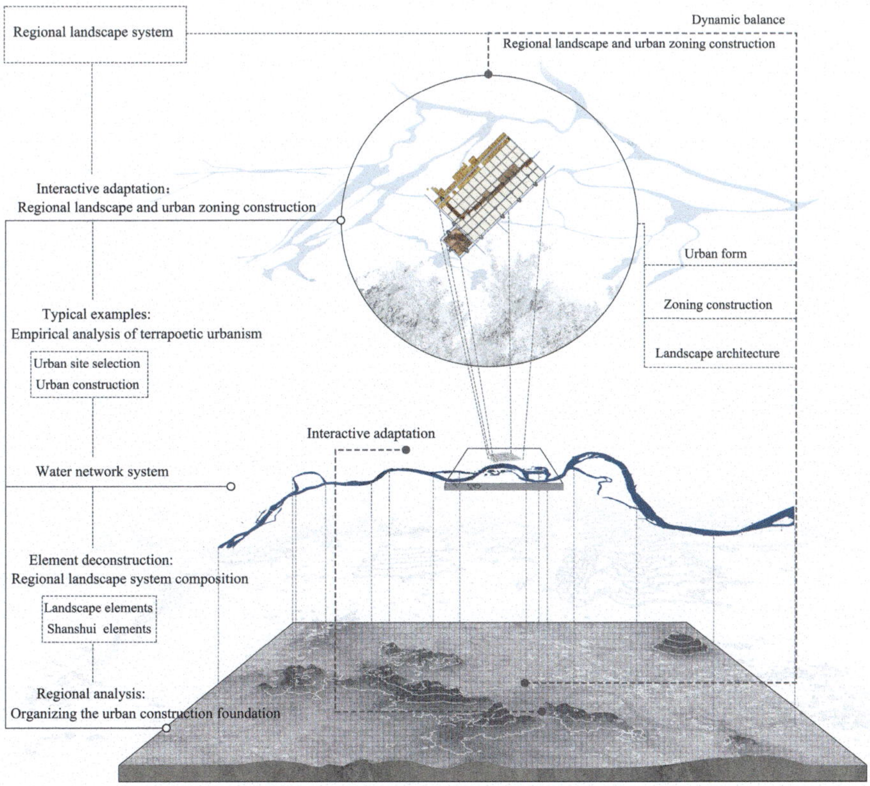

Fig. 2 Schematic diagram of the regional landscape system analysis methodology under the concept of "Terrapoetic Urbanism"

Part II
"Terrapoetic Urbanism" Tradition

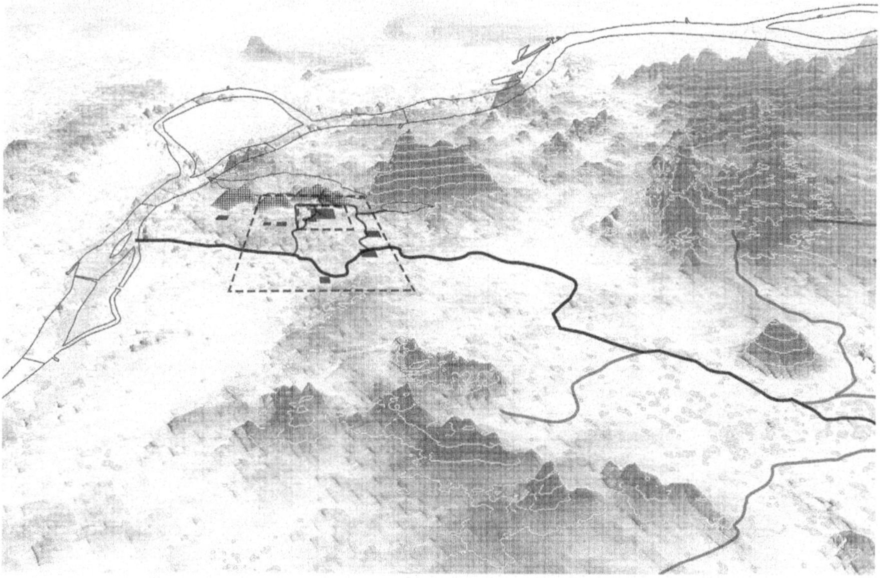

Chapter 3
Cities Through History

Abstract This chapter explores Chinese cities through a historical perspective, focusing on the evolving relationship between Shanshui (mountain–water landscapes) and urban development. It examines how cities were conceived, constructed, and transformed as components embedded within regional landscape systems rather than as isolated entities. Through the analysis of ancient city maps (*yutu*), early water management practices, urban site selection, and spatial organization, the chapter reveals how shanshui conditions shaped urban form, infrastructure, and patterns of expansion across different historical periods. Particular attention is given to the transition from enclosed ward-based cities to more integrated urban–suburban systems, driven by changes in water networks and social organization. At the conceptual level, the chapter discusses key ideas such as context-sensitive design, the unity of humans and nature, and the aesthetic and cultural significance of Shanshui in urban planning. Together, these perspectives demonstrate that the historical evolution of Chinese cities reflects a continuous process of adaptation, negotiation, and balance between urbanization and regional landscape systems.

1 Cities in Ancient Yutu

Ancient city maps or yutu were a way for people in antiquity to perceive and represent cities [1]. In China, maps were also among the earliest practical technical methods to appear. Many Pre-Qin Period[1] (c. 2100–221 BCE) texts, such as the *Shijing* (*(Book of Songs, 1046–476 BCE)),*[2] *Zhou Li*(The *Rites of Zhou, ca. 1046–256 BCE*),[3] contain references to maps [2]. Ancient yutu provide the most direct records of the spatial form and landscape characteristics of ancient cities. Although yutu often cannot match the precision of modern city maps and are more a transmission of imagery and ideas, the visual information—beyond scientific measurement and explanation—allows us to discern the distinctive concepts of the cartographers and the specific subjects they intended to emphasize.

Ancient Chinese cities possessed unique design concepts and methods that integrated mountain-and-water factors and, on this basis, created a great number of

X. Zheng, *Terrapoetic Urbanism*, https://doi.org/10.1007/978-981-95-8676-9_3

Chinese-style urban spaces characterized by "*Tianren Heyi*" (the unity of heaven and humanity) [3]. This is fully reflected in ancient city maps or yutu. The most typical feature of ancient city maps is the use of various graphic symbols to convey geographical information about the region where the city is located, including mountain and river landforms, river–lake water systems, road orientations, and important landmarks, while clearly expressing the relationship of mountain–water–city. This reflects a primitive ecological consciousness and an emphasis on the concept of ingenious integration with nature. The primitive ecological consciousness signifies an awareness of the unity of humans and nature, integrating human construction with natural mountain and water systems and placing them on a single dimension to highlight the harmonious state of *Tianren Heyi*. Ancient city maps represent on paper the Chinese model of mountain–water–city construction and possess significant documentary value (Fig. 1).

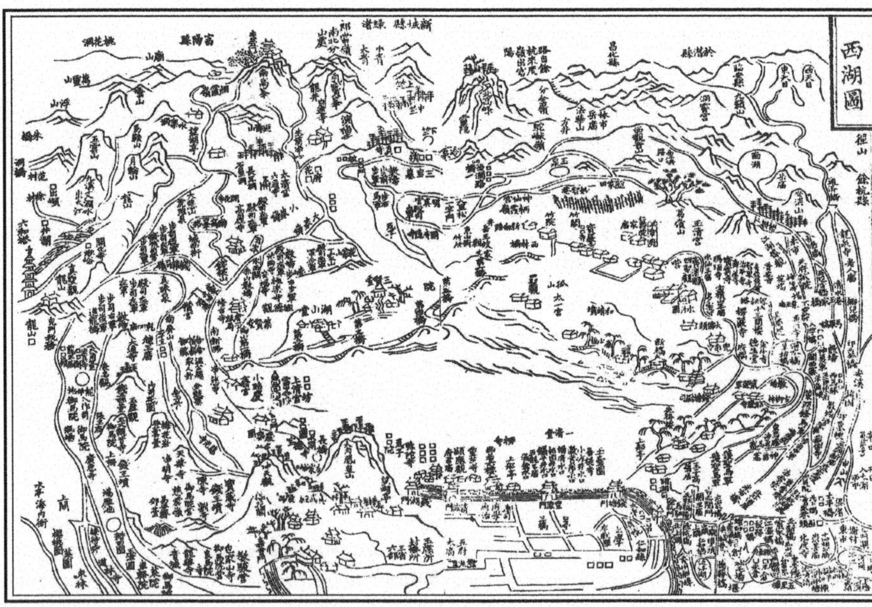

Fig. 1 Relationship among mountain–water–city relationship in the *West Lake Map*. *Image source West Lake Map* (Qing Dynasty, Tongzhi 6th Year, 1867)

2 Historical Timeline

2.1 *Urban Morphological Change Guided by Shanshui*

Urban morphology is the physical manifestation of a city shaped by multiple factors. *Kao Gong Ji* (The artificers' record, ca. 770–221 BCE)[4] records: "The artificers shall construct the capital city, which shall be nine li on each side, with three gates on each side. Within the city, there shall be nine longitudinal and nine latitudinal streets, with nine main avenues. The ancestral temple shall be on the left, the state altar on the right; the imperial palace shall face south, and the market shall be at the rear." Under the constraints of traditional ritual system thought, cities were often constructed in a square form, as exemplified by the Chang'an city of the Sui–Tang Dynasty (581–907 CE). In reality, however, many cities did not possess the geographical conditions to create a perfectly square city during their construction, such as the Ancient Cities of Nanjing, Ningbo, and Shaoxing, as well as the famous mountain city of Chongqing. Chongqing, which originated in the Pre-Qin Period as Jiangzhou, the capital of the Ba state, was not built according to the prescriptions of *Zhou Li* but instead adapted to the natural terrain, resulting in an irregular form [4]. Therefore, following regional environmental conditions and taking nature as a model was an important principle in ancient city building (Fig. 2).

From early on, ancient urban construction reflected planners' high regard for landscape elements: "When establishing a capital or major city, one must choose either at the foot of a great mountain or by a broad river. It should not be placed on arid high ground, so as to ensure an adequate water supply; nor should it be situated in waterlogged lowlands, so as to spare the expense of building extensive ditches and embankments." (the *Cheng Ma* chapter of the *Guanzi*).[5] The natural Shanshui environment thus served as a decisive precondition for urban construction, giving rise to the Chinese model of mountain–water–city urban formation.

As noted in *The Urban History of China*,[6] the study of urban space necessarily considers the site conditions, first, followed by urban form and design. In research on ancient Chinese cities, the Shanshui environment must therefore be treated as a key factor influencing urban development. The concept of landscape-guided urban development has existed since antiquity.

2.2 *Earliest Practices of Water Management*

"When a sage governs the world, its axis lies in water. The work of Yu the Great in controlling floods marks an important milestone in the origin of human civilization [5] and represents China's earliest practice of artificial and proactive water-environment management."

"The great flood spread to the sky, its vast waters surrounding the mountains and submerging the hills, swallowing the common people. I traveled upon four types of

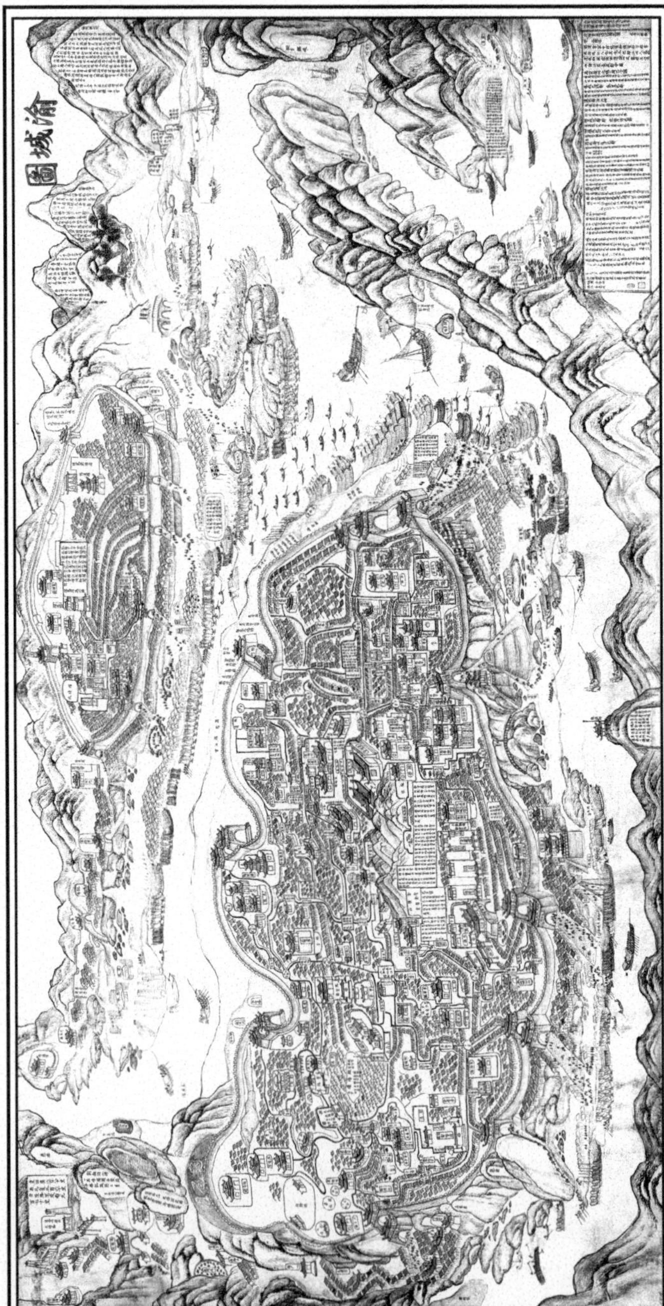

Fig. 2 “Yucheng Map”, the unique urban morphology reflects how site selection and city construction were influenced by the surrounding Shanshui environment. *Image source* “Yucheng Map” (Qing Dynasty, 1860–1886)

conveyances, following the mountain paths to cut down trees as markers, and along with Bo Yi. I gave the freshly hunted birds and beasts to the people. I dredged the rivers of the Nine Provinces, allowing the floodwaters to flow into the Four Seas, and I deepened the irrigation channels to ensure that the water from the fields flowed into the great rivers. Together with Hou Ji, I sowed the grains, providing the people with both grain and meat. We also developed trade, enabling the exchange of goods, thus stabilizing the people's lives." The *Yu Gong* chapter of the *Shangshu*[7] also mentions: "Yu spread the land, following the mountains and cutting down the trees to establish the domains marked by the high mountains and great rivers." "The Nine Rivers were made to flow, and the floods of spring and summer were controlled. The mulberry land began to grow silkworms, and people moved from the hills to the plains, where towns began to form." During the Spring and Autumn Period (771–256BC), thinkers like GuanZi,[8] MoZi,[9] and WuZixu[10] proposed a series of urban disaster prevention theories. In the work the *Du Di* chapter of the *Guanzi*[11] GuanZhong discusses the necessity of addressing natural disasters as a precondition for governing a state. He states: "Those who govern well must first eliminate the five harms." What are these five harms? "Water is one harm; drought is one harm; wind, fog, hail, and frost are one harm; pestilence is one harm; and insects are one harm." Among these, "water is the greatest harm," and thus GuanZi focuses on the prevention and treatment of water-related disasters. He discusses in detail the prevention of flooding, covering aspects from the selection of city sites to the construction, management, and supervision of dikes, drainage canals, and water control systems, forming a comprehensive ancient flood control theory for cities [6].

These theories of water management and agricultural land development and disaster prevention and refuge confirm that the relationship between humans and nature entered a new stage, in which humans could consciously transform and utilize Shanshui environments to achieve the goal of urban construction.

2.3 Disintegration of the Fang–Shi System and Promotion of Integrated Urban–Suburban Development

During the Sui-Tang periods, the development of productive forces and the sharp increase in commodity exchange led to the gradual disintegration of the Fang–Shi system (the ward and market system). Independent and spontaneously formed markets began to appear, and this institutional transformation directly affected the urban pattern. Throughout Chinese history before the Tang Dynasty (618–907AD), the Fang–Shi system had existed, establishing a legally regulated urban commercial space but also restricting the vigorous development of commerce. In this system, bazaar precinct were strictly separated from neighborhood and district system, and the markets were set up and managed by government authorities. Laws and regulations imposed strict control over the time and place of trade.

In the mid-to-late Tang Dynasty, the strict Fang–Shi system began to break down. People were no longer confined to trading only in government-established markets; they also set up temporary trading places along streets and around water systems and canals, such as official street market at city bridges and roadsides, and periodic country fair along canals outside the city. It was precisely because of the presence of these markets that places like Guangling became major hubs for the gathering of goods—"Guangling lies at the great intersection of north and south, where all commodities converge" [7]—and, together with the local custom that "the people of Jianghuai value commerce and neglect agriculture" [8], commercial districts further developed and expanded. As the enclosed neighborhood and district system shifted toward an open street layout, the landscapes of streets and canal-side areas also changed under the influence of commercial activity.

The disintegration of the Fang–Shi system not only brought about the emergence of free markets within cities but also greatly advanced the development of market towns. *A Study of Grass-market Towns in the Song Dyn*asty[12] notes that the periodic country fair described by Song authors were collectively non-official marketplaces mainly distributed in urban–suburban and rural areas. The prosperity of market towns reflected the extension of urban activities into the suburbs, strengthened the connection between urban interiors and exteriors, and represented an expansion of the artificial environment beyond urban boundaries into nature. Connected roads and canals together formed a comprehensive transportation network, enabling the integrated development of natural mountains and waters, landscapes, and settlements within the region of the city.

The disintegration of the Fang–Shi system marked the transformation of cities from a closed to an open development model. Beyond the growth of commerce, it also brought a change in people's thinking, driving integrated urban–rural integration development. At the same time, more public activities and public spaces began to emerge, which in turn fostered the development of gardens and the construction of scenic landscapes.

3 Shanshui Thought and Urban Construction

China has a long history in which urban construction has emphasized an integrated concept, seeking harmony between the artificial environment and the natural environment and embodying distinctive Shanshui characteristics. From the majority of existing ancient capitals and historic cultural cities, one can clearly observe the traditional pursuit of harmonious coexistence between the city and the external environment.

The history of urban construction is in fact also a history of the development of Shanshui culture and Shanshui thought. The *Lai Min* chapter of the *Book of Lord Shang*,[13] records the idea of gathering places shaped by mountains and waters: "Where great mountains and waters converge, a metropolis will be established; where mountains and waters meet on a medium scale, a market town will arise; and where small-scale mountains and waters gather, a village will form."

3.1 The Planning Principle of "Following Natural Advantages and Utilizing Local Resources"

The *Guanzi* is an important classic on ancient Chinese urban planning. In its *Cheng Ma* chapter of the *Guanzi* it states: "Whenever establishing a capital, one must not set it beneath great mountains but upon broad plains; the high ground must not be prone to drought and must have sufficient water; the low ground must not be near water so that ditches and defenses are easy to maintain. Follow heaven's materials, comply with earth's benefits, so the city walls need not be perfectly square, and the roads need not follow strict lines." The principle of follow heaven's materials, comply with earth's benefits underscores a planning approach based on local conditions. The siting and establishment of a capital were to fully consider the Shanshui environment and select locations according to the surrounding Shanshui configuration and their economic practicality.

In the *Du Di* chapter of the *Guanzi* it is further stated: "The sage who establishes a state must choose land that does not tilt and is fertile; dwell with mountains to left and right and with water or marshes nearby; within the city build drainage channels that flow into great rivers. By making use of the natural endowment and the productivity of the land, he provides for the people and nourishes livestock." This passage specifies that the site selection for capitals should favor flat and fertile land, close to mountains and waters, and emphasizes convenient city drainage into major rivers, elaborating in greater detail on how to manage the relationship among mountain–water–city.

In the selection and utilization of the water environment, the *Shui Di* chapter of the *Guanzi*[14] states: "Water is the vital energy of the land, flowing like the veins and arteries of the human body; therefore, it is said that water is a resource that serves all purposes." Here water is likened to the blood of the city—it is both an indispensable factor for the establishment of cities and the foremost source of natural disasters. Thus, the tradition of water management in China dates back to ancient times.

The water environment has a profound influence on urban morphology. The drainage and flood prevention ideas mentioned earlier—"with mountains to left and right and with water or marshes nearby; within the city build drainage channels that flow into great rivers"—exerted deep influence on urban construction through successive dynasties. Consequently, water management and city building were often interlinked. For example, the city walls of Suzhou, Linzi, and Shaoxing also served as flood embankment, and the internal water systems of these cities were connected with the external river–lake networks to facilitate drainage.

Many urban constructions also employed artificially excavated waterways for drainage. In Suzhou, for instance, extensive hydraulic interventions dredged and opened numerous channels such as the Shantang River from the western Changmen

Gate toward Tiger Hill and the Shangtang River linking to the Grand Canal at Fengqiao. Likewise, during the Qianlong reign (1735–1796 AD), the water system of Beijing's western suburbs was renovated: Kunming Lake was widened and the Western Embankment newly constructed, simultaneously achieving water management and enhancing the scenic landscape of the near suburbs. These examples demonstrate that cities flourished by water. In addressing the relationship between cities and the water environment, they not only advanced hydraulic engineering but also created renowned hydraulic scenic sites.

In the selection and utilization of natural mountain ranges, the mountain range orientation is an important factor for macro-level urban site selection. The relationship between mountain bodies, especially low mountains, and cities varies: they may lie in the urban fringe, be located within the city, or even host cities built directly on mountains. Mountains play a dual role for cities. On the one hand, they serve functional purposes, such as providing forest resources, offering military defense through terrain, or protecting against flooding. On the other hand, they create urban and peri-urban imagery and scenic vistas for aesthetic appreciation and cultural association [9]. For example, in the ancient cities of Suzhou and Hangzhou, the western hilly mountain ranges not only provided security and shelter for the cities but also offered natural and landscape resources, constituting an essential factor for their establishment and development.

The site selection of ancient cities was always integrated with the external mountain-and-water environment, following the principle of following heaven's materials, complying with earth's benefits and applying a planning approach based on local conditions. In the *Ba Guan* chapter of the *Guanzi*[15] it is stated: "For a population of ten thousand households, arable land within fifty li (about 25 km) is sufficient. For fewer than ten thousand households, one may supplement from the mountains and marshes; for more than ten thousand households, one should refrain from exploiting them." This reflects such a comprehensive planning concept, which likewise guided the formation and development of ancient Chinese Shanshui cities.

3.2 The Philosophical Concept of Tianren Heyi

As people's intellectual awareness advanced, spiritual life began to guide action. *Zhouyi* (the *book of changes*)[16] summarizes the cosmos: "To establish the way of heaven is called yin and yang; to establish the way of earth is called soft and hard." In urban planning thought, it is said: "The north of water and the south of mountains is called yang; the south of water and the north of mountains is called yin." Accordingly, mountains were usually located to the north of the city and water to the south of the city; mountains represent strength, water represents flexibility; their mutual integration forms the mountains encircling like armrests, waters embracing like sashes essential to the philosophy of Tianren Heyi.

Ancient urban construction in China consistently adhered to the philosophical concept of Tianren Heyi, whose core idea is that heaven, earth, and human beings form a unified whole. Humans cannot separate themselves from nature; they must respect and follow it, seeking coexistence and integration with nature. The pursuit of harmony among heaven, earth, and humans is likewise the pursuit of harmonious unity among mountain–water–city. The application of the tianren heyi concept in ancient urban construction extended through methods of form and number to achieve concordance between humans and heaven, establishing a symbolic tradition in China's urban spatial layout and giving rise to theories such as geomantic soil appraisal, strategic landscape advantages, and fengshui theory [10]. The concept of geomantic soil appraisal has a long history; it is a method of understanding nature that moves from observing objects to abstracting images and then from abstraction back to imitating physical forms. Strategic landscape advantages is a further development of the idea in *Zhouyi* of modeling heaven and imitating earth, emphasizing shanshui environments and elevating urban site selection and mountain range orientation to a macro-regional scale. It stresses the harmony between mountain forms and artistic conception and is, in essence, an overall regional environmental planning theory. The principle in ancient garden design of "Though made by man, it appears as if created by heaven" is likewise an extension of the tianren heyi philosophy.

3.3 The Aesthetic Conception of Attachment to Shanshui

The aesthetic consciousness of Shanshui in ancient China began to awaken during the Wei -Jin Periods (220 AD–420 AD). People admired natural Shanshui, using natural Shanshui as an object of reflection to examine themselves, seeking an inner realm of harmony and tranquility, and taking Shanshui as a carrier to generate diverse cultures. The splendid Shanshui culture has, over the passage of time, profoundly permeated the development of cities and their regional environments.

Ancient literati and scholars sang praises of famous mountains and great rivers, endowing Shanshui with human virtues: mountains with benevolence and steadiness, and waters with tranquility and far-reaching depth. Confucius said: "The wise delight in water, the benevolent delight in mountains; the wise are active, the benevolent are calm; the wise find joy, the benevolent find longevity." Meng Haoran[17] vividly described Dongting Lake: "At eighth month, lake water rises to level with shore. Sky merges with its own reflection off lake into one. Vapor comes up from the Yunmeng Lake. Surging waves strike against the city wall of Yueyang." Rendering the vastness of Dongting Lake so lifelike that the entire Yueyang City seems to tremble, making the city appear exceedingly small within the grand environment.

Wang Xizhi, in his *Preface to the Poems Composed at the Orchid Pavilion*,[18] described the setting: "Here, there are towering mountains and steep peaks, dense forests and tall bamboo groves, along with clear streams rushing rapids, flowing on both sides… Looking up, one sees the vastness of the universe; looking down, one observes the abundance of species. Thus, the eyes are entertained, and the heart is soothed, making it a true pleasure for both sight and sound." By abandoning themselves to nature and pursuing the beauty of artistic conception, literati sought to merge spirit with landscape. Similarly, Wang Bo's *Preface to the Pavilion of Prince Teng*[19] offers the immortal line: "The sunset clouds fly together with the solitary wild duck; the autumn water shares the same color with the vast sky," painting a tranquil and far-reaching masterpiece for the ages.

Influenced by the thought of abandon oneself to nature, poetry, calligraphy, and painting with Shanshui as their theme greatly promoted the construction of the natural environment. When Shanshui was present, people delighted in the pleasures of Shanshui; when Shanshui was lacking, they carried out artificial rock-piling and water arrangement to create Shanshui dwellings. The development of garden-dwelling culture was a prominent chapter in ancient urban construction. Tao Yuanming's[20] phrase, "While picking chrysanthemums neath the eastern fence, my gaze upon the Southern Mountain rests" is a portrayal of this reclusive ideal9. With the advancement of social economy and engineering technology, this Shanshui thought of abandoning oneself to nature reached its peak in the Ming dynasty (1368–1644), characterized by refined arrangements of rocks, flowers, and trees, meticulous construction, and the close integration of cultural connotations with planning and design.

The relationship between Shanshui and the development of ancient Chinese cities reflects the transformation of human understanding of settlement environments in accordance with the progress of social civilization. This evolution moved from an initial stage of passive adaptation to nature, to an active stage of utilizing and reshaping nature after its recognition, and further to a stage of integrating nature with cultural attributes, thereby becoming Shanshui. Such a process of cognition and intervention constitutes the very process of forming the regional landscape system.

With China's long history and its rich and complex natural environment, the land of Huaxia witnessed the emergence of various ancient civilization forms related to Shanshui, and generated multiple planning ideas associated with natural Shanshui. Among them, the concept of context-sensitive design/planning encapsulated in "follow heaven's materials, comply with earth's benefits" formed the basis of planning and site selection; the philosophy of tianren heyi emphasized the harmonious integration of humans and nature as the core of planning; while the aesthetic conception of abandoning oneself to nature represented the incorporation of humanistic attributes into the regional environment. These examples demonstrate that urban construction was inseparably connected with the regional landscape system (Fig. 3).

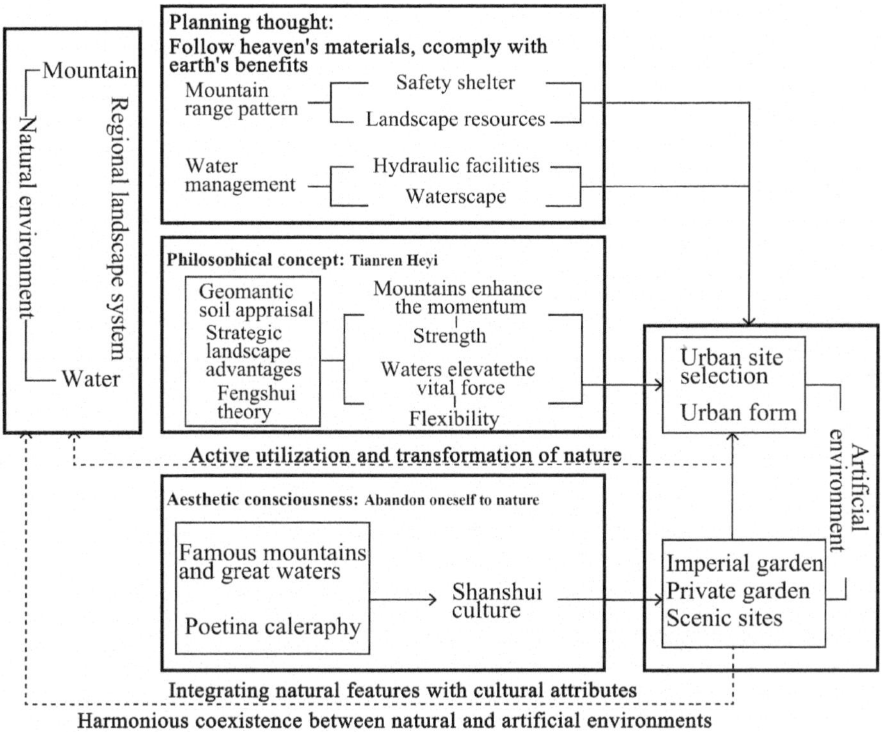

Fig. 3 Diagram of Shanshui thought and urban construction

Notes

1. Pre-Qin Period (c. 2100–221 BCE): The period from the early dynasties (Xia, Shang, Western Zhou) to the end of the Warring States before Qin unification, marked by feudal states, early agriculture, and initial management of rivers, mountains, and settlements, laying foundations for Chinese landscape, urban planning, and political geography.
2. *Shijing* (Book of Songs, 1046–476 BCE): The earliest Chinese poetry anthology, documenting human life, agriculture, settlements, rivers, and mountains during the late Western Zhou and early spring and autumn periods.
3. *Zhou Li* (The Rites of Zhou, ca. 1046–256 BCE): A classical Chinese text outlining political, social, and administrative systems of the Zhou dynasty, including urban layouts, water management, gardens, and farmland organization.
4. *Kao Gong Ji* (The Artificers' Record, ca. 770–221 BCE): A classical Chinese technical text, detailing standards for architecture, bridges, roads, gardens, waterworks, and crafts.
5. The *Cheng Ma* chapter of the *Guanzi* (ca. 400–350 BCE): A text on water management, land allocation, and agriculture, emphasizing river works, irrigation, and settlement planning in relation to natural landscapes.
6. *The Urban History of China* (2019) by Fu & Cao: A chronological study linking Chinese urban forms, public space, and landscape traditions with long-term settlement and planning.
7. The *Yu Gong* chapter of the *Shangshu* (compiled c. 1046–771 BCE): Describes ancient Chinese rivers, mountains, and provinces, linking landscape, resources, and governance—foundational for historical spatial and planning insights.

8. GuanZi, Guan Zhong (c. 720–645 BCE): Qi state chancellor whose reforms in land, water, and settlement planning influenced early Chinese landscape and urban management.
9. Mozi (c. 470–391 BCE) was a Chinese philosopher and social reformer during the Warring States period (475–221 BCE). He criticized wasteful rituals and lavish construction, promoting functional, efficient, and socially beneficial projects.
10. Wu Zixu (559–484 BCE) was a statesman, military strategist, and engineer during the late spring and autumn period (770–476 BCE) of China. Born in the State of Chu, he later served the State of Wu, where he played a pivotal role in urban planning, hydraulic engineering, and fortification projects.
11. *Du Di* chapter of the *Guanzi* is traditionally attributed to Guan Zhong (720–645 BCE) but reflects contributions from multiple scholars over time. It focuses on measuring and managing land, covering terrain, water, and soil fertility for agriculture, taxation, and territorial administration.
12. *A Study of Grass-Market Towns in the Song Dynasty* is a modern Chinese scholarly monograph published in the late twentieth to early twenty-first century. The book examines grass-market towns (caoshi)—informal commercial settlements that emerged outside formal city walls during the Song Dynasty (960–1279)—analyzing their spatial structure, economic functions, and relationship to transportation networks and rural hinterlands.
13. *The Lai Min chapter of the Book of Lord Shang* was compiled during the Warring States period (ca. 350–300 BCE) and is traditionally attributed to Shang Yang (ca. 390–338 BCE), a leading Legalist reformer of the State of Qin, though the text is recognized as a composite work. The chapter outlines policies for attracting and settling population through land distribution, agricultural incentives, and territorial control, linking governance to the management of land, labor, and settlements.
14. The *Shui Di* chapter of the *Guanzi* was compiled during the Warring States period (ca. 475–221 BCE) and is traditionally attributed to Guan Zhong (ca. 720–645 BCE). The chapter examines the relationships among water systems, landforms, soil, and human settlement, emphasizing how rivers, wetlands, and terrain condition agricultural productivity, urban location, and state administration, making it an early theoretical source for environmental planning and regional landscape management.
15. *The Ba Guan chapter of the Guanzi* was compiled during the Warring States period (ca. 475–221 BCE) and is traditionally associated with Guan Zhong (ca. 720–645 BCE). The chapter sets out a systematic framework for observing population, land use, resources, infrastructure, and social order as the basis of effective governance.
16. The *Zhouyi* was formed between ca. 1046–771 BCE, with its core hexagram system traditionally attributed to King Wen of Zhou (d. 1047 BCE) and explanatory line texts associated with the Duke of Zhou (eleventh century BCE), while later interpretations were added over time. As a foundational cosmological and philosophical text, it articulates principles of change, balance, and the dynamic relationship between heaven, earth, and human action.
17. Meng Haoran (689–740 CE) was a major poet of the Tang dynasty, closely associated with the shanshui (mountain-and-water) poetic tradition. His poetry centers on rivers, fields, rural settlements, and seasonal change, articulating an ideal of harmony between human dwelling and natural landscapes.
18. Wang Xizhi (303–361 CE), the most celebrated calligrapher of the Eastern Jin dynasty, was a scholar-official deeply engaged with literati culture and landscape appreciation. His *Preface to the Poems Composed at the Orchid Pavilion* (353 CE) records a gathering beside winding streams and wooded hills, where natural setting, water management, and social ritual were deliberately integrated.
19. Wang Bo (650–676 CE), a prominent poet of the early Tang dynasty, was known for his literary talent and official career in southern China. His *Preface to the Pavilion of Prince Teng* (675 CE) celebrates the famous riverside pavilion in Nanchang, vividly describing the surrounding landscape, river vistas, and human-built structures.

20. Tao Yuanming (365–427 CE), also known as Tao Qian, was a poet and recluse of the Eastern Jin dynasty (317–420 CE). He is celebrated for his pastoral poetry and essays that depict rural life, gardens, and natural landscapes, emphasizing harmony with nature and simplicity over official or urban life.

References

1. Chen, W. (2013). An exploration of methods for historical city conservation: Discoveries from Ancient Urban Maps—A case study on core issues in the comprehensive protection plan of the Nanjing Ming City Wall. *Architect, 3*, 75–85.
2. Dong, W. (2009). The theory of Ancient Chinese cartography and its modern significance (I): Reflections on Pei Xiu's "six principles of cartography." *Architect, 6*, 29–34.
3. Zhang, G. (2006). *Study on landscape constraints in ancient Chinese urban design*. Tsinghua University.
4. Zhu, L. (2015). An analysis of ancient Chinese urban morphology. *Urban Geography, 6*, 275.
5. Research Group on "Improving China's Water Governance System". (2015). Historical Evolution and Experience of China's Water Governance and Institutional System. *Water Development Studies, 15*(8), 5–8.
6. Fu, L. (2004). *Studies on landscape cities*. Hubei Science and Technology Press.
7. Wang, P. (2012). *Tang Huiyao* (p. 4). Shanghai Ancient Books Press.
8. Liu, S. (2013). *New Anecdotes of the Great Tang* (Vol. 6). Zhejiang Publishing Group Digital Media Co., Ltd.
9. Wang, D. (2002). *Chinese landscape culture and urban planning*. Southeast University Press.
10. Chen, Y. (2006). A study on the evolution of Suzhou's contemporary urban morphology. *Urban Planning Forum, 3*, 36–44.

Chapter 4
Cognition and Inheritance

Abstract In this Chapter, Cognition and Inheritance, examines how ancient cities developed enduring regional landscape systems through a deep understanding of nature and strategic human intervention. In periods marked by limited technological capacity and frequent challenges, such as flooding, warfare, and population growth—city builders reorganized regional Shanshui (mountain–water) conditions to create resilient landscape systems that provided ecological support, disaster mitigation, and long-term spatial stability. Through detailed analyses of Suzhou's ancient water network and the construction of Kunming Lake and Beijing's urban water system, the chapter demonstrates how water management functioned as a critical medium shaping urban form, resource circulation, and cultural identity. It further highlights traditional Chinese concepts such as Tianren Heyi (unity of humanity and nature) and learning from nature as foundational principles linking urban space with regional landscapes. In response to contemporary challenges of rapid urbanization and cultural homogenization, the chapter argues for the renewed recognition and inheritance of regional landscape systems as a strategic framework for resilient, culturally grounded, and sustainable urban development.

From the scale of a national capital to that of local prefectures and counties, the construction of ancient cities was invariably a vast undertaking. Particularly in eras when science and technology were far from advanced, urban construction faced formidable challenges arising from natural disasters, military conflicts, and demographic growth. After acquiring a comprehensive understanding and analysis of nature, ancient city builders intervened through human agency to reorganize the natural conditions of the region, thereby forming a resilient regional landscape system capable of withstanding disasters and serving defensive and protective functions—becoming both the cultural substratum and the ecological foundation for urban development. Within the process of dynamic adaptation, cities and the Shanshui environment of their regions continued to evolve, sustaining themselves over millennia. In this process, the regional landscape system served as a crucial medium for urban development, guiding and facilitating the emergence and evolution of distinctive urban spatial forms.

X. Zheng, *Terrapoetic Urbanism*, https://doi.org/10.1007/978-981-95-8676-9_4

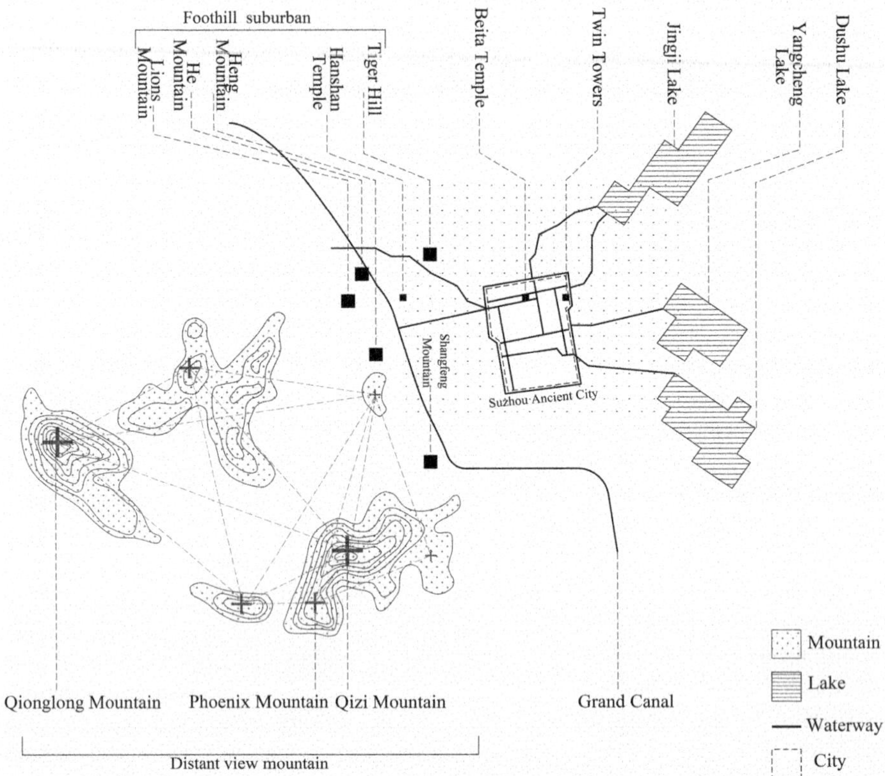

Fig. 1 Schematic diagram of the "mountain–water region" symbiotic relationship in Suzhou Ancient City mediated by regional landscape systems. Base map adapted from: Chen Yong, *Research on the Structural Evolution of Suzhou's Ancient City*, October 2006

Taking the construction of the ancient city of Suzhou as an example: since antiquity, flooding has continuously threatened this "watertown amid marshes." Confronted with the trials of nature, inhabitants undertook extensive water conservancy works, excavated waterways, and diverted water into the city, initiating a measured degree of human intervention in order to adapt to the natural geographical environment. This process of reciprocal interaction between internal and external systems established the urban environmental support framework that sustained the ancient city over a millennium, while also giving rise to its distinctive landscape system. The entire urban operation and resource allocation were fully founded upon this system. The crisscrossing waterways not only fulfilled multiple city functions—including water diversion, drainage, flood discharge, transportation, defense, and pollution purification—but also engendered a distinctive and outstanding regional culture, ensuring the enduring prosperity of Suzhou. Ultimately, this forged a symbiotic relationship of mountain–water–city in Suzhou (Fig. 1).

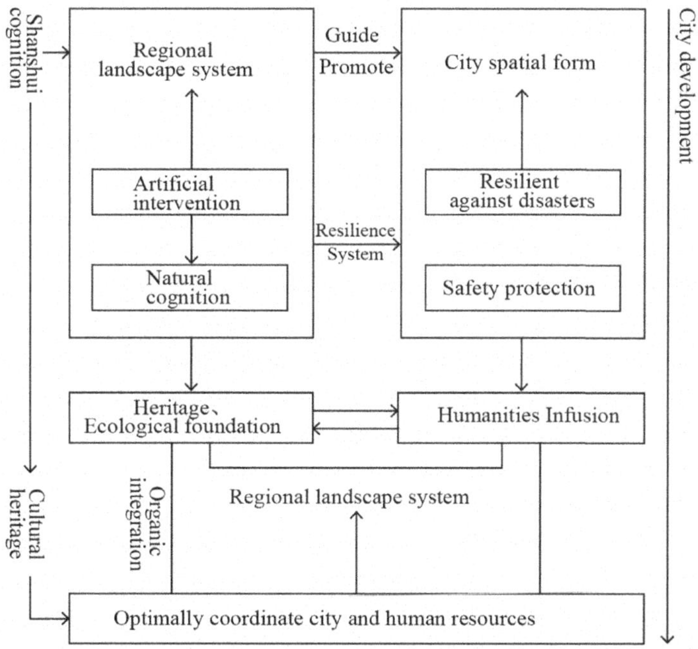

Fig. 2 Influence of Shanshui perception and cultural heritage on regional landscape systems and city spatial form

Another example is the construction of Kunming Lake in the Summer Palace and the improvement of Beijing's water system. At the initial stage of city establishment, the inner city water network drew its source from the springs within Zizhuyuan. Subsequently, in the western suburbs, Kunming Lake was constructed as a reservoir to regulate and store runoff from the Western Hills, eliminate flood hazards, and support irrigation agriculture. In addition, the Kunyu River and the Chang River were excavated to connect with the inner city water system, while to the east, the Tonghui River was opened to link with the northern terminus of the Beijing-Hangzhou Grand Canals in Tongzhou. In this way, goods from the south, transported via waterborne grain transport along the Grand Canal to Tongzhou, could reach the inner city through the Tonghui River. Before the completion of Kunming Lake, however, the west-high/east-low topography of Beijing, combined with unstable upstream water sources, often caused the Tonghui River's water level to fall short of the requirements for waterborne grain transport, resulting in frequent grounding of vessels. It is thus evident that the foremost function of Kunming Lake's construction was to regulate stormwater and flood flows, irrigate surrounding farmland, provide a water supply for the city, and ensure a stable water level for downstream waterborne grain transport. Though located in the western suburbs of Beijing, its significance for urban development was profound (Figs. 2 and 3) [1].

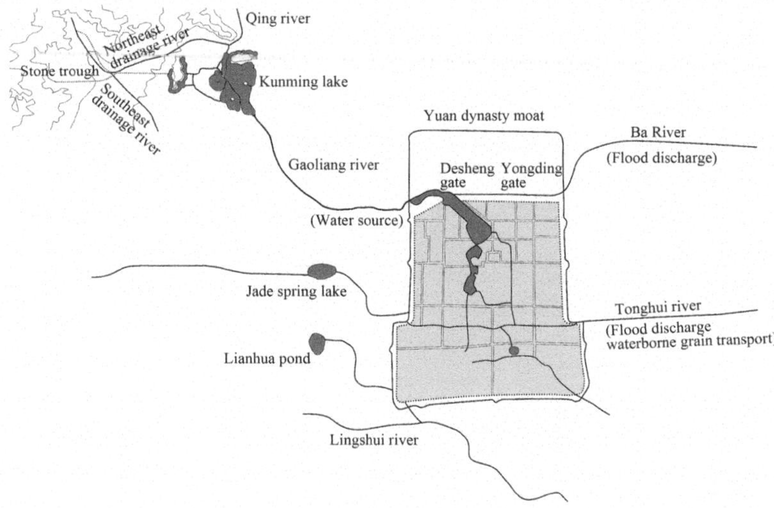

Fig. 3 Schematic diagram of major water systems in Ming-Qing Dynasties (1368–1912) Beijing

Since ancient times, the Chinese philosophical concept of Tianren Heyi (unity of heaven and humanity) and the design principle of following (or learning from) nature have both emphasized the interactive relationship between the city proper and its surrounding natural environment, regarding the two as an integrated and inseparable whole. When combined with the humanistic dimension of the city, mountain-and-water landscapes are thus endowed with cultural and spiritual value. Therefore, it is of critical importance to explore the historical evolution of the human settlement environment, to examine the methods and experiences of artificial intervention in nature, and to analyze the formation of regional characteristics—comprising both cultural and material heritage that must be preserved. On this basis, it is essential to undertake in-depth research into the series of issues facing contemporary cities with regard to urban culture, distinctiveness, and heritage continuity.

In contemporary China, city construction is undergoing a period of rapid development. Influenced by a series of factors—including the acceleration of urbanization, the convergence of global cultures, external shocks, and the limitations inherent in its own development—the relationship between the city and the regional landscape system has become distorted. This has led to a neglect of the integrity, coordination, and mediating role of the landscape system in the creation of the human settlement environment, reflecting, to a certain extent, a rupture in current planning practices with regard to the understanding and inheritance of traditional mountain-and-water thought. As times have evolved, numerous urban visions have been successively proposed, such as the "millennium city," "low-carbon city," and "green city." These more human-centered, harmonious, diverse, and organically developed urban themes call for the construction of sustainable ecosystems to address global challenges and regional pressures. With the introduction of the strategy for building a life community integrating mountains, waters, forests, farmlands, lakes, and grasslands, along with the implementation of ecological civilization construction and the "Beautiful China" initiative, the regional landscape system has once again been elevated to an important strategic component at the planning stage. The thousand-year blueprint for city development cannot be separated from the resilience provided by the regional landscape system formed by landscape elements such as shanshui. For the city, the regional landscape system represents the rational integration and coordination of urban resources—both natural and artificial—to address the social and urban issues arising from global challenges and regional pressures in contemporary development. By organically integrating the regional landscape system with the city, the renewal and growth of the urban fabric can be driven forward, enabling it to confront diverse challenges while achieving cultural continuity and sustainable development.

To study the present, one must first learn from the past. In an era when transformations in the regional landscape system, as well as the scale and speed of urbanization, are accelerating, it is imperative to fully recognize the historical principles governing the construction of regional and city landscape systems. The wisdom and refinement embodied in the harmonious coordination between humanity and nature remain profoundly significant for contemporary landscape and city construction.

Reference

1. Hou, R. (2009). *The life imprints of Beijing City*. SDX Joint Publishing Company.

Part III
Constituent Elements and Research Framework of "Terrapoetic Urbanism"

Chapter 5
The Figure–Ground Relationship Between Shanshui and the Urban

Abstract This chapter introduces Terrapoetic Urbanism as a framework for understanding urban development through the long-term interaction between cities and regional landscape systems. Rather than treating Shanshui (mountain-and-water landscapes) as a scenic backdrop, the book demonstrates how regional landscape systems function as the spatial, ecological, and cultural substrate of urban formation. Integrating regional and dynamic analysis with historical interpretation and empirical case studies, it traces how natural landforms, water networks, and human interventions jointly shaped urban site selection, spatial structure, and resilience across different historical periods. Through analyses of cities, districts, and river basins, the book reveals a process of continuous adaptation between landscape systems and urban growth. By reinterpreting traditional landscape wisdom within contemporary planning discourse, this work offers critical insights for sustainable urbanization, resilient infrastructure, and landscape-based urban design in rapidly changing environments.

The focus of Terrapoetic Urbanism lies in analyzing the regional landscape system as the medium that promotes urban development. The constitutive elements refer to the various factors that embody the primary content of the regional landscape system, as well as the set of processes defined by their interactions under certain guiding principles. The analysis of the constitutive elements of terrapoetic urbanism encompasses two main aspects: first, the examination of the regional landscape system and its elements as the foundation of urban development; second, the summarization of the traditional principles governing the interactions between these elements and urban construction, while exploring the general processes of urban construction and development under the influence of traditional ecological concepts (Fig. 1). Terrapoetic Urbanism thus studies shanshui as existing in the form of a regional landscape system, which serves both as the foundation of urban construction and as the medium facilitating the evolution of urban development. The relationship between the two is one of "figure" and "ground:" the regional landscape system constitutes the "ground," while the city—including government offices, markets, residences, temples, gardens, and scenic construction—represents the "figure."

X. Zheng, *Terrapoetic Urbanism*, https://doi.org/10.1007/978-981-95-8676-9_5

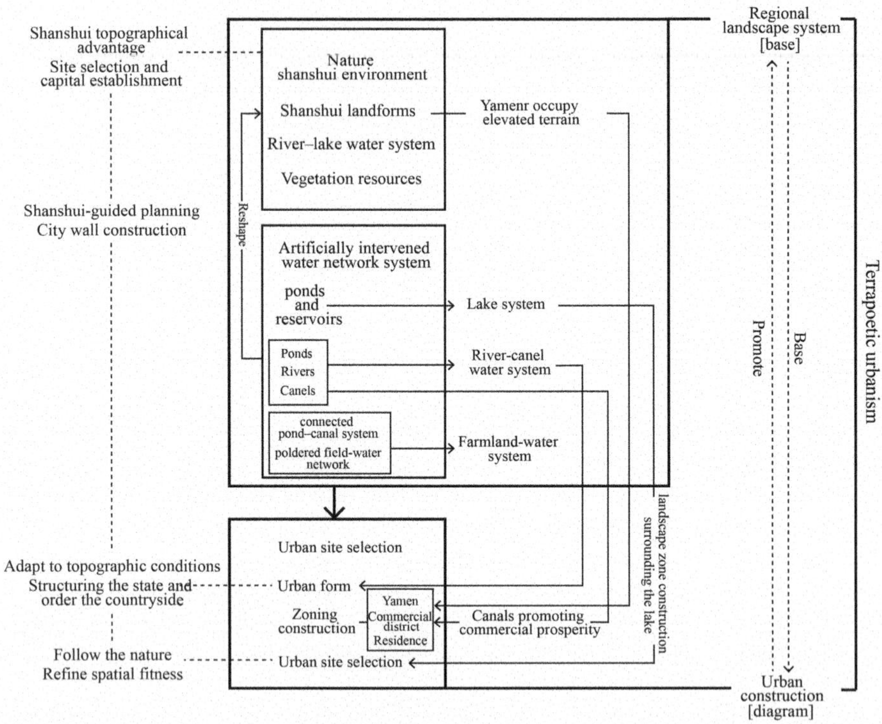

Fig. 1 Diagram of the constituent elements of Shanshui urbanization

Illustrating the composition of the regional landscape system and the interrelationships among various components of urban construction.

1 The Regional Landscape System as the Foundation of Urban Construction

The regional landscape system refers to the totality of the natural Shanshui environment and the artificial water network environment within a defined scope of settlement. It is a comprehensive entity composed of mountain ranges, water systems, vegetation, and a series of hydraulic facilities including ponds, canals, and waterways. The regional landscape system (i.e., the Shanshui environment) contains two aspects: first, the natural Shanshui conditions of the area where the settlement or city is located, along with its surroundings, which provide the essential basis for forming a settlement pattern, including topography, river–lake systems, and vegetation resources; second, the artificial interventions carried out by humans in the pursuit of a livable environment, which reshaped the regional water network environment. The integration of these two aspects jointly constitutes the foundation for urban development—namely, the regional landscape system.

2 Diagram of the Elements of the Regional Landscape System

2.1 Natural Environment

The natural environment is the foundation and framework for the formation of the regional landscape system and constitutes an essential component in its construction. It primarily includes Shanshui landforms, river–lake water systems, and vegetation resources.

(1) Mountain–river landforms

China is located in the eastern part of the Eurasian continent and on the western coast of the Pacific Ocean. With its vast territory, it possesses extensive and unique natural geographical landforms, making it one of the countries with significant regional differences in the human–land system. The terrain of China is diverse, with a general topography that is higher in the west and lower in the east, descending in a stepped pattern. Undulating mountain ranges, majestic plateaus, expansive plains, continuous hills, and fertile basins—these five basic landform types are all distributed across China, with mountains, plateaus, and hills accounting for about one-third of the total land area. The wide extent of mountainous areas is a distinctive feature of China's topography, creating varied conditions for agricultural development, providing abundant forest, mineral, and tourism resources, while also serving as a shelter for human habitation.

As a crucial factor in the geographical environment, Shanshui landforms serve as the spatial foundation for the distribution of regional resources. As expressed in the saying, "Only with land can landscape emerge [1]," landforms are the basis of landscape architecture. The morphology of land not only influences the evolution of natural ecosystems but also shapes the modes of human activities and behaviors [2]. Each regional land unit has its own distinctive geomorphological characteristics. People adaptively modify according to the geomorphic environment of the region, following and utilizing nature, ultimately producing landscape forms such as urban site selection in accordance with mountain terrain, housing construction, farmland reclamation, mountain aggregation, and water diversion for irrigation.

(2) River–lake water systems

China, as one of the countries with the greatest number of rivers and lakes in the world, has more than 1500 rivers with drainage areas exceeding 1000 km^2, and over 24,800 lakes in total, among which more than 2800 are natural lakes larger than 1 km^2. Although China possesses abundant water resources overall, their distribution is uneven across regions, with significant regional differences in the density of river networks. The nation's water systems can generally be divided, according to the circulation patterns of river runoff, into two categories: exorheic rivers that flow into the sea, and endorheic rivers that disappear within inland areas. These two categories of rivers delineate two major hydrological zones: exorheic regions and endorheic regions. Generally, exorheic regions are larger than endorheic ones, with the overall trend showing a gradual decrease from southeast to northwest. Marked hydrological differences exist between the rivers of the south and the north. Southern rivers are characterized by abundant water volume, low sediment content, dense vegetation, and high coverage, whereas northern rivers contrast significantly in these respects. Consequently, northern rivers present fewer obstacles to transportation than southern rivers. However, their shallow channels create difficulties for navigation, and in the rainy season they are prone to flooding. From ancient times until the Western Jin Period (265–316 AD), the economic center of China was located in the north, primarily because the Yellow River Basin provided more favorable natural conditions. After the Western Jin, the economic center gradually shifted southward, largely due to the abundant rainfall, fertile land, and favorable agricultural conditions of the Jiangnan region, coupled with the development of water transport, which stimulated economic growth.

In the *Shui Di* chapter of the *Guanzi* it is stated: "Water is the vital essence of the land, flowing as the veins and arteries; thus it is said: water is the material of substance." Here, water is metaphorically described as the lifeblood of the city, an indispensable factor in its establishment. The formation of most human settlements has been inseparably linked to water. On the one hand, water is the source of life, essential to both daily living and production; on the other hand, in ancient times when overland transportation was underdeveloped, water transport held significant advantages and served as a crucial condition for promoting regional economic development. The tradition of settling by water is reflected in the names of many cities. For instance, Linqing, Linxiang, Lintong, and Jiangpu contain the meaning of "beside a river." Place names such as Guidong, Fenxi, Weinan, Weiyuan, Luoyang, Heyin, and Heyuan indicate relative positions to rivers. Meanwhile, Hejiang, Hankou, Shanghai, Shuangliu, Sanshui, Hequ, Qujiang, Longquan, Ganquan, and Jiangning, as well as many other urban centers, were named after their relationship to rivers or springs [3].

In addition, water systems have also played an important role in the formation and evolution of city forms, especially as natural river–lake networks exerted strong guiding influence at the initial stages of city development. Hangzhou City is a typical city shaped by the impact of the water environment. Located in the Jiangnan watertown region, the urban form of Hangzhou has, from its very inception, been constrained by its geographical environment, with the most significant natural geographical factors being the Qiantang River and the West Lake.

(3) Vegetation resources

Vegetation is a precious resource upon which humanity relies for survival and development. It serves as the support system for global life and as a repository and refuge of biodiversity. With its vast territory, complex terrain, and diverse climate, China provides exceptionally favorable natural environments for plant growth. Consequently, it possesses an abundance of plant species, constituting a valuable genetic repository of vegetation resources.

Vegetation resources have three principal functions: ecological, ornamental, and biological. The ecological function refers to the capacity of vegetation to improve the ecological environment and regulate climate. The ornamental function denotes the inherent aesthetic qualities of vegetation, making it an indispensable element in the landscape. The biological function emphasizes its role as a natural resource available for human use, since virtually all plants, animals, and microorganisms exist within specific vegetation systems. To a certain extent, the richness of vegetation resources was a crucial criterion for the initial site selection of settlements. Chinese civilization originated in the Yellow River Basin, where the ancient natural environment was favorable: the climate warm and humid, the land fertile, with verdant mountains and abundant waters, and diverse plant life—all of which provided advantageous conditions for the survival of our ancestors.

With diverse landform changes, abundant water resources, and a great variety of vegetation, China's natural Shanshui resources have, overall, been rich, providing a favorable natural foundation for the development of ancient cities. However, the natural environment has never been entirely suitable for human habitation. Since ancient times, people have judged the habitability of a region based on the configuration of mountains and waters, taking into account factors of fengshui, safety, military strategy, and economy. This reflects the dual character of the natural environment: while it affords habitability, it also conceals potential risks.

a. The natural barrier function

The natural environment served as a natural defensive system in ancient times. From the perspective of urban safety, the orientation of mountain ranges was a crucial basis for macro-level site selection in city planning. Mountain bodies might be situated in the suburban areas, at the edge of the city, or the city might even be built directly upon the mountain itself, functioning as a natural protective barrier and a shelter shaped by nature. Similarly, water systems, regarded as the lifeblood of the city, played an equally important role in ensuring security within ancient urban construction.

b. Abundant material resources

The natural environment was also the primary source of material and forest resources. Timber, birds and beasts, minerals, medicinal herbs, and foodstuffs were common resources found in the mountains and forests. In the eras when people "cultivated along ridges and tilled the land, or pursued deer and fowl for sustenance," mountain dwellings relied heavily on forest and material resources from the mountains. Likewise, water resources were closely tied to the origins and development of humanity, with rivers serving as crucial cradles of civilization.

c. The natural linkage

The development of water transport greatly stimulated regional economic prosperity. In eras when transportation depended solely on human or animal power, large-scale long-distance transport was nearly impossible, whereas water transport was the only feasible means to achieve this goal. Water systems strengthened exchanges between different regions, serving as a natural linkage. Regions with abundant water resources provided favorable conditions for the development of grain transport, which in turn fostered regional prosperity.

d. Natural landscape resources

The natural environment possesses both ecological and ornamental functions, making it a form of natural landscape resource. The interwoven settings of Shanshui are complex, exhibiting marked regional differences across various areas. Combined with unpredictable climatic conditions, these environments not only provide favorable conditions for human habitation but also present challenges. Floods, droughts, pest infestations, earthquakes, and wild animals were unavoidable survival difficulties for humanity. As the ancient saying goes: "When winds and rains come in due season, the nation prospers and the people are at peace; when disasters strike frequently, the people suffer and the nation finds no respite." The natural environment is thus a system in which habitability and risk coexist. Human survival depends on the natural environment, which in turn exerts reciprocal influence upon humanity. People could only sustain their existence by continually adopting measures to ensure stability. With the deepening of human understanding of nature, the relationship between humans and nature evolved from one of simple dependence and utilization into one of proactive adaptation. This transformation reflects the inherent driving force of nature itself, which induced measured human intervention and thereby fostered the harmonious development of humans and nature.

2.2 *The Water Network System After Human Intervention*

Human intervention in nature was most prominently manifested in the management of the water environment. Marx pointed out in 1930 that a defining characteristic of Eastern society was that hydraulic works consistently functioned as state-led public projects [4]. The American sinologist Karl Wittfogel[1] also observed that "Eastern and Western societies represent two entirely different social formations, with the formation and development of Eastern society being inseparable from water management [5]."

From ancient times, China emphasized agriculture, and "hydraulic irrigation and river defense and dredging" were consistently prioritized throughout successive dynasties. Furthermore, water systems, as the lifeblood of the city, were indispensable conditions for urban development. Across the world, clusters of economically and culturally advanced cities are mostly concentrated in alluvial plains and delta regions rich in water resources.

As noted by scholar Deng Tuo[2] in *A History of China's Relief from Famines*, prior to the 1940s, "throughout more than three millennia, scarcely a year was without disaster, and seldom a year without famine; some Western scholars even referred to China as a 'land of famine,' which was by no means an exaggeration." Among the three major disasters of ancient Chinese society, two—floods and droughts—were directly tied to water, indicating that China's preoccupation with the governance of the water environment stretches back to antiquity.

The saying "If rivers lack waterways, if streams lack embankments, and if marshes lack barriers, then heavy rains bring floods in destructive torrents" exemplifies this condition. The legend of Yu the Great's flood control is, in fact, a symbolic representation of humanity's protracted struggle with floods across a long span of history [6]. The ancient maxim declared: "To govern the state, one must first govern the waters; to govern the waters is to govern the state."

Through hydraulic works, people continuously regulated the regional water system, reshaping diverse river–lake networks and, together with natural Shanshui, constructing the regional landscape system as a vital ecological and cultural carrier upon which cities developed.

Human regulation of the water environment mainly comprised the following types: storage, diversion, drainage, irrigation, and dredging for navigation. Among these, water-conservation measures focused on "storage" were ultimately manifested in the form of lakes and ponds; measures emphasizing "diversion, drainage, and navigation" were expressed in the linear forms of canals and channels; while measures centered on "irrigation," combined with farmland development, together constituted the agricultural landscape system.

(1) The system of artificial lakes

In the *Guoyu, Zhou Yu Xia (discourses of Zhou, lower section)*[3] it is recorded: "Ponds and reservoirs, though low and damp, accumulate their beauty." Wei Zhao annotated: "To store water is called 'pei,' which means pond." Hydraulic works focusing on "storage" were primarily realized through ponds and reservoirs. Such ponds collected surrounding water sources by utilizing low-lying terrain, forming pools mainly intended to resolve the temporal and elevational contradictions between water availability and water use. Their purpose was to conserve water for better utilization of water resources. This was typically achieved by constructing river-blocking dams at mountain gullies or river gorges, thereby forming artificial lakes.

China's earliest pond–reservoir hydraulic works can be traced back to the mid-spring–autumn period (722–481 BCE), with the construction of the large-scale Shaopei hydraulic project in the Huainan region. Subsequently, many such projects were built in the Jianghan, Jianghuai, and regions south of the Yangtze River. In essence, these pond–reservoirs were equivalent to what we call reservoirs today. According to their construction sites, pond–reservoirs can generally be divided into two types: plain reservoirs and valley reservoirs. Early examples such as the Shaopi in Huainan City, the Hongxi Pond in Runan County, the Liumen Pond in Nanyang City, Jianhu Lake in Shaoxing City, and Lianhu Lake in Danyang City all belonged to the category of plain reservoirs; whereas the Maren Pond in Biyang County, the Chengong Pond in Yangzhou City, and Chishan Lake in Jurong City, were representative of valley reservoirs [7].

The construction of pond–reservoirs was carried out in accordance with the principle of adapting measures to local conditions, making ingenious use of existing topography, water sources, and other natural conditions. Full consideration was given to utilizing natural low-lying areas and marshlands prone to water accumulation, or to artificially constructing ponds to intercept natural rivers or mountain runoff in order to form storage capacities for flood control and irrigation. The development of pond–reservoir hydraulic works began early and proved highly effective, and the engineering techniques of pond–reservoirs also made significant progress. There were both large-scale reservoir projects organized by national and local governments, and small-scale, diverse groups of ponds and lakes constructed by households making use of favorable regional water resources. These projects greatly transformed the regional water network environment (Fig. 2).

(2) River–canal water network system

a. Pond–rivers (tanghe)

   The pond–river refers to an artificial river, a watercourse formed by embanked levees, and later more generally used to describe rivers constructed by humans. Unlike canals, which serve functions of navigation, irrigation, water supply, and diversion, the primary purpose of the pond–river was to resist flooding disasters and tidal incursions. The term pond–river

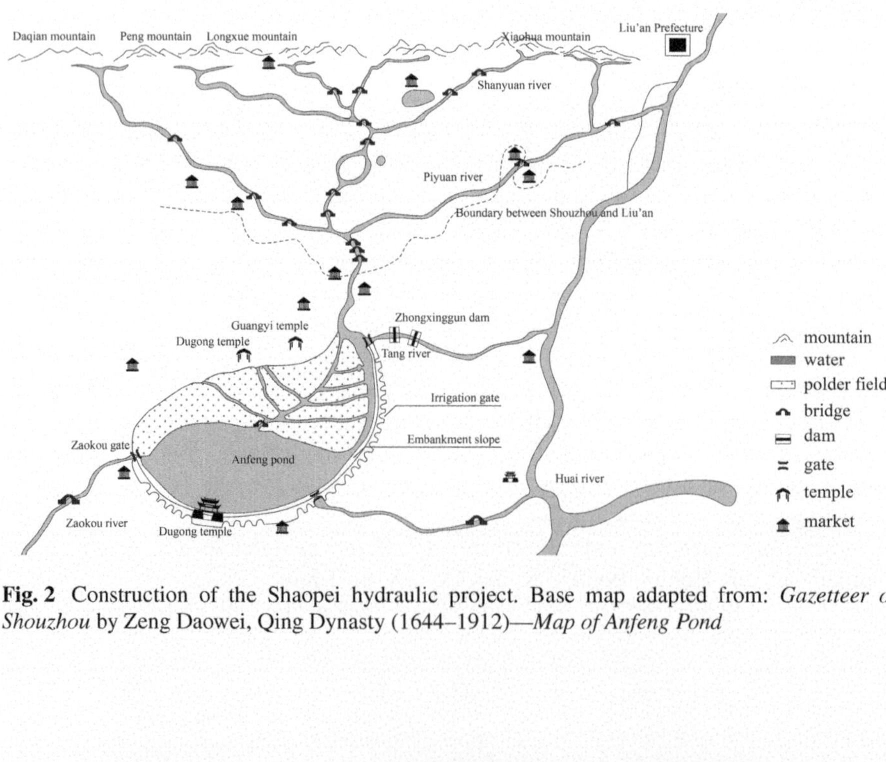

Fig. 2 Construction of the Shaopei hydraulic project. Base map adapted from: *Gazetteer of Shouzhou* by Zeng Daowei, Qing Dynasty (1644–1912)—*Map of Anfeng Pond*

appeared mostly in the eastern coastal regions. The early construction of pond–rivers originated among local populations of the coastal areas, who, based on the varying topography of tidal flats, built earth embankments to resist tidal flows. Before the tenth century, in Cixi, such embankments were referred to as "scattered ponds" (san tang), small in scale and lacking systematization. Large-scale pond–river engineering works began with the construction of the Dagutang in the Northern Song period. Traditional pond–rivers were built with earthen levees, sometimes surfaced with stone slabs, thereby creating the picturesque scene of parallel land and water routes: rows of willows and camphor trees along the banks, and boats passing each other upon the waters.

b. Ditches and channels

Ditches and channels is a general term for excavated waterways designed for irrigation and drainage. In ancient cities, the moat was a type of ditch–channel system. In many cases of urban construction, river courses were dug or modified for drainage purposes. For example, in Suzhou City, the base water system was dredged and cut through to form multiple waterways, thereby creating a dual grid pattern of land and water. The saying "cities thrive because of water" illustrates that, in the process of negotiating the relationship between the city and the water environment, waterworks were constructed not only to control flooding and supply irrigation, but also to transform water into a vital component of the urban landscape system, making waterscapes an indispensable element of the city's spatial and aesthetic composition.

c. Canals

In ancient China, the improvement of rivers and the construction of artificial waterways were primarily undertaken for the purpose of irrigation, and in the earliest periods such works were generally referred to as "ditches" or "channels." If the excavation of the Han Gou is regarded as the starting point of the Chinese canal tradition, its history extends back more than two millennia. The concept of the canal itself, however, first appeared in writing in Ouyang Xiu's *New History of the Tang* (*Xin Tang Shu*) [8]. During the Spring–Autumn Period, King Fuchai of Wu[4] initiated large-scale excavation near present-day Yangzhou in order to secure the transport of military supplies. This project later developed into the Grand Canal, a monumental waterway of approximately 1,700 km in length, running from north to south and linking more than twenty cities and regions across the six provinces of Beijing, Tianjin, Hebei, Shandong, Jiangsu, and Zhejiang.

The earlier-excavated section of the Grand Canal in Wuxi[5] gave rise to the city's renowned four major wharfs, the Rice Market Wharf, the Cloth Wharf, the Silver and Currency Wharf, and the Silk-Cocoon Wharf of which more than thirty cultural heritage sites remain to this day. The ancient canal also accumulated Wuxi's distinctive Wu cultural traditions and local customs, such as folk arts including Wuxi Opera, Wu songs, and Jiangnan silk-and-bamboo music, as well as folk crafts like Huishan clay figurines, paper-horses, and Wuxi embroidery.[6] Moreover, folk customs such as river lanterns, temple fairs, and lantern parades flourished along the canal. This cultural vitality further gave rise to a number of representative traditional historical neighborhoods in the Jiangnan region, such as Jieguan Pavilion Lane near Sanliqiao, Rihui Lane at the foot of Renmin Bridge, and Taosha Lane beneath the South Gate Drawbridge.[7]

The canal, while strengthening navigation, simultaneously connected regions and water bodies through artificial waterways, thereby accelerating the circulation of goods and the movement of people, and to a certain extent influencing the development of cities and market towns along its route—particularly in shaping urban form and determining the distribution of bazaar precinct. In addition to fulfilling the function of waterborne transport, the opening of canals also served purposes of irrigation, flood diversion, drainage, and water supply. This network of artificial channels, excavated streams forming semi-artificial waterways, and regulated natural rivers greatly improved the habitability of regional environments and facilitated the formation of a more integrated water system network.

(3) Combined farmland irrigation and water conservancy system

a. Development of farmland irrigation and water conservancy

Ancient Chinese civilization was splendid and sustained by a large population, and therefore agriculture was consistently emphasized, with "water conservancy irrigation" and "river defense and flood diversion" regarded as the foremost tasks of successive dynasties. From the advent of agrarian society, organized farmland irrigation practices emerged. Broadly speaking, the development of farmland irrigation and water conservancy underwent the following stages:

During the slave society period (Shang Dynasty (1600–1046 BC) to the Zhou Dynasty (1046–256 BC), irrigation adhered to the well-field system, where the construction of agricultural water facilities primarily relied on ditches laid out within the well-field.

By the Warring States Period (475–221 BC), with the consolidation of the feudal enfeoffment system,[8] agriculture prospered, and large-scale canal systems replaced ditches, bringing rapid advancements in hydraulic engineering. Examples include the earliest large-scale canal system in China, the "Twelve Canals of the Zhang River," and the world-renowned Dujiangyan Irrigation System.[9]

In the Eastern Han Period (25–220 AD), the focus shifted to the construction of ponds and weirs, beginning in the Huai River system and gradually extending to other river basins. After the Eastern Han, pond-and-weir water conservancy experienced accelerated growth.

By the Tang (618–907 AD) and Song (960–1279 AD) dynasties, with relative political stability and after the groundwork laid during the Six Dynasties Period (220–589 CE), water conservancy in the Jiangnan region developed rapidly. In addition to diversion canal systems, major achievements included reservoir ponds and dams, brine-resisting and freshwater-preserving projects, and polder fields along lakeshores. Among these, the most remarkable was the development of Taihu polder fields, which became highly advanced by the late Tang dynasty [9].

b. Major models of farmland irrigation and water conservancy

① Pond–canal connection projects

China's pond–canal connection projects originated as early as the Spring–Autumn Period and became prominent from the Warring States to the Han Dynasty, especially in the hilly basins and plains along the middle and upper reaches of the Huai River and the Han River. This type of water diversion irrigation project was largely influenced by topography and climate, resulting in distinct differences between the north and the south. In North China, where plains dominate, rainfall is scarce, and river density is low, irrigation was primarily achieved through canals that diverted water. In contrast, in the south, with abundant rainfall and hilly terrain, irrigation relied more on the construction of dams and weirs to store water for agricultural use. The Huai and Han River regions, located in the transitional zone between the north and the south, developed a hybrid type of farmland irrigation, typically characterized by the interlinked use of ponds and canals.

Pond–Canal Connection Projects also encompassed two distinct types of interlinking methods, which together created expansive irrigation networks that regulated the distribution of regional water resources. The first type involved constructing dams in accordance with topography to stratify water levels, thereby forming storage zones that were linked to multiple ponds through canals. This system created a "long-vine-and-melon" type of irrigation network. The second type diverted water from multiple points along a source river to connect with different ponds at various water levels, while upstream canals could also replenish downstream canals. This resulted in a mesh-like irrigation system characterized by interlinked ponds and canals.

② Polder farmland projects

Polder fields represent the principal form of cultivating farmland in low-lying areas, essentially reclaiming land from water. Their construction method involved enclosing low-lying lands, riverine sandbanks, ponds, marshes, river channels, and lakes with embankments, thereby converting them into agricultural fields. As Fan Zhongyan[10] once remarked: "In Jiangnan, there ought to be polder fields, each extending for dozens of li like a large city, with canals inside and sluice gates outside; in times of drought, the gates are opened to draw upon the benefits of the Yangtze River, and in times of flood, the gates are closed to guard against the river's hazards. Thus, neither drought nor flood prevails, bringing great agricultural prosperity" (Fig. 3).

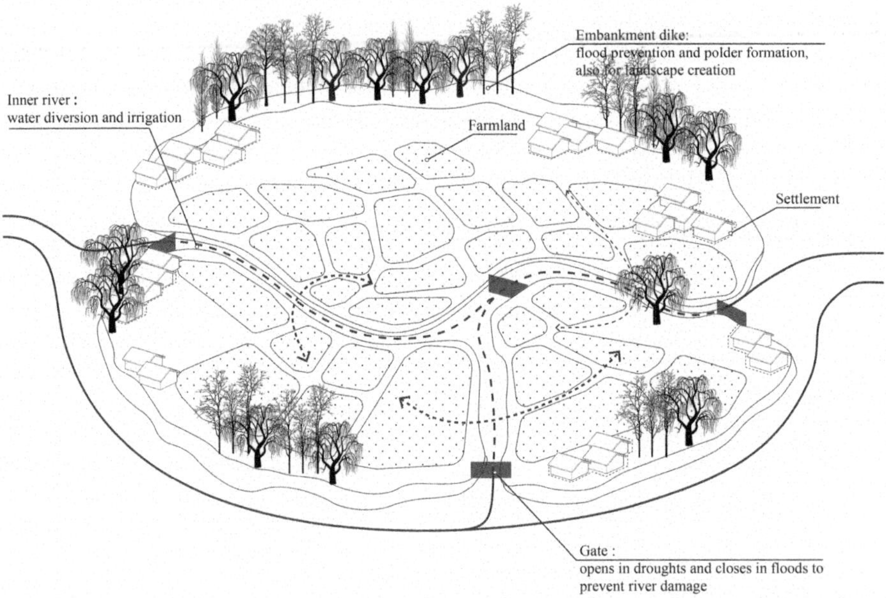

Fig. 3 Schematic diagram of ancient polder engineering. Base map adapted from Zhu Xuexi, *History of Farmland Water Conservancy in the Taihu Basin, in Famous Ancient Hydraulic Projects of China* (Chap. 5)

During the Spring–Autumn and Warring States Periods (770–221 BC), there were already signs of developing shallow marshes and constructing embankments to enclose fields around Taihu Lake. By the Five Dynasties period under the Wuyue Kingdom (907–978 CE), the Taihu Plain had already formed a systematic pattern of polder farmland. This represented an indigenous agricultural engineering technique in China's low-lying plains, which significantly transformed the water-network environment of flatlands crisscrossed by rivers, improved the living conditions of marshy lands, and promoted the development of cities.

Shaoxing and Ningbo City, located on the southern plain of Hangzhou Bay, are two major cities that serve as cases of the coordination between urban spatial form and farmland water conservancy construction The Ning-Shao Plain is one of the oldest areas of polder farmland reclamation. Over approximately two thousand years of cultivation, the Ning-Shao Plain transformed from a barren marshland into a prosperous Jiangnan watertown settlement. The water conservancy system composed of embankments, ponds and sluices was the primary reason for changing the settlement environment of this region. The construction of such water conservancy facilities created a land pattern densely interwoven with waterways, within which cities developed and evolved into Jiangnan water network cities. The renowned Jianhu project, which created an expansive lake stretching hundreds of li, became a famous scenic site. Generations of literati visited and composed poetry there, inaugurating the origins of China's lake culture.

The linear water channels formed through water diversion for irrigation and excavation of canals for flood discharge, the surface ponds and lakes created by flood interception and water storage, as well as the irrigation water networks formed by the interconnection of canals and ponds, were integrated with natural water bodies, extending across the entire region. These systems supported daily life, navigation, irrigation, and landscape construction, thereby reshaping the regional water network environment. The combination of natural mountains, natural waters, and the water networks resulting from human intervention jointly constituted the foundation for urban development, namely the regional landscape system. Upon this foundation, urban construction reflected a spatial pattern coordinated with the broader Shanshui system.

Notes

1. Karl A. Wittfogel (1896–1988) was an American sinologist and historical sociologist known for his theory of "hydraulic civilization," which links large-scale water-control systems to state power and urban organization in ancient China.
2. Deng Tuo (1912–1966) was a Chinese historian and intellectual who authored *A History of Famine Relief in China* (Zhongguo Jiuhuang Shi, first published in 1956), a systematic study of disaster relief from antiquity to the late imperial period. The book analyzes how water control, land use, granary systems, and state intervention shaped settlement resilience.
3. *Guoyu, Zhou Yu Xia (Discourses of Zhou, Lower Section)* was compiled during the late Spring and Autumn period (475–403 BCE) as part of Guoyu, an ancient Chinese historical text recording speeches, events, and political strategies of the Zhou dynasty.
4. King Fuchai of Wu ruled the State of Wu from 495 BCE to 473 BCE during the late Spring and Autumn period. He is remembered for urban and military projects, including canal construction and city fortifications.
5. The Grand Canal (Wuxi section, Wuxi City, Jiangsu Province) functioned not only as a transportation corridor but as a structuring element of the city's historical urban form. The emergence of specialized wharfs reflects a canal-based commercial zoning system, in which waterborne trade directly shaped land use patterns, occupational clustering, and waterfront morphology.
6. The concentration of Wu regional performing arts, handicrafts, and ritual activities along the canal illustrates how hydraulic infrastructure acted as a cultural landscape, sustaining intangible heritage through everyday spatial practices.
7. The formation of traditional Jiangnan neighborhoods, such as Jieguan Pavilion Lane near Sanliqiao Bridge demonstrates a canal-oriented settlement pattern, characterized by narrow lanes, mixed residential–commercial use, and close integration between water, streets, and social life.
8. The feudal enfeoffment system, implemented during the early Zhou dynasty (ca. 1046–256 BCE), organized territory by granting land to nobles in exchange for loyalty and military service.
9. The Dujiangyan Irrigation System (built in 256–251 BCE) in Sichuan, China, was engineered by Li Bing and his son during the Warring States period. It exemplifies early integrated water management, flood control, and agricultural planning, shaping both the local landscape and urban settlement patterns.
10. Fan Zhongyan (989–1052 CE) was a Northern Song dynasty statesman, reformer, and scholar-official. He promoted public works, flood control, and regional planning, linking governance with landscape management and urban development in the Yangtze-Huai river regions.

References

1. Wang, Y. (2014). *Principles of landscape ecological planning*. China Architecture & Building Press.
2. Li, L. (2011). *The humanization of nature*. Beijing Forestry University.
3. Shibo, Y. (2013). *A history of Chinese cities* (p. 52). Peking University Press.
4. Marx, K., & Engels, F. (1976). *Selected works of Marx and Engels:* (Vol. II, p. 64). People's Publishing House.
5. Wittfogel, K. A. (1989). *Oriental despotism: A comparative study of total power* (p. 58). China Social Sciences Press.
6. Gui, M. (1997). An outline of the history of natural disasters in ancient China. *Agricultural Archaeology, 3*, 230–242.
7. Zhang, F. (2009). *A history of irrigation engineering technology in ancient China*. Shanxi Education Press.
8. Cheng, Y. (2008). The formation, development, and prosperity of the grand canal of China. *Journal of Liaocheng University (Social Sciences Edition), 3*, 1–7.
9. Wang, J., & Zhang, F. (1990). *A history of farmland water conservancy in China*. China Agriculture Press.

Chapter 6
The Processual Theory of Terrapoetic Urbanism

Abstract This chapter presents an analytical and empirical examination of representative cases of urban formation in ancient China, elucidating the role of mountain–water urbanization across multiple spatial scales. Through examples from Suzhou, Shaoxing, and Hangzhou, the chapter demonstrates that urban form emerged not from isolated design intentions or formal aesthetics, but from integrated responses to landscape structure, hydrological conditions, and institutional frameworks. The analysis shows that regional landscape systems shaped through long-term human intervention constituted the ecological foundation and cultural substratum of urban development, revealing a dynamic and adaptive relationship between urbanization processes and their supporting natural environments.

The construction of ancient cities fully utilized the natural environment together with the water network environment, both in their original form and after artificial intervention, to jointly shape a coherent regional landscape system. From the siting of capitals to zoning schemes, from garden-making to the creation of scenic and historic sites, every stage reflected an interdependence and integration with the Shanshui environment. In this process, the artificial city environment and the regional landscape system together formed a resilient mountain–water–city structure. By examining the interactive relationship between the regional landscape system and the development of urban construction, the following wisdom and experience have been distilled.

1 Picturesque Mountain-and-Water Landscape and the Siting of Capitals

"The way of geography lies entirely in Shanshui," the picturesque mountain-and-water landscape has always been a guiding principle in traditional construction. Capitals were sited through comprehensive consideration of the Shanshui environment, taking the natural environment as the foundation for urban construction. As mentioned above the *Cheng Ma* chapter of the *Guanzi*,[1] On Establishing the Capital,

X. Zheng, *Terrapoetic Urbanism*, https://doi.org/10.1007/978-981-95-8676-9_6

in any case of establishing a state capital, situate it either at the foot of a great mountain, or above a broad river. Avoid placing it so high as to approach the drought. Avoid placing it so low as to approach the flood. Follow heaven's materials, comply with earth's benefits, hence the inner and outer city walls need not accord with the compasses and square, nor its roads with the level and marking line. Follow heaven's materials, comply with earth's advantages emphasizes a planning philosophy that fully considers the Shanshui environment in the siting of a capital, selecting locations based on the surrounding Shanshui pattern and on economic practicality. The *Du Di* chapter of the *Guanzi*[2]: When a sage sets the site for a state, he invariably chooses ground that will not shift or subside, and he selects a terrain configuration that is rich and fertile. With a mountain to the fore, and watercourses or marshland running to left and right, he lays out interior runoff channels as conduits and leads them to discharge into the great river. Thus, taking advantage of heaven's endowment and what the earth produces, he benefits and sustains the people and rears the six kinds of domestic animals. The siting of a capital should be on terrain that is level and fertile, backed by mountains and close to water, with convenient internal drainage channels discharging into a major river. This reflects the principles governing capital siting in ancient China: mountains encircling like armrests, waters embracing like sashes, level terrain, and water transport convenience. For example:

1.1 Beijing—Girded by the Cang Sea and the Taihang Mountains, Resting Against Juyong Pass and Clasped by the He and Ji Rivers—Truly, a Heaven-Bestowed Realm

Beijing is situated at the northwestern edge of the North China Plain, at the confluence of the Yanshan Mountains and the Taihang Mountains. The junction of these two mountain ranges forms a semicircular, eastward- and southward-opening mountain bend, known as the "Beijing Bend." From a comprehensive view of Beijing's topography, it is characterized by mountains to the back and facing the sea, exhibiting the strategic formation of "dragon coiling and tiger crouching," with a majestic and commanding landscape configuration. The ancients said: "the land of Youzhou—on the left, it is girded by the boundless sea; on the right, it is embraced by the lofty Taihang Mountains; to the north, it is cupped by Juyong Pass; and to the south, it opens toward the Heji River. Truly, it is a realm bestowed by Heaven…"

1.2 Lin'an—Backed by Mountains and Bordered by Waters, Endowed with Convenient River Transport, and Attaining the Height of Prosperity and Splendor

The city of Lin'an (present-day Hangzhou City) was characterized in its overall spatial layout by being backed by mountains and bordered by waters. It possessed two prominent peaks—South High Peak and North High Peak—along with several smaller hills arranged around West Lake: to the north stood Baoshi Hill and Solitary Hill, while to the south rose Xizhao Hill and Phoenix Hill; in the northwest lay Feilai Peak. To the west, the city was directly adjacent to West Lake, within which several small islands were constructed. Together with embankments and bridges, these elements integrated the mountains, West Lake, and the urban fabric into a unified whole. Within the city, waterways such as the Middle River and East River traversed the entire urban area, connecting to the Grand Canal and enabling convenient water transport. The development and construction of Lin'an took place on the basis of making the best use of the natural mountain-and-water conditions while continuously improving the natural environment [1]. The superior geographical setting played a decisive role in the city's growth, and it enjoyed a period of exceptional prosperity. The spatial arrangement of Lin'an, extending from west to east, vividly demonstrated the distinctive Chinese model of mountain–water–city spatial configuration.

1.3 Yangzhou, Situated to the Southeast of Shugang, Is Characterized By a Network of Interlacing Waterways and Enjoys a Uniquely Poetic Geographical Advantage—Possessing Exclusively the Two Halves of the Bright Moon

Yangzhou is located at the confluence of the Grand Canal (the Han Canal[3] section of the Sui–Tang Grand Canal) and the Yangtze River. The local terrain is higher in the west and lower in the east, with the hilly area in the northwest—known as the Shugang region—being the highest. From west to east, the land slopes gradually in a fan-shaped pattern. Beneath the Shugang lies an alluvial plain formed by the deposition of the Yangtze River, while the river itself flows along the southern edge of this plain, together constituting a mountain encircling like armrests, waters embracing like sashes natural mountain–water configuration. Superimposed upon this natural mountain–water configuration, through centuries of anthropogenic modification—such as excavating canals outside the city (including the Han Canal, Shanyang Ditch, and Yilou River) and within the city (the Market River and the Official River), as well as constructing reservoirs (such as Chengong Pond, Goucheng Pond, Upper Leitang, Lower Leitang, and Xiaoxintang) a complete regional landscape system was created

[2]. This system provided a favorable natural background for the construction and development of the city.

1.4 Suzhou City, with Its Western Embrace of Lake Taihu, Its Northern Reliance on the Yangtze River, and a Dense Network of Waterways Interwoven Throughout the City

Located within the Taihu Lake Basin in the lower reaches of the Yangtze River, Suzhou City is backed by Lake Taihu to the west, bordered by the Yangtze River to the north, and facing the sea to the east. Low hills encircle its western suburbs, while a constellation of shallow lakes—including Yangcheng Lake, Jinji Lake, and Dushu Lake—is distributed across its eastern suburbs. The city is interlaced with a dense network of waterways and supported by an expansive hinterland. The surrounding mountain and water landforms have provided a solid natural and ecological foundation for the emergence and development of the city.

1.5 Shaoxing, Founded upon the Nine Hills and Structured Along the Five Rivers, Was Ingeniously Constructed Above Marshy Terrain

The ancient city of Shaoxing (Yuecheng) was established in the transitional zone between the Kuaiji Mountain Range and Hangzhou Bay, known as the Shanhui Plain. At the time of its founding, the Shanhui Plain was largely marshland with a harsh environment. The people of the ancient Yue region chose to construct settlements upon isolated hills (guhills) that rose above the wetlands, and Shaoxing's ancient city was no exception. The "Zhong Hill" (also known as Wolong Mountain, now called Fushan), Ji Hill, Peng Hill, Guai Hill (also known as Feilai Hill, Guishan Hill, or Baolin Hill, now called Tashan Hill), Baima Hill, Baolang Hill, Emei Hill, Huozhu Hill, and Huangzhuo Hill were the nine isolated hills situated around the key water transport route of the Ruoye Creek in the Shanhui Plain. Following the advice of Fan Li[4] (a political, economic, and military strategist of the Spring–Autumn Period (770–476 BCE) from the State of Chu), the city was constructed here, making the Nine Hills the very foundation of its urban construction. In addition to the mountain enclosure, five natural water systems traversed the city, laying the groundwork for the eventual formation of a water city with extensive hydrological networks and interlaced waterways (Fig. 1).

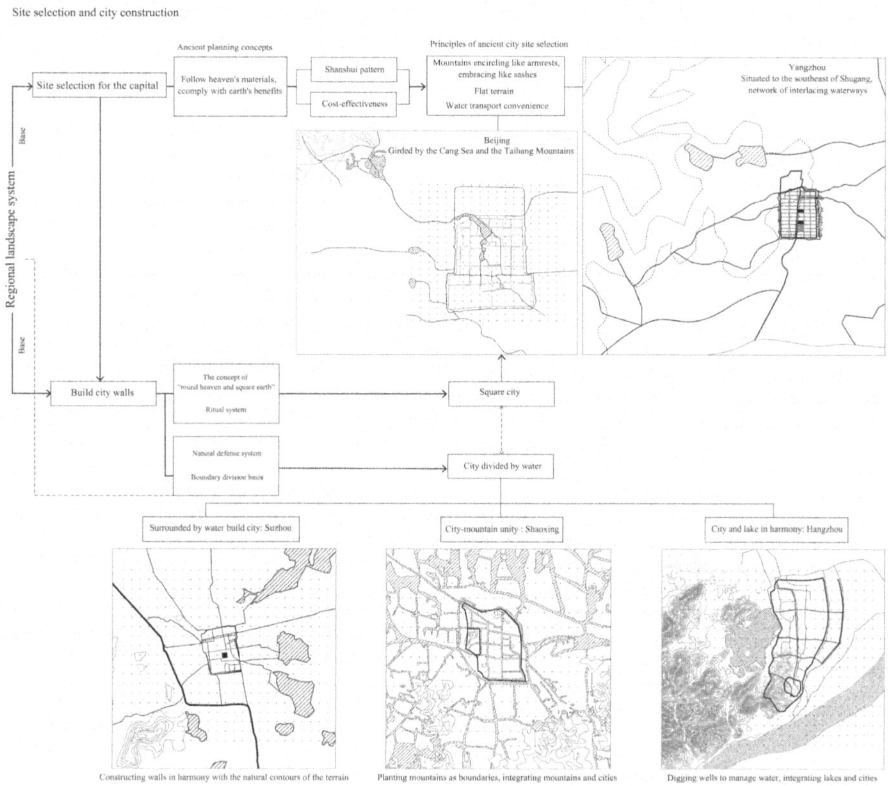

Fig. 1 Analysis of urban site selection and relationship with Shanshui

2 Harnessing the Advantages of Shanshui for Guidance, Constructing the City and Erecting Defensive Walls

In ancient China, the city was enclosed by the city wall, most commonly in a square or rectangular plan. Some cities assumed a regular square configuration, while others were notably irregular—because urban site selection often followed the logic of mountains and waters, the city's form was inevitably influenced by the Shanshui environment. On the one hand, mountains and waters functioned as a natural defensive system consonant with the protective role of the city wall; on the other, they long served as bases for boundary demarcation. City founders chose sites in relation to mountains and water and, through context-sensitive design/planning, improved the setting to create distinctive urban forms and spaces: some cities were enclosed along the water's course, and the city wall accordingly became irregular; in other cases, the city wall was tightly integrated with mountain masses so that the mountains themselves formed part of the fortifications; in still others, the city was combined with extensive water bodies to produce a lake city ensemble. The city's morphology was

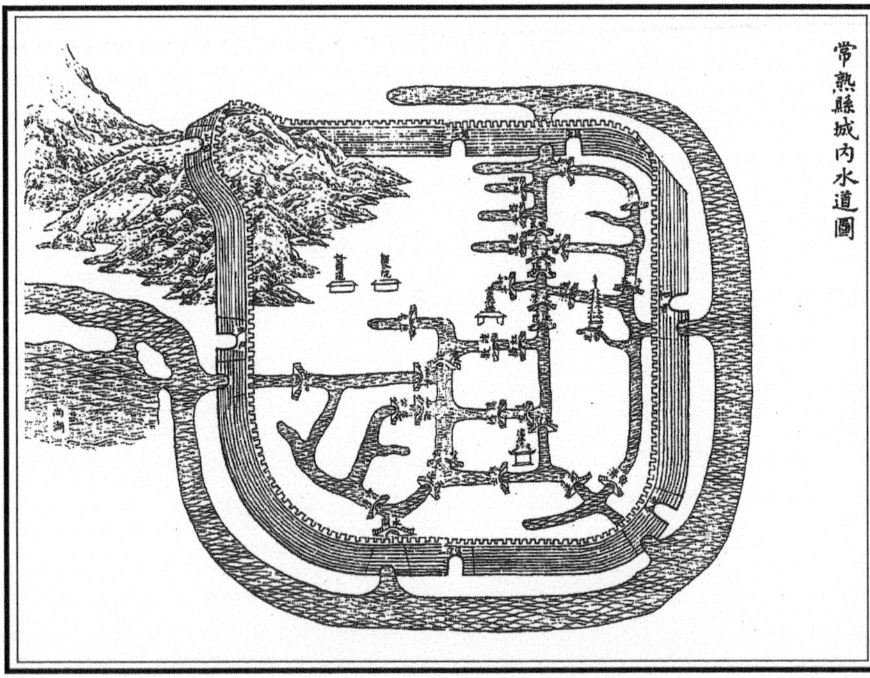

Fig. 2 Map of waterways within Changshu County Town. This illustration depicts the relationship between the natural landscape and the urban site, showing the city integrated into the regional landscape system. Portions of the low hills were incorporated into the town, while an artificially organized network of waterways extended both inside and outside the city walls. Image Source: Zhang Guowei (Ming Dynasty 1368–1644 CE), History Section, Geography Category, Waterways Subcategory, *Wuzhong Water Conservancy Compendium*, Volume 1, Explanatory Notes on the Map of Waterways Within Changshu County Town

inextricable from the development of the regional landscape system upon the foundation of this regional landscape system, comprehensive coordination meant that the Shanshui environment and artificial intervention jointly shaped the evolution of urban form (Fig. 2).

2.1 The Ancient City of Suzhou City: Aligning with the Water's Course and Constructing the City Wall

In the ancient city of Suzhou, the corners of the city wall were not constructed at right angles in accordance with conventional urban form in ancient China, but were instead designed in an arc. The primary reason for this design was to take into account the outer river channels surrounding the city, thereby avoiding damage to the city wall from the rapid currents. At the same time, this curvature facilitated the smooth flow of water and enabled the passage of boats.

In the ancient city of Suzhou City, the construction of the city gates differed from other, more regular cities: their locations were coordinated with the alignment of the waterways, each placed on a principal channel and equipped with both water and land gates, yielding an asymmetrical arrangement.

2.2 Ancient City of Shaoxing: Zhongshan as Boundary, Mountain–City Integration

Constructed during the Spring–Autumn Period, the ancient city of Shaoxing was laid out upon the "Nine Hills of Yue." To the northwest, Zhong Hill—the highest of the nine served as the boundary, and no city wall was erected there, yielding a "one-round, three-square" boundary configuration in the city's urban form.

The boundary of the city, while constituting a defensive system, also created an enclosed, inward-looking urban space. In Shaoxing, the mountain mass was adopted as the boundary and no city wall was erected along that edge, thereby integrating nature into the city. Drawing upon the mountain's landscape resources for the construction of governmental offices and garden precincts, this approach articulated a boundary-making concept that privileged scenery and embodied the ruler's principle of following (or learning from) nature.

2.3 Ancient City of Hangzhou: Well-Digging for Water Management, with Lake–City Integration

During the reign of Emperor Dezong of the Tang (779–805 CE),[5] Li Mi[6] established six wells,[7] bringing water from West Lake into the city so that residents were freed from the hardship of brackish water. In the second year of Changqing (822),[8] Bai Juyi[9] dredged West Lake and its surrounding channels and enhanced the lake environment. From the Southern Song (1127–1279) through the Qing (1644–1912), the lake was repeatedly dredged and regulated as part of sustained water management, thereby achieving a durable urban water supply; the water environment was improved, and the city and the lake came to exist in integrated symbiosis (Fig. 3), through a series of targeted artificial intervention measures.

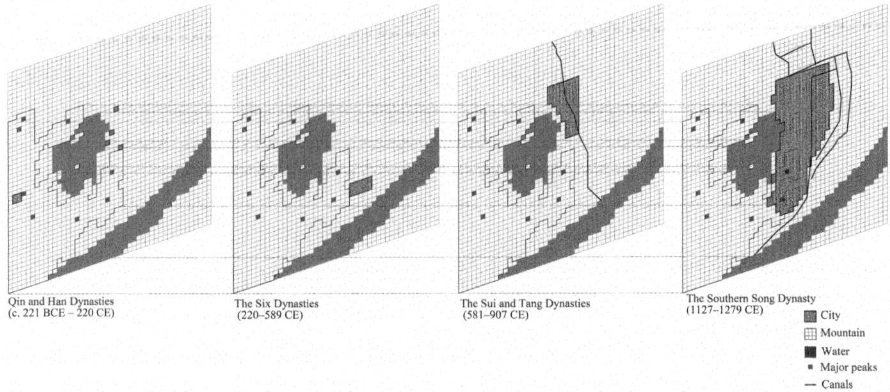

Fig. 3 Schematic diagram of the evolutionary relationship between Hangzhou City and West Lake

2.4 *Ancient City of Nanjing: Following the Shanshui Pattern, Urban Construction Produced a "Pin"-Shaped City Structure Defined by the City Wall*

Nanjing is likewise a paradigmatic city with an irregular boundary. Celebrated as the "land of the Six Dynasties and imperial Jinling," it lies in the Ning–Zhen–Yang hilly region at the lower reaches of the Yangtze River, adjoining the Yangtze River Delta to the east, backed by the hills of southern Anhui to the west, facing the Taihu water network to the south, and linking with the Jianghuai Plain to the north; the Yangtze cuts across the city's northwest. The Ming-period Nanjing, following the shanshui pattern, developed an outer enceinte approximating the "pin" character. To the northeast, the boundary was defined by Xuanwu Lake near the southwestern foot of Zhong Mountain, while to the north it stretched to the Yangtze's bank, reaching Lulong Mountain; within the city wall were enclosed Mount Bazi, Mount Wutai, Mount Shiren, Mount Jilong, and Mount Fuzhou. To the southeast, the perimeter incorporated the Qinhuai River, thereby taking in the Southern Tang capital of Jinling together with the Six Dynasties' Jiankang and the Eastern Prefectural City, forming the largest and most robust city in Nanjing's history. This sinuous and markedly irregular urban form and city structure has persisted to the present, exemplifying the principle advocated in the *Guanzi*[10]: "Follow heaven's materials, comply with earth's benefits, hence the inner and outer city-walls need not accord with the compasses and square, nor its roads with the level and marking line [3]."

3 Following Natural Endowments and Harnessing Strategic Landscape Advantages, Integrating Statecraft with Urban–Rural and Regional Landscape Systems

In ancient cities, the internal spatial structure was primarily composed of three main functional zones: yamen, commercial markets, and residential quarters. Outside the city, there were market towns, settlement, farmland, and temples. The construction of each functional zone often relied on the natural environment and the water network environment that had been modified through human intervention, and these zones influenced one another. Inside and outside the city, the artificial environment and the natural environment coexisted harmoniously. Each functional zone was organically integrated into the regional landscape system, forming a Shanshui city (mountain-and-water city) that realized urban–rural integration, thereby promoting the long-term prosperity of the city (Fig. 4).

3.1 Adapting to Mountainous Terrain and Constructing the Yamen

In ancient China, towers, halls, and official premises (referring to large-scale buildings such as multi-story towers, main halls, official halls and offices, which served administrative, ceremonial, and public functions within the city) were represented by the yamen [4]. The yamen (referred to in the *The Rites of Zhou (Zhouli)*[11] as guanfu, in the Han dynasty as guansi, and after the Tang dynasty as yamen, gongshu, gongxie, or yamen gate) was the place where officials conducted administrative and judicial affairs. It was the highest-ranking building complex within a city. In the construction of ancient cities (unrecorded in glossary), yamen were often built by relying on mountain bodies, hills, or other elevated and dry terrains, or adjacent to lakes. From a defensive perspective, mountains and water served as natural barriers (unrecorded in glossary) for the city. From a landscape perspective, mountains and water possessed high-quality scenic resources: elevated and dry mountain terrain offered expansive views, facilitated the formation of control axes and focal points, and guided the development of the urban spatial form. In Chinese history, government office districts in cities were often located adjacent to lakes—examples include the Xiyuan Sanhai area in Beijing, the foothills of Phoenix Mountain beside West Lake in Hangzhou, and the vicinity of Yuehu Reservoir in Ningbo—where landscape and hydrology were integrated to foster water-oriented city development.

(1) Shaoxing Ancient City: Occupying Zhongshan to Guide the City Spatial Structure

The Shaoxing Ancient City had its yamen district consistently located in the Zhongshan area at the northwest corner. Because Zhongshan resembles a giant dragon coiled above the plain—with steep terrain on the north side and a broad, gentle slope

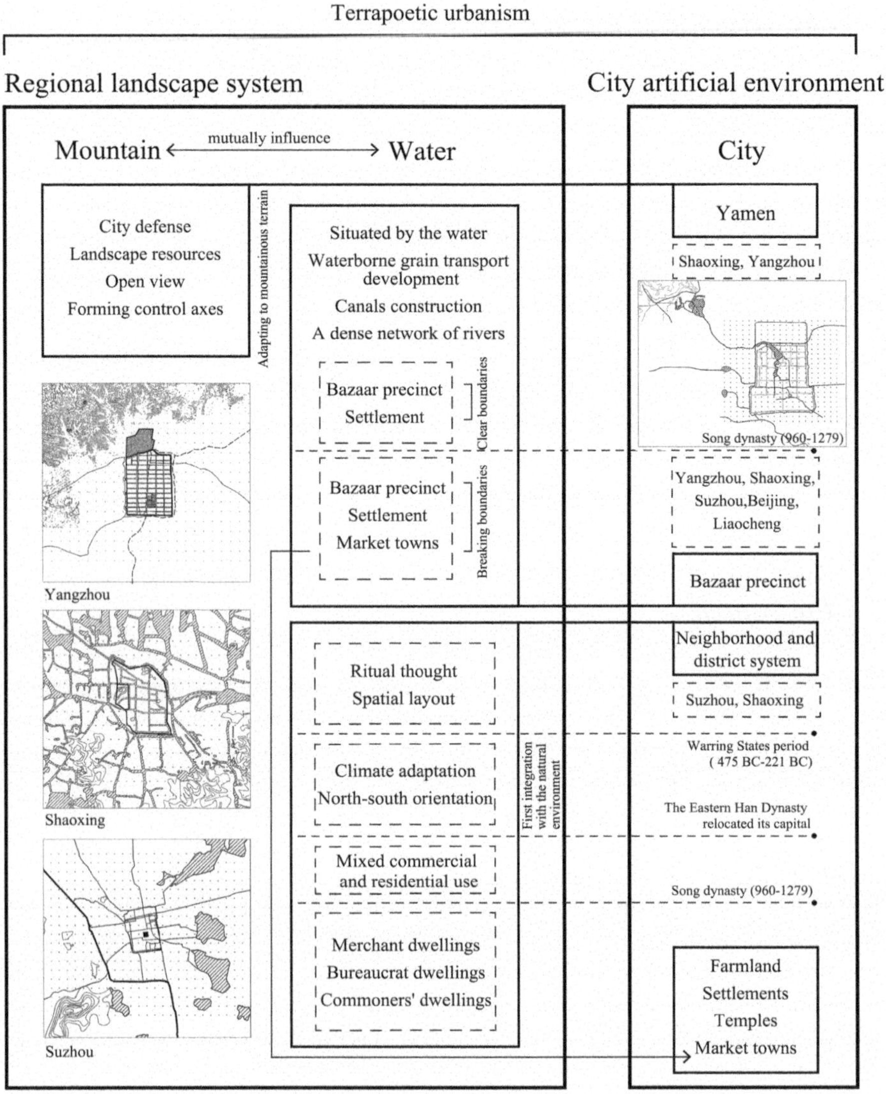

Fig. 4 Schematic diagram of the relationship between regional landscape systems and urban zoning construction

on the south—it was regarded as the optimal site for constructing a palace. Fan Li[12] built a small city to the south of the mountain, erecting a palace platform on the gentle foothills. Since then, the Zhongshan area has served as the administrative center of Shaoxing across successive dynasties, housing the seats of prefecture and state administrations.

(2) Yangzhou Ancient City: Utilizing the Terrain of Shugang, Constructing a Zicheng and Building the Yamen

From the construction of the Jiangdu Palace by Emperor Yang of the Sui[13] to the establishment of the Yangzhou Grand Commandant's Office, the Huainan Jiedushi yamen, and the prefectural and county offices in the Tang dynasty, the Yamen District of Yangzhou was consistently located atop Shugang. The formative pattern of the Zicheng of Yangzhou during the Sui (581–618) and Tang (618–907) Periods began in the tenth year of the Kaihuang era (590 CE), when Emperor Yang served as Governor-General of Yangzhou and stationed at Jiangdu, constructing the Jiangdu Palace on Shugang and placing the yamen within it. In the early Tang Dynasty (618–712), the administrative seat of Yangzhou continued to use the former site of Jiangdu Palace, thereby laying the foundation for the subsequent development of the city [5].

3.2 Commercial District, Situated by the Water, Benefiting from the Convenience of Waterborne Transportation

The development of a city is inseparable from commercial and economic activities, the growth of commerce depends on the prosperity of waterborne grain transport, and the advancement of waterborne grain transport also stimulates urban prosperity. In ancient China, the evolution of commercial districts can roughly be divided at the Song Dynasty (960–1279): before the Song Dynasty, the market and residential quarters were strictly separated, with fixed trading venues; during the Song Dynasty, the boundaries between market and residential quarters were broken, allowing open trading venues to form along streets, and the grass market (caoshi)[14] became widespread. This was accompanied by the rise of market towns, closely related to the construction of the Grand Canal and the southward shift of the economic center. In cities with extensive river and water network systems, the distribution of commercial districts was often concentrated on both sides of waterways, as seen in the city structure of Luoyang City and Yangzhou City during the Sui and Tang periods, and Shaoxing City in the Southern Song Period.

(1) Yangzhou: Canal Rerouting and Transformation of the Commercial District

During the Sui and Tang Periods, Yangzhou City was equipped with a Great Market and a Small Market,[15] both located at the central position of the Luocheng[16] and established along the city river. The reshaping of the water network system prompted the emergence of a second type of commercial form: people, relying on sites adjacent to the canals, would temporarily set up trading venues—such as the official street market at bridgeheads and along streets within the city, and the periodic country fair by the canals outside the city.

In the second year of Baoli (826 CE), when Wang Bo served as the Salt and Iron Transport Commissioner, due to the fact that "the official river within Yangzhou City was shallow, and during drought seasons grain-transport ships became stranded," a canal was excavated at Qili Port to the west of Changmen Gate in the city's south, meandering eastward to connect with the old official river at Chanzhi Temple Bridge. This newly dug waterway was deeper and more navigable, leading to a flourishing of waterborne traffic. From then on, the official river within the city gradually lost its shipping function, while the canal encircling the southeast and eastern side of the Luocheng became the main navigation channel. Along the canal east of the city, a periodic country fair gradually emerged.

(2) Shaoxing, Situated at the Confluence of the Fu River and the Canals, Witnessed Remarkable Commercial Prosperity

As a water network city, Shaoxing City experienced remarkable prosperity in the development of commercial districts driven by waterborne grain transport, and on multiple occasions served as an economic and political center. During the Southern Song period, historical records indicate that more than a dozen commercial districts were distributed within Shaoxing City, bringing commerce to an unprecedented peak. These numerous commercial districts formed an extensive commercial network inside the city, with most located along the Fu River and the eastern Zhejiang canals.

The prosperity of waterborne grain transport and commercial districts greatly stimulated the development of surrounding landscapes. Initially, the practice of constructing imperial roads and planting willows along both sides of river channels created land-based transportation corridors with roadside greenery. Subsequently, privately built yet publicly accessible public-oriented commercial gardens associated with tea and wine establishments emerged, integrating leisure and social functions into the urban economic and cultural fabric.

This series of landscape greening improvements carried out around the canal and water network significantly improved the habitability of the surrounding environment and strengthened the connections among various functional districts within the city [6]. Canal-front cities established wharves along the canal to facilitate communication between the interior and exterior, serving purposes such as cargo loading and unloading, passenger embarkation and disembarkation, commercial trade, and provision replenishment. Scenic landscaping was developed along the riverbanks, making them attractive venues for visitors to gather and socialize. Meanwhile, wharf areas were equipped with facilities for recreation, commerce, services, and storage, thus forming in antiquity a well-functioning urban hub centered on water bodies.

(3) Suzhou City Changmen Gate: Adjacent to the Canal, Creating the "Prosperous Changmen Gate"

Suzhou City, historically known as a watertown within an extensive wetland region, relied fundamentally on water transport as an indispensable mode of communication and exchange. Following the opening of the Grand Canal, its waters were directed into the city, making Changmen Gate a crucial nodal point for the entry and exit of canal traffic. As the western city gate of the ancient city of Suzhou City, Changmen Gate capitalized on its proximity to the canal and its advantageous transportation accessibility. The surrounding Changmen district consequently developed into one of the most prosperous commercial districts in the country, celebrated in historical records with the epithet "Prosperous Changmen Gate," symbolizing the zenith of economic vibrancy and cultural flourishing in relation to the canal-based urban economy.

During the Ming (1368–1644) and Qing (1644–1912) Periods, the re-dredged and reconnected Grand Canal once again became the principal artery of water transport. Flowing past the western side of the ancient city of Suzhou City, the canal passed through Changmen Gate in the northwest, a strategically significant location where five waterways converged, ensuring highly convenient water–land transportation. Most goods entering and leaving Suzhou City were concentrated at the Nanhao Wharf outside Changmen Gate, where myriad merchants anchored their vessels. This thriving exchange not only stimulated the prosperity of the Changmen district but also contributed to the overall economic and cultural flourishing of the entire city of Suzhou City.

(4) Shichahai in Beijing, Directly Connected to the Grand Canal, Served as a Vital Hub Where Both Grain Transport and Commercial Activities Flourished

The Shichahai area originated as a chain of lakes left behind after the course of the Yongding River was altered. In the Yuan Dynasty (1271–1368), under the leadership of Guo Shoujing,[17] a thirty-year program of hydraulic works was undertaken, which connected the Grand Canal and laid the foundation for the present-day river system and waterside landscape of Beijing. This also transformed Jishuitan into an inland port of the Grand Canal within Dadu City. In the urban planning of Dadu, following the principle of "palace in the front, market in the rear" from *Kao Gong Ji (The artificers' record)*,[18] the Shichahai district was designated as the commercial center. Consequently, the Shichahai area developed into a dual hub for both grain transport and commercial activities [7].

During this period, river transport and maritime transport operated in parallel, with Jishuitan functioning as a major port of Dadu. Grain transport vessels and merchant ships from across the country converged at the grain transport wharf in Shichahai, forming a vital hub of exchange. At the same time, the Xiejie market on its northeastern shore, together with the bustling area in front of the Bell and Drum Towers, constituted the most prosperous commercial district of the entire city, where merchants and traders gathered in great numbers. For a time, the Shichahai district experienced an unprecedented peak of economic vitality and urban vibrancy.

(5) Liaocheng: The Grand Canal Encircling the City, Shaping the "Metropolis North of the Yangtze River"

Liaocheng is located in the western plain region of Shandong, and its ancient city is also known as Dongchang Ancient City. Due to repeated breaches of the Yellow River and the resulting floods, the ancients excavated earth to form channels, creating a distinctive pattern in which lake waters encircled the city. In the Sui Dynasty, the Yongji Canal flowed past the western side of the ancient city. After the Grand Canal was re-excavated and dredged in the Yuan Dynasty, the Huitong Canal section ran north–south along the eastern side of Liaocheng Ancient City. Once the canal was opened to navigation, it became connected with the surrounding lake waters, forming the integrated water system around the ancient city.

During the Ming-Qing Dynasties, the Grand Canal served as the principal artery of north–south water transport, and the Dongchang Prefectural City was one of the nine major commercial ports along the ancient canal. Outside the city, the canal wharfs varied in size and were connected in a continuous belt. Among them, both the large and small wharfs were the most important sites for the aggregation and distribution of urban goods. Vessels plied back and forth, masts stood in dense array, merchants thronged like the tide, and merchant ships awaiting unloading shuttled continuously, stretching for several li, creating a scene of remarkable prosperity. Owing to the flourishing trade, a new, vibrant urban district emerged outside the city, closely dependent on the Grand Canal for its commercial vitality (Fig. 5).

3.3 Dwelling by Water: Constructing a New Residential Pattern

In ancient Chinese cities, residential districts for common citizens first appeared during the Warring States period, in the Qin capital of Xianyang, where the ideology of ritual system profoundly influenced the spatial distribution of dwellings. The relationship between residential space and the natural environment was first clearly manifested when the capital was relocated to Luoyang during the Eastern Han dynasty. To adapt to the northern climate, the residential space adopted a north–south spatial orientation, which effectively shielded against the northwesterly winds in winter and facilitated heating. In the Song Dynasty, the spatial layout of cities underwent a transformative change: commercial and residential functions were no longer strictly segregated, resulting in a more open urban environment and an unprecedented flourishing of urban civilization. People began to pursue more suitable living environments, and mountain-and-water environments gradually became significant factors influencing the spatial organization of residential areas.

Within the urban area, elevated and well-drained sites with convenient transportation and pleasant surroundings were predominantly occupied by the private residences of bureaucrats and members of the gentry. Along the Grand Canal and other major water transport routes, the dominant residents were merchants, while the dwellings of ordinary people were scattered throughout the city. In the flat regions crisscrossed by dense water networks, the influence of these networks was particularly evident in shaping the subdivision of urban streets and in defining the spatial boundaries of residential courtyard wings.

(1) Uzhou Ancient City: Dwellings Resting upon Rivers, Abundant Waterfront Ports, and Small Bridges

During the Tang-Song Dynasties, the dual checkerboard pattern of adjacent waterways and roadways (water–land adjacency, with rivers and streets running parallel) further refined the water network framework, ultimately forming a waterway system

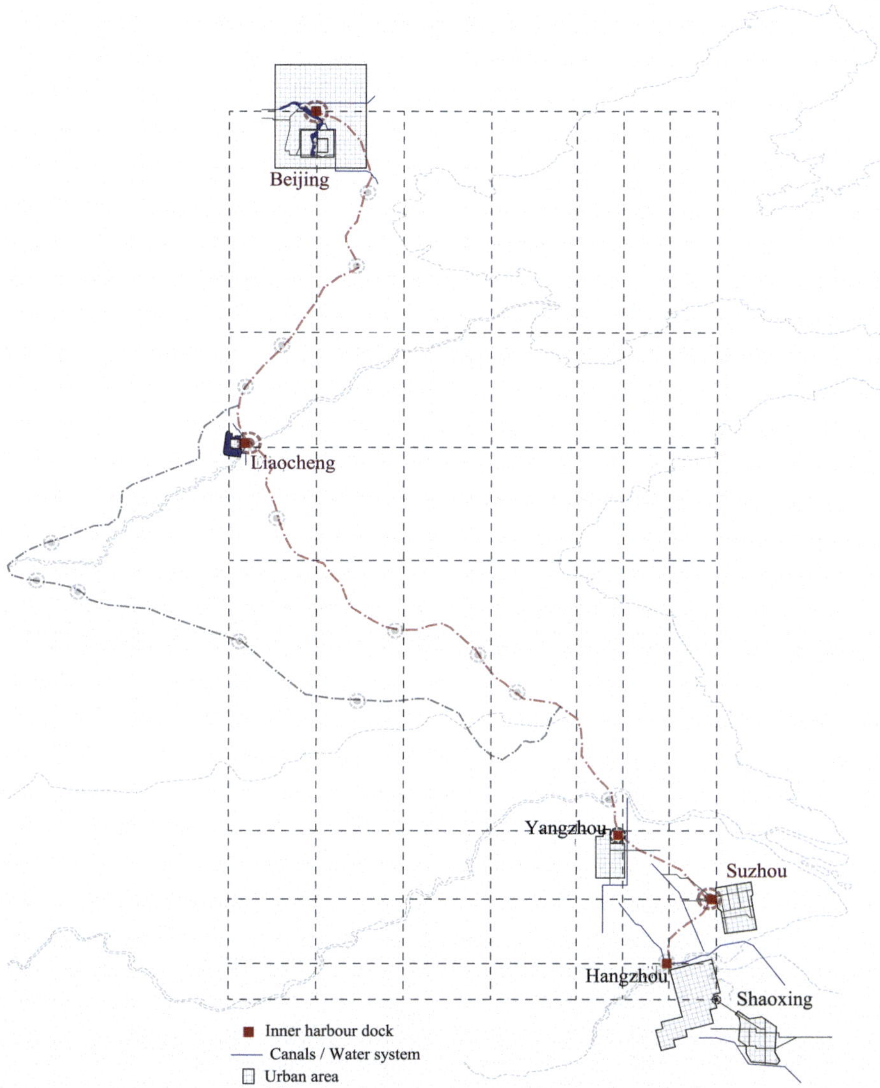

Fig. 5 Schematic diagram of the relationship between the city commercial district and the canal water network

characterized as "three horizontal, four vertical, and one ring." This configuration created the distinctive Jiangnan watertown landscape where "dwellings resting upon the river, with abundant waterfront ports and small bridges (Fig. 8) [8]."

(2) Shaoxing Ancient City: Where There Is a Temple, There Is Always a Hill Nearby; Where There Is a Household, Water Is Always Accessible

During the Southern Song period, Shaoxing exhibited a distinctive water network pattern comprising spatial configurations such as "one river flanked by one street," "one river flanked by two streets," and "a river without streets." The pervasive presence of the water network meant that almost every household was located along the water, with boats serving as a common means of daily transportation. This arrangement gave rise to a characteristic residential pattern, poetically captured by the adage: "Where there is a temple, there is always a hill nearby; where there is a household, water is always accessible [9]." Such a pattern became a defining feature of Shaoxing vernacular dwellings, in which the iconic scene of "small bridges, flowing water, and residential houses" vividly embodied the harmonious integration of urban spatial form, hydrological infrastructure, and the cultural landscape.

4 Following the Principles of Nature, with Refinement Lying in the Appropriateness of Form and Function

The construction of gardens and landscapes does not exist in isolation, but is closely interlinked with the shanshui environment and city functions. The prosperity of the city provided the economic and cultural foundation for the creation of gardens, while the Shanshui environment constituted the material basis for their formation (Fig. 6).

4.1 Garden Construction with City Functions as the Carrier

Yangzhou City: Three Parts of the Moon's Brightness, Two Parts Belong to Yangzhou

"Hangzhou City excels in its lake-and-mountain scenery, Suzhou City excels in its commercial districts, and Yangzhou excels in its garden pavilions." The origins of Yangzhou gardens can be traced to the Han Dynasty (206–220 CE), with renewed flourishing during the Tang Dynasty, and their zenith in the Ming-Qing Periods, earning the enduring acclaim that "the excellence of Yangzhou gardens is unrivaled under Heaven [10]." Just as the poetic imagery suggests that even the moon favors Yangzhou, during its peak the city boasted over two hundred private residential gardens. The immense wealth of the Yangzhou salt merchants provided abundant financial and material resources for garden construction. Among these were the lavish private residential gardens of the salt merchants, the public-oriented commercial gardens associated with tea and wine establishments designed primarily for leisure and entertainment, and the academy gardens devoted to reading and scholarly instruction. Architectural features such as thatched pavilions over water and perfumed, shaded corridors formed refined compositions; tea houses were often constructed along the rivers, with their south-facing facades supported on stilts overhanging the water—surpassing even the willow-lined riverbanks in charm.

Fig. 6 Relationship between residential forms and water networks in Suzhou City. *Image source* (Qing Dynasty, 1644–1912) Xu Yang, "Scene of Prosperity in Gusu," Sect. 11, Shantang Street

4.2 Landscape Construction of Scenic and Historic Sites Dominated by Low Mountains

In traditional China, scholars and literati often revered nature, especially picturesque mountain-and-water landscapes. Consequently, locations endowed with such natural beauty invariably became the most favorable sites for the construction of scenic landscapes and gardens. The landscape resources, material resources, and relative ease of human development in low mountain areas determined their status as the preferred shelters for early human habitation, thereby fostering the emergence of early settlement civilizations. The integration of artificial construction with the surrounding landscape can be regarded as the prototype of the earliest urban scenic sites. Many of these low-mountain and hilly regions subsequently became the sites of historically renowned detached imperial palaces and gardens, and in some cases, even served as the administrative centers of the state [11].

(1) Suzhou City: Exiting the Western Gate of the Ancient City, the Scenery Here is Uniquely Superb

The suburban areas of Suzhou City are endowed with abundant Shanshui resources, where diverse geological conditions have shaped a variety of distinctive Shanshui landforms. In particular, the low hill and hilly areas stretching westward from the ancient city toward Taihu Lake—although of modest elevation—are richly vegetated and often situated adjacent to bodies of water, thereby offering excellent visual and scenic effects. In the western low hill and hilly zones of the city, Hu Hill (Tiger Hill), He Hill, Lion Hill, Heng Hill, Shangfang Hill, Lingyan Hill, and Tianping Hill form an "encircling" configuration, spatially embracing the western urban periphery. Each of these sites, in combination with deliberate human development, has evolved into a renowned mountain scenic and historic area, infused with profound Suzhou City cultural characteristics, thereby creating peri-urban scenic and historic areas where scenery and culture are seamlessly integrated. Meanwhile, Qionglong Hill, Dongting East Hill, and Dongting West Hill, situated along the eastern shores of Taihu Lake, provide a vast, verdant backdrop linking the city to the foreground mountains.

Hu Hill (Tiger Hill) is a quintessential example of scenic area development in the suburban periphery of Suzhou City, located in the northwestern sector of the ancient city. Celebrated as the "foremost scenic and historic site in Wuzhong," it encompasses an area of just over 300 mu and rises to a height of slightly more than 30 m, yet it embodies the grandeur described as "the finest manifestation of the natural ravines east of the Yangtze." With its precipitous cliffs and varied landforms of remarkable Shanshui character, the site offers a multitude of visual impressions. The renowned literatus Su Dongpo[19] once remarked: "To come to Suzhou City without visiting Hu Hill is a matter of regret [12]." Within the context of the ancient Shanshui environment of the Suzhou City region, Hu Hill held extraordinary significance—not only as a model for peri-urban scenic and historic area development, but also as a critical spatial mediator linking the urban fabric with the surrounding natural environment, thereby deepening the integrative connections between city and hinterland.

(2) The Garden Complex of the Western Suburban Low-Mountain Areas in Beijing

The "Three Hills and Five Gardens"—a collective term in the history of classical Chinese gardens referring to the imperial palace and garden complex in the western suburbs of Beijing during the Qing dynasty—were successively constructed from the Kangxi reign (1662–1722) to the Qianlong reign (1736–1795). Since the Liao (907–1125) and Jin (1115–1234) Dynasties, the western suburbs of Beijing have been regarded as a scenic and historic area, with the eastern slopes of the Western Hills characterized by undulating peaks, a series of lakes, abundant springs, and a Shanshui environment evocative of the Jiangnan water towns. The Three Hills specifically refer to Fragrant Hill, Jade Spring Hill, and Longevity Hill, the latter two being branches of the Western Hills. These hills host Jingyi Garden, Jingming Garden, and the Summer Palace, along with the surrounding Changchun Garden and the Old Summer Palace, together constituting the scenic area of Beijing's western suburbs. The Three Hills and Five Gardens also fulfill an equally important role as an element of urban functions. Their enduring charm lies not only in the extraordinary peaks and ridges, but also in the cultural landscape formed through the integration of natural resources with historical and cultural heritage over the passage of time.

4.3 Scenic Landscape Construction Centered upon Lakes

In Chinese history, cities were often provided with lakes or reservoirs at the upper reaches of catchment areas, such as Kunming Lake in Beijing, West Lake in Hangzhou, and Moon Lake in Ningbo. Within the urban fabric, lakes served as crucial sources of domestic water supply and as environmental carriers: they were not only essential supports for human existence and urban development, but also key scenic elements in traditional gardens and scenic landscapes. Taking the lake body as the center, integrating scenery—that is, city development by water—perpetuated the city structure, thereby shaping urban landscapes where scenery and culture are intricately interwoven (Fig. 7).

Since ancient times, China has undertaken landscape construction around lakes. The earliest recorded large-scale project of excavating an extensive water body within a garden can be traced to the Pre-Qin Period (about 2100–221 BCE), as documented in historical records: "The water flows east into Li Lake [13]. Beside the lake stood the Zhanghua Terrace." The Zhanghua Terrace, encircled by mountains and embracing the lake, formed a waterside scene [14]. Subsequent landmarks in the history of lake landscape construction include the Taiye Pool of the Jianzhang Palace in the Han dynasty, the Qujiang Pool of the Furong Garden in the Tang dynasty, and the Kunming Lake of the Qingyi Garden in the Qing dynasty—each serving as a milestone in the evolution of lake-centered landscape design.

Among them, the lakes of multiple cities bear the name "West Lake." The widespread emergence of the West Lake as an urban landscape symbol is the crystallization of the evolution of Chinese ecological wisdom and the transformation of

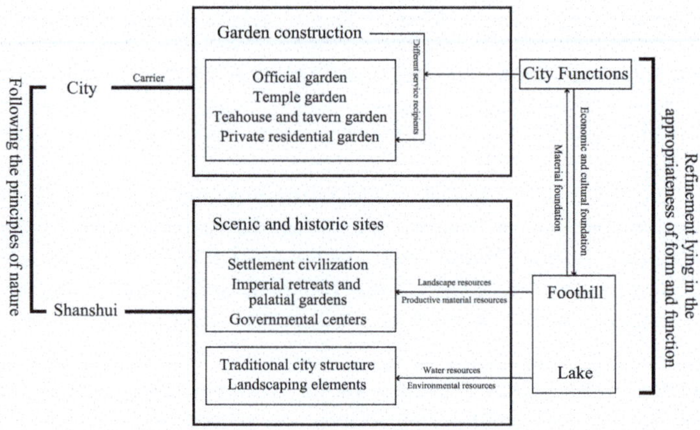

Fig. 7 Schematic diagram of mutual utilization and dynamic adaptation between artificial and natural environments supported by regional landscape systems

Fig. 8 Panoramic view of West Lake. *Image source* Dong Bangda (Qing Dynasty), Scroll of Ten Scenes of West Lake

urban civilization, embodying a holistic environmental view of "mountain–river–object–human." Historical records document more than eighty instances of the West Lake. An analysis of the similarities in the spatial relationship between city and lake, as well as the causes and processes of their formation, reveals that it represents a typical urban spatial pattern of "lake–city" integration characteristic of China's hydraulic society period. Professor Mao Huasong of Chongqing University has conducted in-depth and insightful research on West Lake culture; here, his existing findings are drawn upon to enrich this book's discussion of landscape construction around lakes [15].

The widespread presence of the West Lake began during the Wei and Jin Dynasties (220–420 CE), initially constructed with the lake body serving a protective function, under the premise of ensuring urban security and defense. After the Tang Dynasty, the construction of hydraulic facilities became the primary focus of West Lake projects, during which phenomena of landscape construction gradually emerged. Beginning in the Song Dynasty, the motivations for West Lake construction shifted toward landscape construction for the urban topographical scenery, a trend that continued into the Qing Dynasty examples include the formation of Zhangpu West Lake [16], Baoqing West Lake, Damingfu West Lake, Fuzhou West Lake, and Hangzhou's Ten Views of West Lake and Eighteen Views. Over the course of its development and consolidation from the Wei and Jin through the Two Song Dynasties, West

Lake culture established a relatively universal pattern in aspects such as diversified construction objectives, urban water system organization, and landscape architecture elements, reflecting the ecological wisdom of harmonious integration among human and nature, human and society, and human and landscape. This, in turn, gave rise to a typical urban landscape paradigm in Chinese history [15].

Hangzhou West Lake stands as the representative of West Lake culture, located in a shallow inlet near the estuary of the Qiantang River. In its early period, it was a marine bay, and the present-day site of Hangzhou's urban area was still a shallow tidal flat, with the earliest settlements situated at the foothills on the west side of the lake. Due to flooding of the Qiantang River, this inlet gradually transformed into a lake. With the combined effects of flooding and sedimentation, by the Eastern Han Dynasty (25–220 CE) West Lake had become separated from the river and the sea, forming a lagoon that gradually shifted from brackish to freshwater. The sedimented and reclaimed area eventually became the urban foundation site of Hangzhou. During the Tang Dynasty, Prefect Li Mi excavated six wells within the city, channeling water from West Lake into the city via bamboo tubes and ceramic pipes as a form of urban water-supply infrastructure, thereby meeting the drinking water needs of residents [17]. In the Changqing reign, Bai Juyi constructed a bridge–causeway between Solitary Hill and the city, which not only played an important role in urban hydraulic works but also contributed to the final shaping of West Lake's form. In the Sui-Tang period, with the flourishing of religion, religious activities became frequent around West Lake, leading to the construction of numerous gardens and buildings dominated by religious culture. Gradually, West Lake began to assume the character of a scenic site for leisure and sightseeing. By the Song dynasty, with political and intellectual transformations, urban civilization became increasingly sophisticated, and public gardens were widely constructed, furthering the landscape construction of the West Lake Scenic Area. From the Yuan through the Ming and Qing dynasties, Hangzhou West Lake remained a popular destination for recreation and tourism. Several dredging projects were carried out, with the dredged silt used to build causeways, thereby perfecting the construction of the West Lake Scenic Area. During the Qing dynasty, the emperors Kangxi and Qianlong visited West Lake multiple times; their efforts in both the governance of the lake and the construction of gardens greatly promoted its development. Their inscriptions and poems further enriched the cultural connotations of the West Lake landscape (Fig. 8).

Yuehu Lake in Ningbo, also known as West Lake, is situated in the southwest of the urban area of Ningbo City. It was excavated during the Zhenguan reign of the Tang Dynasty (627–649). In the seventh year of the Taihe reign (833 CE), Wang Yuanwei,[20] the magistrate of Mou County, undertook hydraulic works, "diverting the waters of Ta Mountain and constructing a dam at Jiangxi Stream," channeling water into the city to create the Sun Lake and Yuehu Lake, from which the populace greatly benefited. By the northern and southern Song dynasties, Ningbo had gradually developed into a prosperous city center and a strategic metropolitan hub near the capital. The city's water conservancy systems were successively dredged and improved, forming an urban water network system with Yuehu Lake as its core. Around the lake were extensively built pavilions, terraces, and towers, with seasonal flowers and trees planted throughout the year, creating the celebrated "Ten Islets Scenic Views of Yuehu Lake." Since the Song and Yuan Dynasties, Yuehu Lake has also been the academic center of eastern Zhejiang, attracting gatherings of literati and poets. On its sandbank islands and surrounding lands, a profound accumulation of cultural layers has been deposited, constituting a wealth of brilliant traditional culture (Fig. 9).

Since antiquity, the construction of cities in China has, in its own distinctive way, expressed profound respect for the Shanshui environment, pursuing sustainable development by respecting nature, transforming nature, and adapting to nature. From a regional perspective on urban construction, natural mountains and waters together with the water network system shaped through artificial intervention jointly constitute the regional landscape system upon which the city depends for development. Within this system, the natural environment and the artificial environment permeate and integrate with each other; the zoned construction of different urban functions and the region's natural elements after artificial intervention are mutually utilized and dynamically adapted. The urban form and the Shanshui pattern influence one another, shaping a livable environment with distinctive local characteristics.

The construction of a city is inseparable from the regional landscape system upon which it depends—its Shanshui environment. Therefore, urban construction should take into account the relationship between urban functions and the regional landscape system, and, through rational zoning, ensure that the city operates efficiently while achieving harmonious integration with the natural environment. During the process of urban expansion, it is necessary to maintain the integrity of the regional landscape system that integrates urban areas and suburbs, in order to realize resilient and sustainable development.

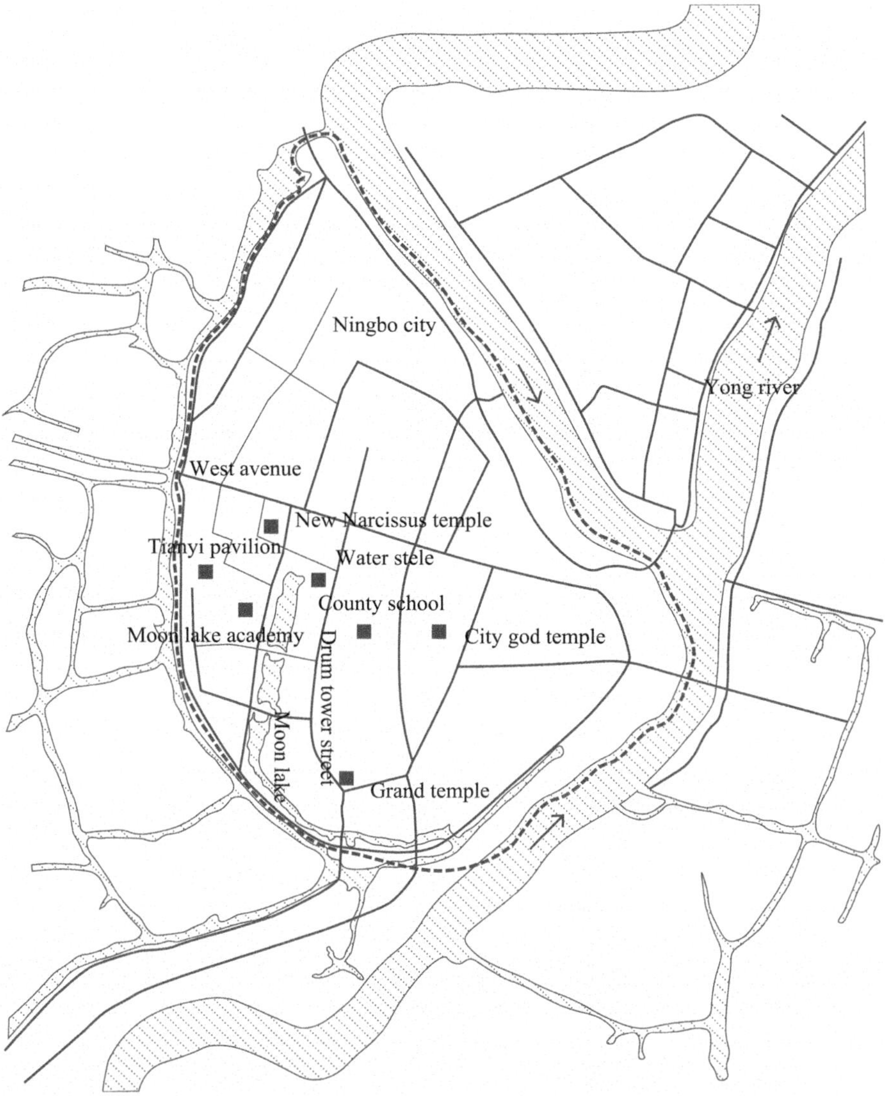

Fig. 9 Yue Lake, constructed through water conservancy development in Ningbo, served as the center of the city's water network system. Surrounding it were commercial districts, government offices, academies, and other establishments. Continuous garden landscaping transformed the area into the city's most prosperous region. Map adapted from: Yoshikazu Shibata, translated by Bu He, A History of Chinese Cities, Peking University Press, 2014 edition

Notes

1. *The Cheng Ma chapter of the Guanzi* (c. 350–26 BCE) discusses transport capacity and logistics as foundations of state governance, highlighting the role of mobility networks in territorial and economic management.
2. *The Du Di chapter of the Guanzi* (c. 350–26 BCE) addresses land measurement and classification as foundations of territorial and resource governance.
3. The Han Canal (Hanguo) is an early artificial canal constructed in the Spring and Autumn period to connect the Yangtze River with the Huai River system. As a precursor to the Grand Canal, it facilitated military movement, grain transport, and the emergence of canal-based urban and waterfront landscapes.
4. Fan Li (c. 536–448 BCE), a statesman of the Spring and Autumn period, is remembered for his strategic governance and later economic practice. His legacy reflects an early understanding of adapting human activity to terrain, waterways, and regional resources, linking political decision-making with environmental and spatial conditions.
5. The reign of Emperor Dezong of the Tang Dynasty (779–805 CE) was marked by fiscal reform, military decentralization, and pressures on urban and hydraulic systems. These conditions influenced the maintenance of cities, waterways, and administrative landscapes during the mid–late Tang period.
6. Li Bi (722–789) was a Tang-dynasty statesman and strategist whose ideas emphasized adaptive governance and coordination between central authority and regional conditions. His political thought provides context for understanding territorial management, infrastructure maintenance, and landscape governance in the mid–late Tang period.
7. The Six Wells (Liu Jing) were a set of public freshwater wells built mainly during the Tang–Song periods to supply urban drinking water and complement surface water systems.
8. The second year of the Changqing era (822 CE) falls in the mid-Tang period, when urban and water-management systems were still functioning but increasingly shaped by local administration.
9. Bai Juyi (772–846) was a Tang-dynasty poet and local official whose writings and administrative actions engaged closely with urban life, waterways, and gardens. His work provides valuable insight into the management and cultural perception of cities and landscapes in mid-Tang China.
10. *The Guanzi* (c. 350–26 BCE) is a Warring States text that discusses governance through land, water, and resource management.
11. The Rites of Zhou, compiled in the late Warring States period (c. 300–50 BCE), outlines an idealized bureaucratic system that includes principles for city planning, infrastructure, water management, and spatial order. It is a foundational text for understanding early Chinese concepts of institutionalized landscape and territorial governance.
12. Fan Li (c. 536–448 BCE) was a statesman of the Spring and Autumn period whose political and economic practices emphasized adapting human activity to terrain, waterways, and regional resources. His legacy provides early insight into integrated land–water management and territorial governance.
13. Emperor Yang of the Sui Dynasty (569–618; r. 604–618) initiated large-scale canal construction that unified north–south waterways, reshaping transport networks, urban centers, and regional landscapes. His projects laid the infrastructural foundation for later imperial water-based urban systems.
14. A grass market (caoshi) is an informal market located outside city walls, typically along roads or waterways, linking urban and rural exchange.
15. Great Market and Small Market form a hierarchical market system, organizing commercial activity at citywide and neighborhood scales.
16. An outer city wall (luocheng) refers to an additional defensive and administrative wall built outside the inner city, enclosing expanded urban areas such as suburbs, markets, waterways, and agricultural land. It reflects stages of urban growth and the integration of defense, circulation, and peri-urban landscapes in premodern Chinese cities.

17. Guo Shoujing (1231–1316) was a leading Yuan-dynasty scientist and hydraulic engineer. Beyond calendar reform, he played a key role in urban water supply and river management for the Yuan capital, exemplifying integrated planning of astronomy, hydrology, and city infrastructure in premodern China.
18. *Kao Gong Ji (The artificers' record,* ca. 770–221 BCE), compiled in the late Warring States period (c. 350–100 BCE) and incorporated into the Rites of Zhou, is an early Chinese technical treatise outlining standards for city construction, infrastructure, craft production, and spatial order. It provides foundational insight into institutionalized planning and environmental management in early China.
19. Su Shi (1037–1101), a Northern Song scholar-official, articulated an influential view of landscape that integrates natural scenery, urban space, and human emotion. His writings and civic activities provide key insights into cultural landscape perception and urban–water relationships in Song China.
20. Wang Yuanwei was a Tang-dynasty local official active in the mid–late Tang period (eighth–ninth century). Figures like him were responsible for everyday territorial administration, including water management and urban maintenance, and thus played a practical role in shaping local landscapes beyond imperial planning.

References

1. Wang, D. (2002). *Chinese landscape culture and urban planning.* Southeast University Press.
2. Wang, H. (2016). *Yangzhou canal world heritage.* Nanjing Normal University Press.
3. Yang, J. (2004). The unique charm and preservation of the ancient urban pattern of Nanjing. *City Planning, 12,* 41–46.
4. Guo, J. (2013). The Ming–Qing bureaucratic custom of "officials not repairing their offices". *Evening Digest, 23.*
5. Li, T. (2002). *Studies on the history of Yangzhou in the Tang dynasty* (p. 10). Jiangsu Ancient Books Press.
6. Du, M. (2014). *The transformation of Yangzhou gardens: People and landscape.* Tongji University Press.
7. Hou, R. (1997). On the planning and design of the Yuan Dynasty Capital (Dadu). *City Planning, 3,* 10–13.
8. Xu, S., Lei, Q., & Zhu, J. (2010). *Suzhou geography* (p. 11). Guwuxuan Press.
9. Zhang, J. (Tang Dynasty). *Seeing Zhu Qingyu off after passing the imperial examination: Complete Tang poems* (Vol. 384).
10. Li, D. (2014). *Records of Yangzhou's painted boats.* China Pictorial Publishing House.
11. Zhu, G. (2009). *Research on the narrative urban space of Suzhou's old city.* Central South University.
12. Huang, W. (2010). Tiger Hill—The first scenic wonder of Wuzhong. *China Place Names, 1,* 26–27.
13. Li, D. (Northern Wei). (2013). *Commentary on the water classic.* Collated by Chen Qiaoyi. Zhonghua Book Company.
14. Chai, S. (2016). *Research on the landscape art of traditional lakes in Zhejiang.* Zhejiang A&F University.
15. Mao, H., Du, C., & Chen, X. (2014). The ecological wisdom of "West Lake Culture" and its contemporary significance. *Landscape Architecture, 6,* 59–63.
16. Wu, X. (2014). *The historical development and aesthetic interpretation of "West Lake" culture.* Chongqing University.
17. Lü, Y. (1987). A study on the historical evolution of Hangzhou. *Journal of Hangzhou University, 17*(4), 10–18.

Part IV
City and Regional Development within the "Regional Landscape System"

Chapter 7
Region: Integration of Lakes and Cities—The Landscape Evolution of the Three Major Lakes in the Lower Reaches of the Qiantang River and Urban Development

Abstract This chapter examines the three major lakes in the lower reaches of the Qiantang River Jianhu, West Lake, and Xianghu and analyzes their landscape evolution in relation to the urban development of Shaoxing, Hangzhou, and Xiaoshan. By comparing their shared motivations for lake construction, modes of hydraulic intervention, lake–city scale relationships, and divergent trajectories of management and transformation, the chapter demonstrates how sustained historical interventions in hydrological systems reshaped the regional landscape system and, in turn, structured urban form and regional organization. The differing outcomes of the three lakes reveal that moderate and continuous human intervention was central to maintaining lake–city coexistence, highlighting the role of the regional landscape system as a mediating framework integrating ecological, natural, and cultural processes in traditional Chinese urban development.

The landscape evolution of the three historically renowned lakes in the lower reaches of the Qiantang River—Jianhu, West Lake, and Xianghu—has been closely related to the morphological formation of the three cities of Shaoxing, Hangzhou, and Xiaoshan. By analyzing both the commonalities and differences in the dynamic development processes of these three lakes and the three cities, it is possible to demonstrate how the intervention of hydraulic infrastructure in historical periods altered and reshaped the regional landscape system, thereby advancing the patterns of urban development.

The construction of the three major lakes shared similar motivations, namely the response to challenges of the living environment, and all belonged to large-scale hydraulic projects led by the government. Each was integrated with its respective city to form the distinctive mountain–lake–city landscape pattern. However, their subsequent development and transformation diverged, due to differences in the types of lakes themselves, their methods of construction, and the varying scales and degrees of interdependence between the lakes and the cities. These factors became critical to shaping the city–lake coexistence system. The landscape evolution of the three major lakes, together with urban development, illustrates how moderate and appropriate human intervention gave rise to the regional landscape system, integrating ecological,

X. Zheng, *Terrapoetic Urbanism*, https://doi.org/10.1007/978-981-95-8676-9_7

natural, and cultural elements. This system functioned as a principal medium for city building, thereby advancing regional urbanization and sustainable development.

1 Harmonizing Humans and Heaven—Hydraulic Infrastructure and the Lake Landscape Evolution in the Lower Reaches of the Qiantang River

The alluvial fan in the lower reaches of the Qiantang River, namely the Northeastern Zhejiang Plain (including the Hangjiahu Plain on the north bank and the Ning-Shao Plain on the south bank), is one of the most renowned Jiangnan watertowns and fertile "lands of fish and rice" in China. Historically, this region has been characterized by economic prosperity, cultural flourishing, urban vibrancy, and picturesque scenery. In earlier times, the area was extensively interlaced with rivers and lakes, among which the three major lakes became nationally celebrated for their landscape, culture, poetic references, and legends, as well as for their close interdependence with their respective cities—Shaoxing's Jianhu, Hangzhou's West Lake, and Xiaoshan's Xianghu (Fig. 1).

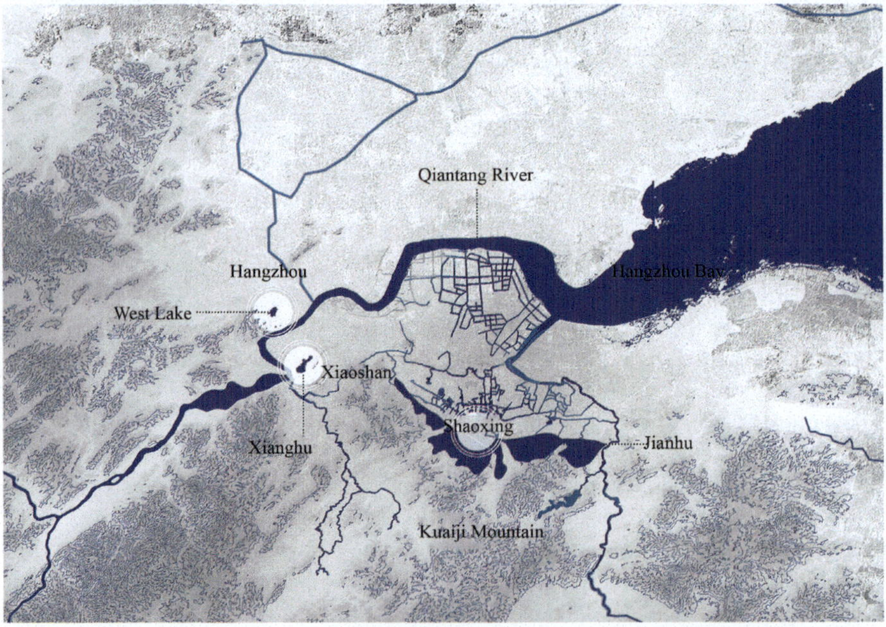

Fig. 1 Diagram of the geographical environment of the lower reaches of the Qiantang River and the locations of Shaoxing, Hangzhou, and Xiaoshan

In the Ming Dynasty (1368–1644 AD), *Zhang Dai's*[1] *West Lake Dreaming*[2] *(Xihu Mengxun)* opens with a vivid metaphor concerning the three great lakes, illustrating their broad renown and cultural influence throughout history: "Since Ma Zhen first opened Jianhu, from the Han dynasty (206 BC to 220 AD) to the Tang dynasty (618–907 AD) it was earliest to gain fame. Later, by the Northern Song dynasty (960–1127 AD), West Lake arose and eclipsed it; people all flocked to West Lake, while Jianhu's reserved elegance could no longer compare with the dazzling allure of West Lake. As for Xianghu, it lay secluded in desolation, with few boats or carriages arriving, thus the refined literati scarcely spoke of it." Here, West Lake, Jianhu, and Xianghu are not merely geographical entities but historical-cultural landmarks, their prominence reflecting the dynamic interplay of landscape, literary imagination, and urban cultural identity.

"My younger brother Yiru often compared West Lake to a beautiful woman, Xianghu to a recluse, and Jianhu to an immortal. I do not quite agree. I regard Xianghu as a maiden, lying in graceful shyness, as if still seen in her unmarried years; while Jianhu is like a noblewoman from a distinguished family, worthy of admiration but not of familiarity; whereas West Lake is like a celebrated courtesan of the pleasure quarters, resplendent in both voice and appearance, yet standing at the doorway offering a smile, to be flirted with by everyone. Precisely because all may trifle with her, all also yearn after her; and precisely because all yearn after her, all at the same time treat her with a certain levity."

"The immortal of reserved elegance, the courtesan of dazzling beauty, the recluse of desolate seclusion, or the maiden, the noblewoman, and the celebrated courtesan—these personifications of the three great lakes are all vivid, lifelike, and strikingly expressive. They also, from another perspective, reflect the historical periods of formation, the geographical location conditions, as well as the varying degrees of preference among different social strata for appreciating the lakes."

"While the three great lakes attracted numerous men of letters to compose celebrated poems, their processes of formation and transformation were closely intertwined with regional development. The lakes were not entirely natural in origin but were gradually perfected through continuous human interventions—constructing dikes, building dams, installing sluice gates, erecting bridges, dredging and desilting—while also adding pavilions, terraces, towers, vegetation, and trees along the shores to engage in landscape construction. They thus emerged as the outcome of sustained coordination between artificial intervention and natural processes."

"At the same time, Jianhu West Lake, and Xianghu, as important hydraulic infrastructures, enabled this region to possess, at least a millennium ago, a well-developed agricultural system that sustained high-yield rice cultivation to support the cities. This nurtured the development of Shaoxing, Hangzhou, and Xiaoshan (then situated in the hinterland between the two), creating an astonishing and enduring regional prosperity. It profoundly reflects the great works and wisdom of humankind in transforming and utilizing nature to improve the human settlement environment, and stands as a paradigmatic example embodying the concept of terrapoetic urbanism."

2 Transformation Alongside Water—Tracing the Landscape Evolution of the Three Great Lakes and the Trajectory of Urban Development

The three great lakes were able to stand out in the course of historical development primarily because their construction and existence transcended the mere significance of scenic sites. The evolution of the lake bodies and regional development—including the agricultural system and urban growth—formed a symbiotic relationship of mutual dependence. The three cities nurtured by the three great lakes remain prosperous and flourishing to this day.

2.1 Jian Lake and Shaoxing City

Shaoxing City is one of China's oldest city, serving as a regional center before the twelfth century. Jianhu Lake was also the earliest to be constructed (140 CE), flourishing during the Tang Dynasty (618–907 AD) and the Song Dynasty (960–1279 AD), but later gradually silted up. From the Tang dynasty onward, Hangzhou began to develop, with West Lake rising to prosperity; during the Southern Song Dynasty (1127–1279 AD), it replaced Shaoxing as the regional center, a shift historically referred to as the exchange of Hangzhou and Shaoxing. Xiaoshan, located in the suburban area between the two cities, saw the construction of Xianghu Lake in the Northern Song Dynasty (960–1127 AD). Together with Jianhu Lake, it belongs to the broader geographical setting of Kuaiji Mountain, and has long been regarded as the sister lake of West Lake.

Shaoxing lies to the south of Kuaiji Mountain, facing the Shanhui Plain to the north, extending as far as Hangzhou Bay in the lower reaches of the Qiantang River. Jianhu Lake (known as Jinghu Lake during the Tang Dynasty and as Jianhu Lake during the Song Dynasty was first constructed in the Eastern Han Dynasty (25–220 AD), located in the transitional zone between Kuaiji Mountain and the Shanhui Plain. Throughout the Jin (265–420 AD), Sui (581–618 AD), Tang, and Song Dynasties, and continuing into the Ming (1368–1644 AD) and Qing (1644–1912 AD) periods, the region around Jianhu Lake was recognized as one of the most celebrated scenic sites in the nation. Generations of literati visited the area and composed poetry, among them many distinguished figures of the Tang dynasty, who admired Jianhu and inaugurated the cultural tradition of urban lake culture in China. For instance, Li Bai wrote, "I wish to dream of Wu and Yue, flying across the moonlit waters of Jinghu in a single night."[3] At that time, the influence of West Lake was still less renowned than that of Jianhu Lake, as attested by Yuan Hongdao's remark in *On the Road to Shanyin*[4] *(Shanyin Daoshang)*: "Before the Six Dynasties, people did not hear of the beauty of West Lake."

(1) Seawater Invasion, Brackish Rejection and Freshwater Storage, and City Building upon Isolated Hills (Fig. 2)

In ancient times, the Shanhui Plain was submerged by the seawater surging in from Hangzhou Bay to the north and the floodwaters from Kuaiji Mountain to the south, rendering the area unsuitable for habitation. Early settlements were thus mostly located along the northern foothills of Kuaiji Mountain. By constructing sea dikes at the estuary to block seawater intrusion, the water quality in the plain gradually turned fresh.

During the Spring–Autumn Period (770–476 BC), the State of Yue, near the northern foothills of Kuaiji Mountain, built large and small urban settlements upon four isolated small hills on the plain, known as Isolated Hills, which served as the structural framework. This marked the formation of the ancient city of Shaoxing.

(2) Jian Lake, Stretching Across "Eight Hundred li," Vast and Magnificent, Fostered Urban Prosperity.

In the Eastern Han (25–220 AD), during the fifth year of the Yonghe reign (140 CE), Ma Zhen, the Prefect of Kuaiji Commandery, gathered waters from Shanyin and Kuaiji to form a lake. He constructed an embankment of over seven hundred li (1 li = 500 m), with the ancient city of Shaoxing as its center, dividing the embankment into eastern and western sections. This created a vast artificial lake whose water level stood higher than the surrounding plain, with a total area of about 200 km^2. Complementing the verdant peaks of Kuaiji Mountain, it earned the title of "The Eight Hundred Li of Jianhu" (Fig. 3).

After the construction of the embankment, a series of hydraulic facilities such as sluice gates (doumen), sluices (zha), weirs (yan), and culverts (yinguo) were subsequently installed along the embankment, and the seawall continued to be reinforced [1]. The formation of Jianhu protected Shaoxing city and the Shanhui Plain from the flood hazards of Kuaiji Mountain, restructuring the regional water network system and providing Shaoxing with safe and stable domestic water supply as well as agricultural irrigation for the plain areas. The Shanhui Plain was thus transformed from a vast expanse of open water into fertile land stretching for thousands of li, becoming a developed agricultural region, which further promoted the prosperity of Shaoxing as a water city economy (Fig. 4).

(3) Gradual Decline and Disappearance, Hangzhou-Shaoxing Transition

The construction of Jian Lake helped establish Shaoxing as the regional economic center. However, with the urban development and population expansion, there was an urgent need for newly reclaimed land to sustain the rapidly growing population, leading to a strained relationship between humans and the land. At this time, Jian Lake, covering nearly 200 km^2 in the suburban areas of the city, became an obstacle to urban expansion and agricultural reclamation. After the mid-Tang Period, the trend of filling lakes for land reclamation became more prevalent, and the silt from the mountain floods of Mount Kuaiji continuously obstructed the lake. Due to the

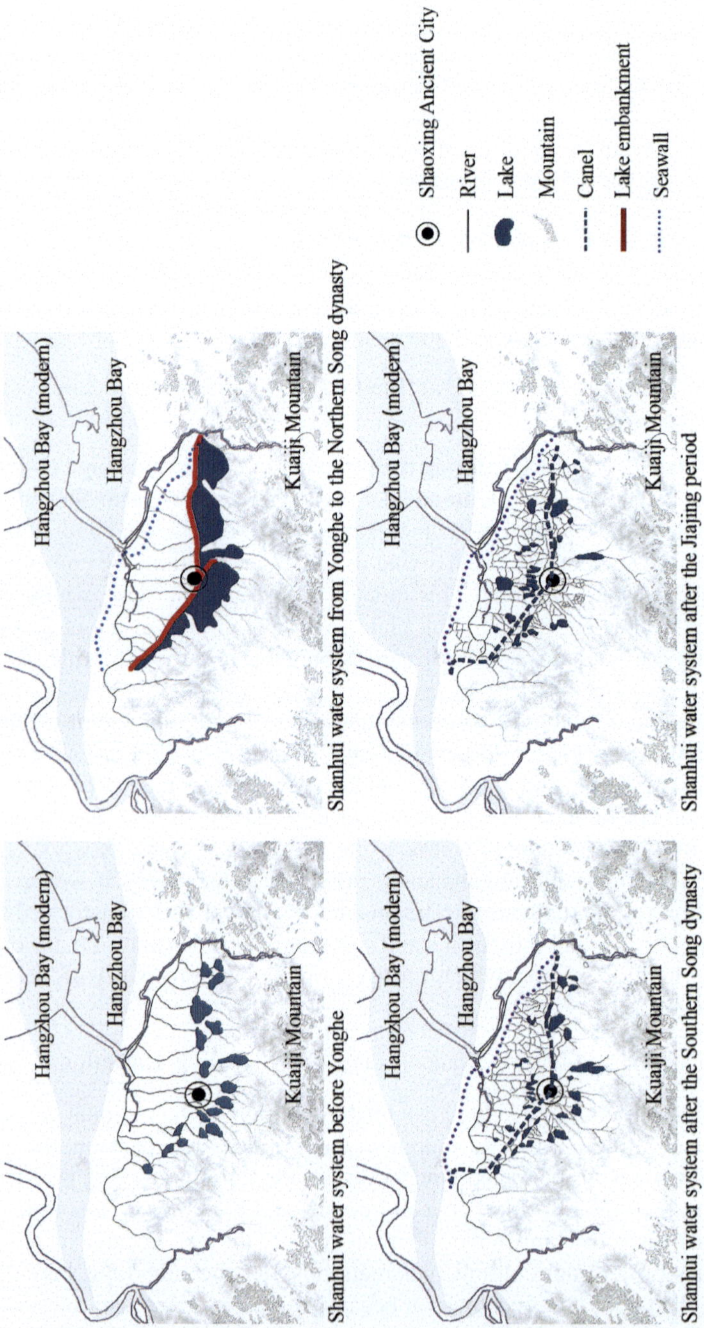

Fig. 2 Schematic diagram of the Jianhu Lake and the Shanhui Plain water system and the evolutionary process of its hydraulic infrastructure

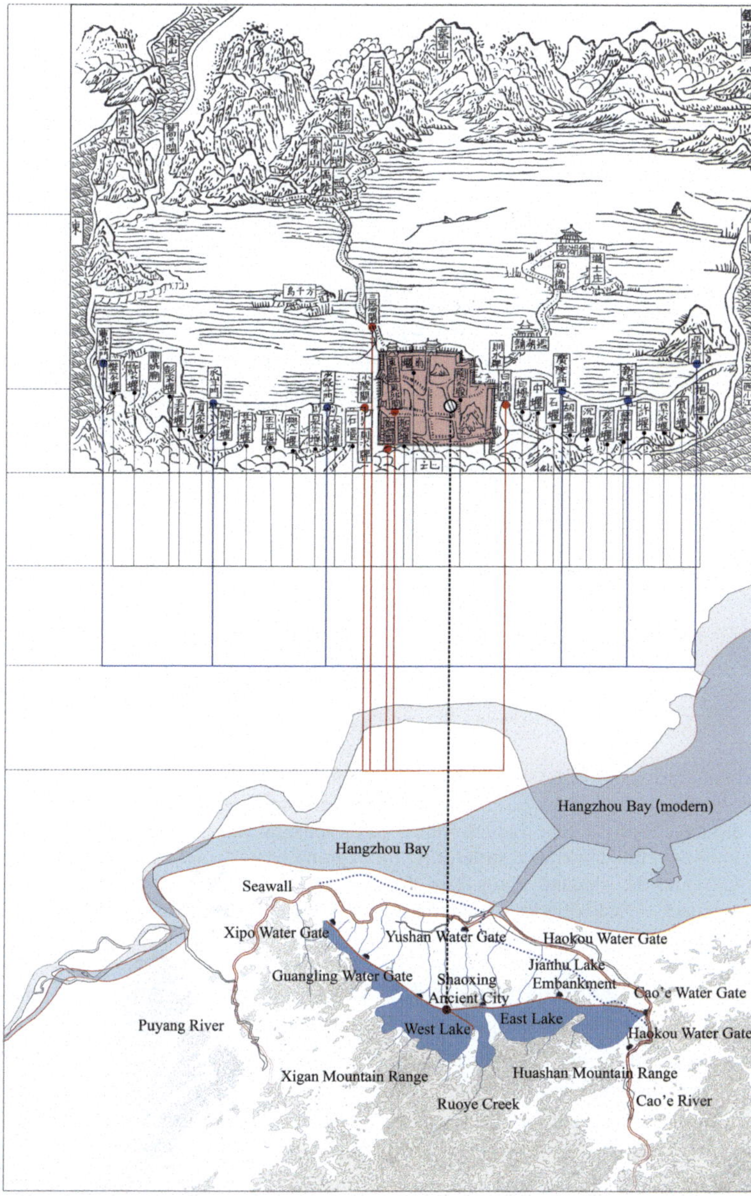

Fig. 3 Schematic diagram of Jianhu Lake and its hydraulic infrastructure during the Yonghe Period (136–141). During the Yonghe period of the Eastern Han Dynasty, embankments were built to impound water and form the large artificial lake Jianhu. Sluice gates, locks, weirs, and underground drains were installed along the embankment, restructuring the regional water network system and protecting Shaoxing City and the Shanhui Plain from flooding from the Kuaiji Mountains. Source: (Ming dynasty, 1368–1644) Gazetteer of Shaoxing Prefecture during the Wanli reign, Illustration of Jianhu Lake

Fig. 4 Schematic diagram of Jianhu Lake and its hydraulic infrastructure during the Yonghe period. During the Yonghe period of the Eastern Han dynasty, embankments were built to impound water and form the large artificial lake Jianhu. Sluice gates, locks, weirs, and underground drains were installed along the embankment, restructuring the regional water network system and protecting Shaoxing City and the Shanhui Plain from flooding from the Kuaiji Mountains. *Source* (Ming dynasty) Gazetteer of Shaoxing Prefecture during the Wanli reign, Illustration of Jianhu Lake

lack of large-scale dredging projects, Jian Lake gradually transitioned from a lake to land, eventually becoming lake fields. By the end of the Song Dynasty only a few small lakes remained, and by the Ming Dynasty, the remaining lakes and marshes had nearly vanished [2].

The renown of Jianhu reached its peak during the Tang and Song Dynasties, but as it became silted up by the end of the Song, the fame of West Lake began to rise. With Hangzhou becoming the capital city of the Southern Song, it gradually replaced Shaoxing as the regional metropolitan center, an influence that has persisted to the present day. Although the reasons for the "Hangzhou-Shaoxing Transition" are highly complex, the decline and eventual disappearance of Jianhu was a significant factor.

Today, to the southwest of Shaoxing, there still remains a relatively broad river channel known as the Keyan–Jianhu Scenic Area.

(4) Transformation into Farmland Through Siltation, Reconstruction of the Water Network System, and Urban Expansion Toward the Plain

After the siltation of Jianhu, the runoff from Kuaiji Mountain flowed directly into the Shanhui Plain where several large retention lakes emerged, thereby reconstructing the regional water network system and facilitating urban expansion toward the northern plain. Today, with Sang Lake in the plain region serving as the central green core, the system integrates three urban clusters, including the ancient city of Shaoxing, to form the greater Shaoxing metropolitan area.

2.2 *West Lake and Hangzhou*

Hangzhou is located on the northern bank of the Qiantang River, at the western edge of the Hangjiahu Plain. Since the Southern and Northern Dynasties (420–589 AD), the city gradually developed, and its growth accelerated especially during its role as the capital of the Wuyue Kingdom (907–978 AD) and later the Southern Song Dynasty. After the successive disappearance of numerous lakes such as Jianhu and Xianghu, West Lake has been preserved to the present day through continuous dredging and governance.

The evolutionary process of West Lake can be summarized as follows:

(1) Natural Lagoon and Shallow Sea Beach

West Lake is situated beside the estuary of the Qiantang River, within a shallow inlet. In its earliest stage, it was originally a bay, while the present-day location of Hangzhou city was still a shallow sea beach. The earliest settlements were concentrated along the western foothills of the mountains bordering the lake.

Due to the floods of the Qiantang River, this inlet gradually transformed into a lake. Moreover, through the combined effects of flooding and sedimentation, by the Eastern Han Dynasty (25–220 AD), West Lake had already been cut off from the river and the sea, forming a lagoon [3]. Over time, it transitioned from saltwater to freshwater, covering an area of approximately 5.6 km^2. The land formed through sedimentation later became the foundation site for the city of Hangzhou.

(2) Diversion of Water into the City, Lake–City Symbiosis

The urban site of Hangzhou was once characterized as "a former land of rivers and seas, with saline and barren soil, and brackish springs," which severely limited the availability of potable water and thus constrained urban development. During the Tang Dynasty (618–907 AD), Prefect Li Mi of Hangzhou excavated six wells within the city and constructed bamboo pipes and ceramic conduits to divert the waters of West Lake into the city, thereby establishing fundamental water-supply infrastructure and meeting the drinking water needs of its inhabitants [4].

As the primary source of potable water for the city, West Lake also functioned in regulating and storing floodwaters from the western mountains, as well as in irrigating farmland. This integration of hydrological management with urban needs provided the material foundation for the prosperity of Hangzhou.

(3) The Enduring and Monumental History of Dredging and Scenic Construction

With the economic and demographic expansion of Hangzhou, the tension in the relationship between people and land intensified, and West Lake was likewise threatened by land reclamation for farmland and by siltation. Successive local officials, fully aware of the significance of West Lake of Hangzhou, attached great importance to its dredging and regulation. The dredged silt was employed to reinforce sea dikes and lake embankments as a means of controlling floods from the Qiantang River, thereby increasing the water-storage capacity of West Lake and ensuring a sufficient potable water supply for the growing urban population.

At the same time, embankments were constructed, and islands were created, thus advancing the practice of scenic construction. Furthermore, with the completion of the Grand Canal, Hangzhou was integrated into the national arterial network of transportation and logistics, ultimately enabling it to surpass its regional counterparts and to emerge as a flourishing metropolis of Jiangnan (Fig. 5).

Continuous dredging has been an important artificial means of intervening in natural processes and maintaining the development of West Lake. Professor Wang Xiangrong of the School of Landscape Architecture at Beijing Forestry University has emphasized that one of the distinctive cultural features of Hangzhou lies in the history of dredging West Lake, which reflects the most essential relationship of enduring mutual adaptation and coordination between humans and nature, and also constitutes the foundation for the sustained development of urban culture.

2.3 *Xiang Lake and Xiaoshan*

Xiaoshan is located between Shaoxing City and Hangzhou City, with Xiang Lake situated to the southwest of Xiaoshan city on the southern bank of the Qiantang River. Its origin can be traced back more than 4000 years when it was formed from a gulf into a lagoon, originally known as Xicheng Lake. Subsequently, due to marshland evolution and human impact, it became silted up and disappeared, until it was restored as a lake through artificial construction during the Northern Song Dynasty, when it was renamed Xiang Lake.

The developmental stages of Xiang Lake can be summarized as follows:

(1) The Disappearance of Ancient Xiang Lake and Its Transformation into Farmland

Xicheng Lake (the ancient Xiang Lake) was gradually reclaimed as farmland due to the increasing population in Xiaoshan and the corresponding demand for cultivated land. In addition, floods from rivers and runoff from surrounding mountains carried sediment that caused progressive siltation, ultimately transforming the lake into a low-lying agricultural field.

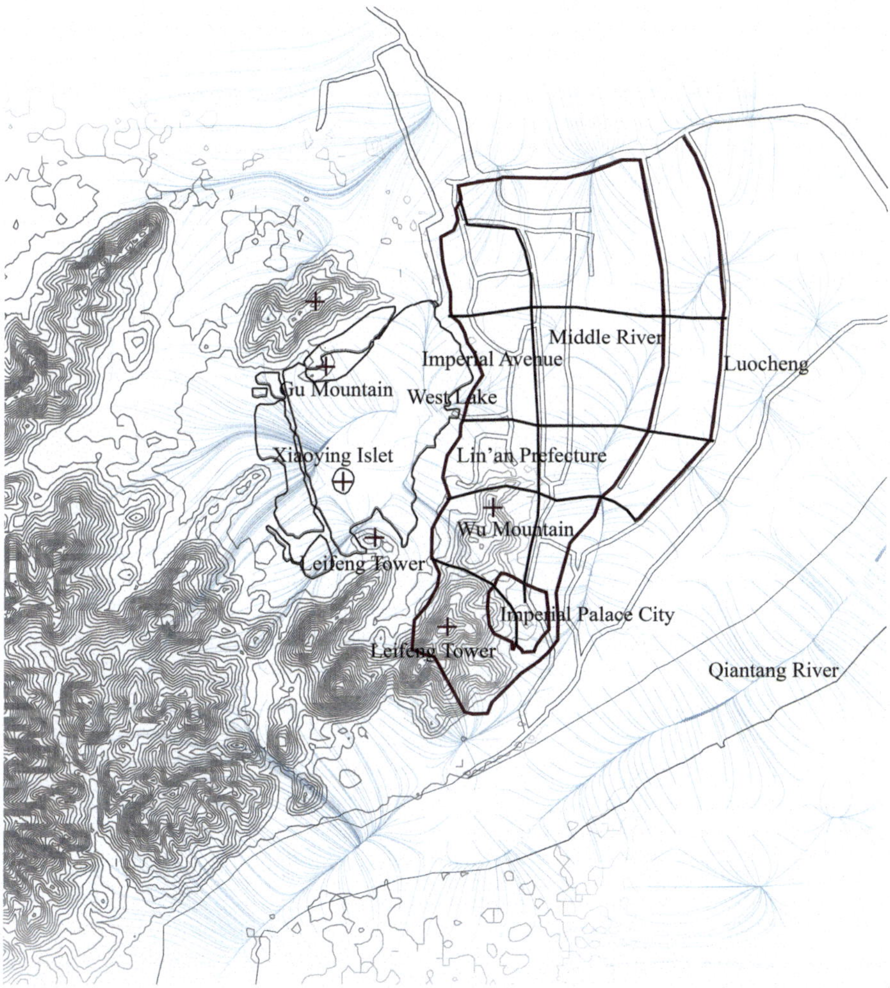

Fig. 5 Diagram of the main water system of West Lake and Hangzhou (Lin'an) during the Southern Song Dynasty (1127–1279)

(2) Artificial Restoration of the Lake and the Elimination of Water-Related Disasters

Before the Song Dynasty, during each flood season, mountain torrents surged rapidly into the city of Xiaoshan due to the topography sloping from south to north. Combined with the surging tides of the Qiantang River, the city often suffered from breaches of the embankments, resulting in repeated flooding disasters. The area around Xiang Lake was particularly unstable, frequently stricken by both floods and droughts. In order to resist these disasters and to ensure sufficient agricultural water throughout the year, the original site of Xiang Lake was restored as a lake, functioning as a reservoir for flood regulation and water storage, which provided the most effective solution to this contradiction [5].

After repeated petitions from the local residents for lake restoration, in the second year of the Zhenghe reign of the Northern Song Dynasty (1112 AD), under the leadership of County Magistrate Yang Shi,[5] Xiang Lake was successfully reconstructed, with its surface area three times that of West Lake [6]. The artificial lake was ingeniously integrated with two mountain ranges oriented southwest to northeast, which served as natural embankments. Additional embankments were constructed along the north–south axis, with sluice gates installed around the periphery to collect runoff from the surrounding mountains. During dry seasons, sluices were opened for irrigation, while in flood seasons, the lake stored and regulated excess water, ensuring that the city of Xiaoshan was protected from water-related disasters. Simultaneously, efforts in landscape construction enhanced the site, making Xiang Lake renowned near and far (Fig. 6).

(3) Eight-Hundred-Year Struggle Between Preservation and Reclamation

The creation of Xiang Lake submerged large areas of original rice fields, fish ponds, and lotus ponds. Although these lands were barren fields frequently subjected to flood and drought disasters, disputes over its preservation or reclamation persisted for more than eight centuries following its reconstruction. This ongoing contention between advocates of lake preservation and proponents of lake reclamation essentially reflects a historical record of conflicts of interest—between local governments and public-minded gentry on one side, and bureaucrats and powerful landlords seeking to appropriate the lake for private gain on the other.

In the late Qing Dynasty (1636–1912 AD), additional embankments divided Xiang Lake into two sections—Upper Xiang Lake and Lower Xiang Lake—with differing water levels. This transverse division impeded the natural circulation of lake water, accelerated siltation, and gradually led to its reclamation into farmland [7]. This transformation was the combined outcome of natural sedimentation and human reclamation activities, with the latter becoming the dominant factor reshaping the regional landscape.

Since 2004, Xiang Lake has undergone a gradual restoration process, particularly as part of its role as the venue for the 2006 World Leisure Expo in Hangzhou, marking a renewed phase in its long history of transformation.

2.4 *Commonalities and Differences*

(1) Commonalities

a. Lake Construction as a Strategic Response to Environmental Challenges

The creation of lakes represented a deliberate strategy to confront challenges associated with flood control, the regulation and storage of rainwater and floodwaters,

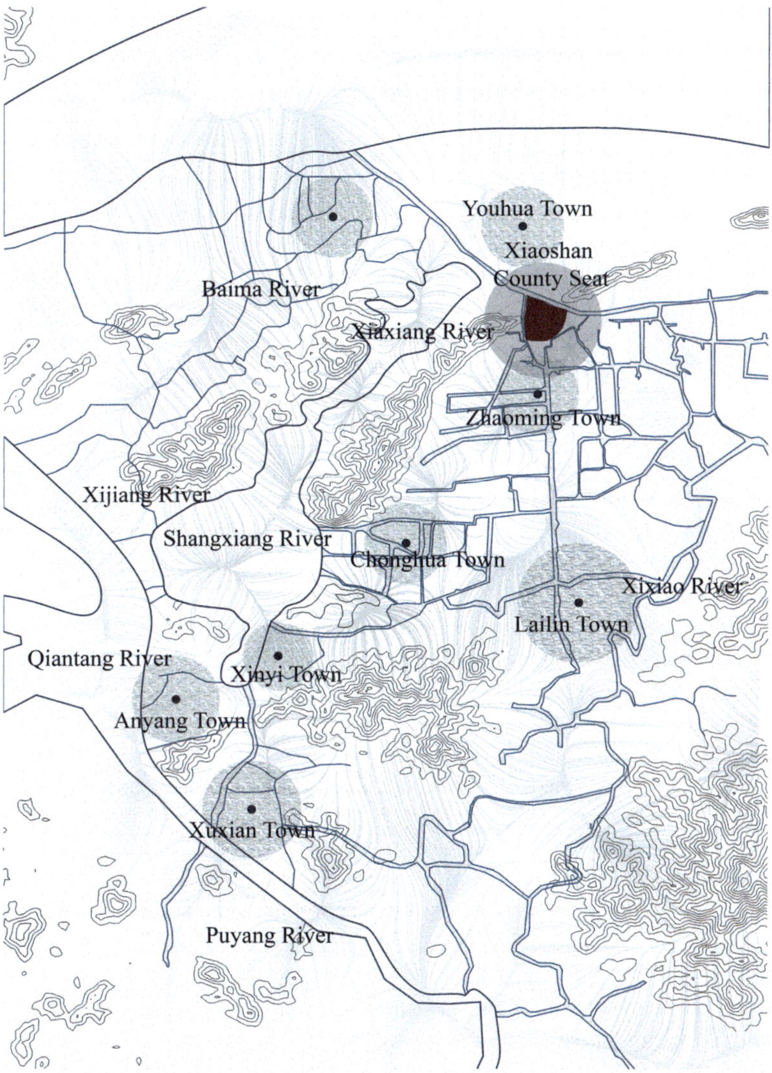

Fig. 6 Diagram of Xianghu Lake hydraulic system and Xiaoshan. Xianghu Lake utilized two mountain ranges oriented southwest to northeast as natural embankments to collect runoff from the surrounding hills. During the dry season, sluice gates were opened for irrigation; during the flood season, water was stored to regulate floods, ensuring Xiaoshan City was protected from water disasters. *Source* Redrawn by the author based on Note 6

the irrigation of agriculture, and the provision of essential water resources for urban development. Through the process of human engineering intervention in natural hydrological systems, new configurations of the regional landscape system were formed.

These interventions facilitated the rapid development of agriculture, resolved recurring flood disasters, and established a more habitable human settlement environment. The resultant material security, ensured by an abundant food supply, promoted the prosperity of commerce and handicrafts, thereby stimulating urban economic growth. Such prosperity attracted significant populations from surrounding regions, leading to flourishing urban marketplaces and further contributing to the emergence of a vibrant and enduring regional culture characterized by both richness and resilience.

b. Government-Led Large-Scale Hydraulic Engineering

All three major lakes were vast in scale, and the process of lake construction required the mobilization of immense human and material resources to build embankments. These were organized and executed as large-scale hydraulic engineering under the leadership of the government, constituting remarkable achievements in ancient human engineering.

In the subsequent processes of lake maintenance and debates over preservation or reclamation, local governments consistently played the role of coordinators and organizers. Within this ongoing human intervention in nature, the concept of "artificial" reflected not only the developmental level of human society but also the advancement of engineering technology. Such historical practices continue to provide meaningful insights for the construction of large-scale urban projects in the contemporary era.

c. The Optimal Landscape Pattern of Mountain–Lake–City

All three major lake are located in the suburban areas of their respective cities. Prior to the construction of the lakes, these areas were originally zones inundated by seawater and mountain runoff, where recurrent flood hazards posed long-term threats to both urban settlements and local inhabitants.

Through artificial embankment construction one zones were transformed into fertile and habitable landscapes. The lakes thereby integrated mountain–water–city, producing an optimal landscape configuration characterized by the harmony of "lake and mountain in mutual reflection, city and lake in unity." This spatial pattern represents not only the synthesis of natural and cultural elements but also the coexistence of scenic beauty and cultural significance.

West Lake is characterized by "cloud-wreathed hills on three sides and the city on one," with Mount Kuaiji to the south, a broad plain to the north, and the eastern and western causeways extending outward from the city. This configuration creates an integrated system in which the city and causeways originate together, and the sequence of mountains, lake, city, and plain unfolds progressively.

Xiang Lake, by contrast, is defined by "a single lake embraced by two mountains." Together, these landscapes generate an artistic conception in which, as Zhu Qiqian[6] observed, "artifice enters nature and thus achieves the marvelous, while serenity tempers richness and thus attains elegance." Over time, this spatial order fostered a region distinguished by scenic beauty, deep cultural heritage, and sustained urban prosperity.

(2) Differences

However, the outcomes of the evolution of the three major lakes were markedly different. Mirror Lake gradually silted up and declined; West Lake, through repeated dredging, was able to achieve a state of coexistence with the city; whereas Xiang Lake remained caught in a persistent struggle between preservation and reclamation, ultimately becoming silted and disappearing.

a. Differences in Lake Types and Modes of Construction

West Lake and Xiang Lake were naturally formed lagoons. West Lake, through continuous dredging, overcame the threats of land reclamation for agriculture, sediment deposition, and urban expansion, with the key aspect of artificial intervention lying in the persistence of dredging. The ancient Xiang Lake, which was naturally formed, became entirely obliterated due to sedimentation and land reclamation; later, it was restored on its original site through artificial reconstruction. In contrast, Mirror Lake was entirely created as an artificial lake through large-scale hydraulic works and embankment construction for water impoundment, with artificial intervention concentrated primarily in the stage of lake creation.

Thus, the transformation through artificial intervention in the natural process constitutes a continuous management process. Its purpose is that, along with the development of the city in terms of society, economy, and population, it is necessary to continuously coordinate the relationship between the artificial and the natural in order to achieve a state of mutual adaptation and dynamic balance, thereby realizing sustainable development.

b. Advantages and Disadvantages of the Scale of the Lake Bodies

In their early stages of construction, the three major lakes promoted the development of regional agriculture and the city. However, with the rapid increase in population and given that all were located in the urban periphery, the lakes faced challenges of land reclamation through infilling, siltation, and the pressures of further urban development.

All of the three major lakes possessed expansive waters and broad surfaces: West Lake covered nearly 6 km^2, while Xiang Lake was three times larger than West Lake, and Jian Lake extended to approximately 200 km^2, nearly ten times the size of Xiang Lake. The greater the area occupied by the lake bodies, the more vulnerable they became to encroachment from urban development, including land reclamation through infilling and increasing siltation. At the same time, excessively vast lake surfaces also heightened the difficulties of organizing comprehensive dredging and centralized management. Among the three, West Lake, being the smallest, maintained a proportion between lake and city that was comparatively suitable, whereas Jian Lake and Xiang Lake occupied areas far exceeding the scale of their respective cities. Thus, scale may have been one of the key factors enabling West Lake to be preserved, while the other two lakes ultimately silted up and disappeared.

c. The Correlation of Lake–City Dependency

The degree of dependency between Xiang Lake and Xiaoshan was not particularly high, as its primary role was to function as a reservoir for flood retention, thereby ensuring greater safety for the city. Its impact was more prominently reflected in the promotion of agricultural development within the surrounding areas. In the contemporary era, Xiang Lake has begun to undergo gradual restoration, yet the process of re-establishing its accumulated history and culture, as well as its former prominence, remains a long and ongoing endeavor.

The degree of dependency between Jian Lake and Shaoxing was relatively high. The Lake Dike was divided into eastern and western sections, with the ancient city of Shaoxing itself forming part of the Dike, directly intervening in the transformation and improvement of the regional water system. The subsequent silting and decline of Jian Lake led to the emergence of several large lake bodies in the plain area, and the city began to expand toward the plains.

The degree of dependency between West Lake and Hangzhou was the highest, being the result of continuous dredging and meticulous construction efforts across successive dynasties. This deeply reflects the historical process of mutual coordination and adaptation between human intervention and nature. The existence of West Lake not only promoted but also defined the urban form of Hangzhou. This close symbiotic relationship between the city and the lake has developed into an even stronger bond today, constituting a successful model of urban development and a paradigm for the advancement of landscape architecture (Figs. 7, 8 and 9).

The upper-right grid illustrates the proportional relationship among the areas of Jianhu, West Lake, and Xianghu, while the lower-right grid depicts the area changes of Jianhu throughout its evolution from artificial lake formation to gradual siltation and abandonment.

With the policy support of local governments for the protection and restoration of lakes, the contemporary development of West Lake, Xiang Lake, and Jian Lake has undergone new transformations. Professor Wang Xiangrong of the School of Landscape Architecture at Beijing Forestry University presided over the Westward Expansion Feasibility Study and Preliminary Planning of West Lake in 2000, which essentially restored the former water bodies of the western part of West Lake. Since 2004, he has also presided over the first and second phases of the Xiang Lake planning and design projects, both of which have been completed, aiming to restore the former landscape of Xiang Lake, while the third phase of planning and design is currently underway. The Hangzhou Institute of Landscape Design undertook specific design work for both the westward expansion of West Lake and the restoration of Xiang Lake. As for Jian Lake in Shaoxing, which has already silted up and cannot be restored, Professor Wang Xiangrong has carried out preliminary conceptual planning research on the development of Central Tea Lake in the plain area, making significant contributions to the contemporary development of the three major lakes.

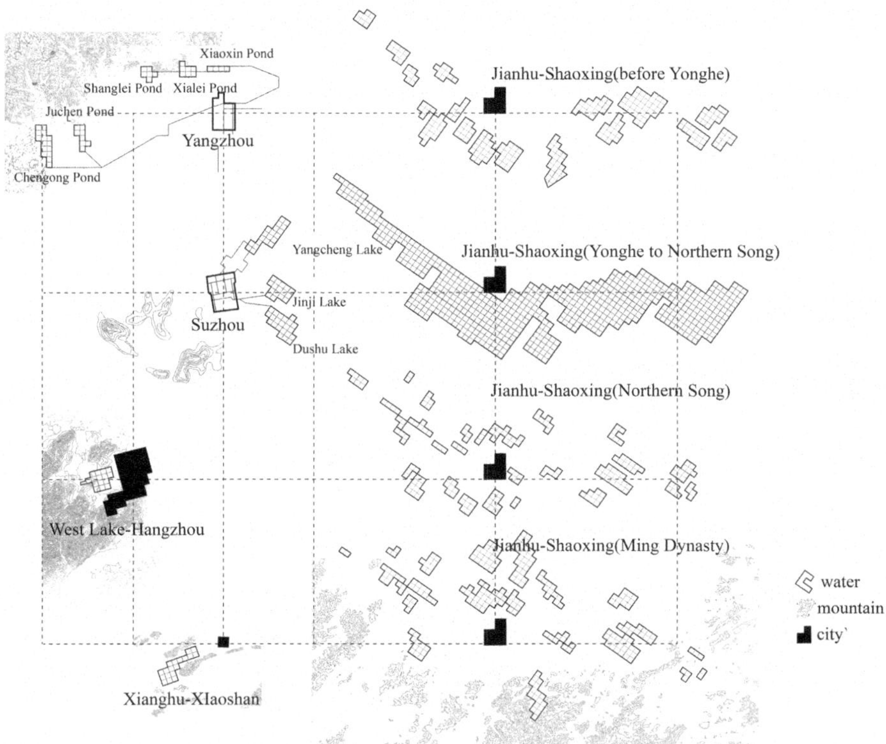

Fig. 7 Lake–city scale comparison diagram of Jianhu, West Lake, and Xianghu

3 Harmonious Coexistence—Human Intervention in Natural Processes, Shaping the Regional Landscape System, and Promoting Regional Development

Through the course of millennia, the formation and evolution of the three great lakes have continuously reshaped the regional landscape system, representing a harmonious integration of human intervention and natural processes, and serving as a paradigm for promoting regional urbanization development.

In global regional development, similar construction methods have consistently held significant importance. These methods can be generalized into similar patterns:

(1) Cities exist within the regional landscape system, and natural resources are the fundamental root enabling their reproduction and growth.
(2) In order to cope with the challenges of the living environment, people have profoundly promoted changes in the regional landscape through continuous artificial intervention in natural processes. These artificial projects achieved remarkable accomplishments in engineering, such as agricultural irrigation systems, water-conveying canals, canals, dams, and road networks.

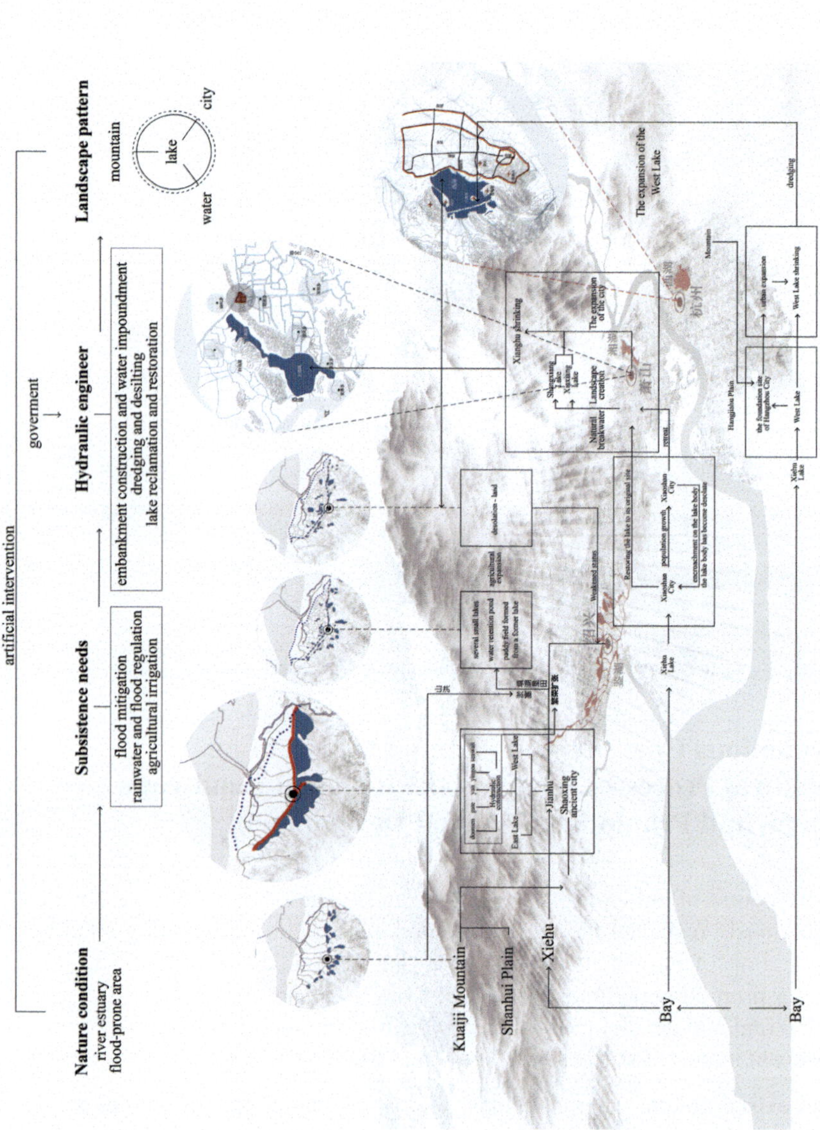

Fig. 8 Evolutionary process of the three great lakes region landscape, reflecting the dynamic equilibrium of mutual adaptation between human activities and nature

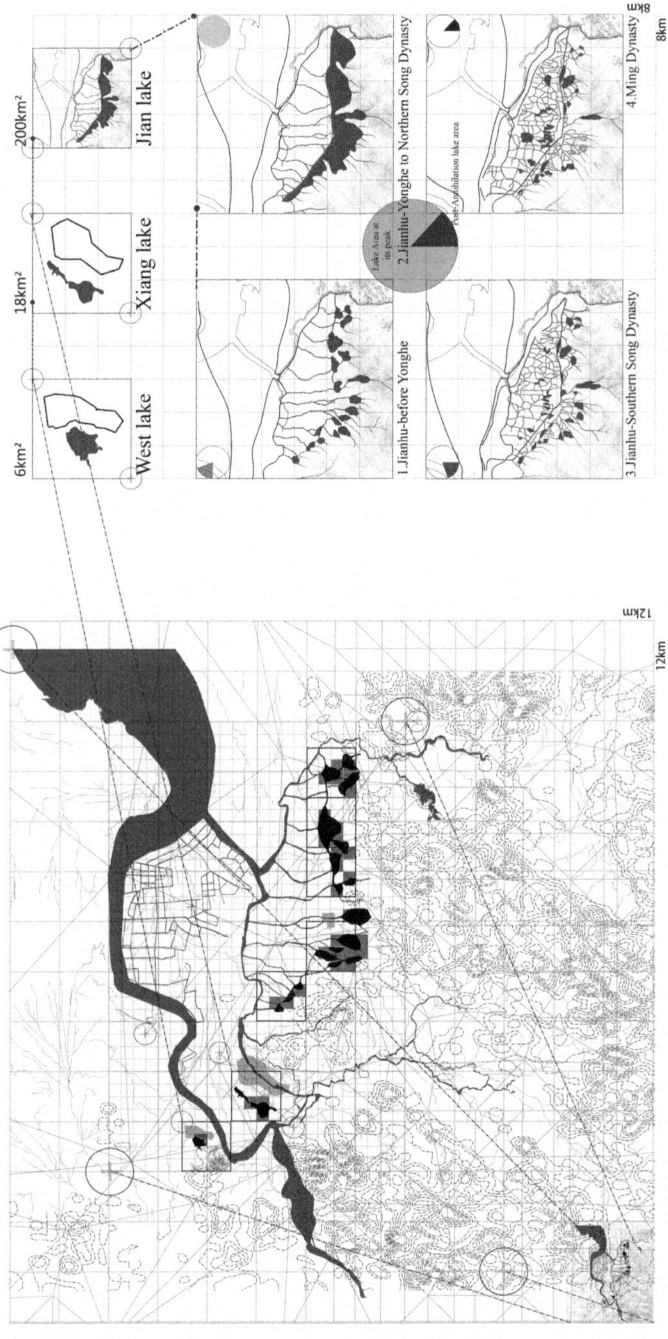

Fig. 9 Schematic diagram of scale comparison and area changes among the three major lakes

(3) The transformation of the regional landscape has, in turn, significantly influenced and guided the region's agricultural development and urbanization. The unique regional landscape formed through the interaction between the artificial and the natural has fostered and defined the distinctive characteristics of cities.

The transformation of the regional landscape has, in turn, significantly influenced and guided the region's agricultural development and urbanization. The unique regional landscape formed through the interaction between the artificial and the natural has fostered and defined the distinctive characteristics of cities [8].

The artificial intervention in natural processes, while improving the human settlement environment, simultaneously gives rise to urban problems. The involvement of the urban landscape system helps to mediate the contradictions that emerge under such interventions, providing an effective operational method and developmental framework for urban construction, and serving as an important medium for shaping a livable environment and nurturing urban culture.

Notes

1. Zhang Dai (1597–c. 1680) was a late Ming writer whose works document urban life, gardens, and scenic practices through memory and personal observation. His writings offer valuable insight into the cultural experience of cities and landscapes in late imperial China, especially around West Lake.
2. West Lake Dreaming written by Zhang Dai in the mid-seventeenth century around the fall of the Ming dynasty, is a memoir-like prose work that reconstructs the gardens, scenic sites, and urban life of West Lake through personal memory. Rather than a gazetteer, it records how landscapes were experienced, used, and culturally remembered in late-imperial China.
3. In this line from Li Bai's poem "Dreaming of Mount Tianmu" (mid-eighth century), the poet imagines traveling in a dream across the Wu–Yue region and over Moonlit Mirror Lake (Jinghu). The verse conveys a fluid, imaginative experience of regional landscapes, where mountains, water bodies, and cultural geography are perceived as continuous and traversable rather than fixed or bounded.
4. "On the Road to Shanyin," written by Yuan Hongdao in the late Ming period, is a travel essay describing a journey through riverine corridors and mountain scenery near present-day Shaoxing. The text foregrounds sequential movement, sensory perception, and everyday landscapes, offering an experiential understanding of regional space rather than a fixed cartographic view.
5. Yang Shi (1053–1135) was a Northern Song scholar-official who served as a county magistrate. In this role, he was responsible for local administration, including water management, settlement order, and everyday landscape maintenance, illustrating how Song dynasty governance shaped county-level environments.
6. Zhu Qiqian (1872–1964) was a modern Chinese scholar and early architectural historian active from the late Qing to the Republican period. His writings and statements on traditional architecture, urban form, and historical construction systems played a foundational role in interpreting premodern Chinese landscapes and cities through modern scholarly methods.

References

1. Che, Y., & Chen, Q. (2001). *Historical geography of Shaoxing*. Shanghai Bookstore Publishing House.
2. Chen, Q. (1962). The rise and fall of ancient Jianhu and the agricultural water conservancy of the Shanhui Plain. *Acta Geographica Sinica, 28*(3), 178–202.
3. Formation of Lagoons: Rivers flowing from mountains into the sea carry large amounts of sediment, which gradually accumulates in ancient bays. Sand spits form at bay mouths, extend continuously, and eventually block the connection between the bay and the shallow sea outside. The bay becomes a lagoon; with further development, the sand spits connect completely, cutting off the lagoon from the sea. The lagoon gradually retreats inland and its water becomes fresh. During the widespread appearance of lagoons, the Hangjiahu and Ning–Shao plains also gradually formed as low, lake-filled plains due to falling sea levels and sediment deposition. (Reference: Hou, H. (1989). The physical geography and evolution of Xianghu. *Journal of Hangzhou University, 16*(1), 89–95.)
4. Lü, Y. (1987). A study on the historical evolution of Hangzhou. *Journal of Hangzhou University, 17*(4), 10–18.
5. Hou, H. (1989). The physical geography and evolution of Xianghu. *Journal of Hangzhou University, 16*(1), 89–95.
6. Shibo, Y. (1985). "Xianghu Water Conservancy Records" and "Xianghu Studies": The beginning and end of the Xianghu water conservancy in Xiaoshan County, Zhejiang. *Journal of Chinese Historical Geography, 2*, 220–248.
7. Wang, S. (2006). Studying the historical facts of Xianghu to recreate the "Pearl" of Xianghu. *Journal of Agricultural Library and Information Science, 18*(6), 13–14.
8. Steiner, F. (2010). Conservation as catalyst: Lady Bird's urbanism. *Topos: The International Review of Landscape Architecture and Urban Design, 71*, 74–79.

Chapter 8
City: The Integration of "Mountain–Water–City"—The Foothill Hydrological Network in the Western Suburbs of Suzhou City and the Development of the Ancient City

Abstract This chapter takes the ancient city of Suzhou and the foothill hydrological network of its western suburbs as the research object, examining the integrated "mountain–water–city" model within traditional Chinese urban development from a regional perspective. By analyzing the formation and evolution of four key elements natural substratum, river–lake hydrological networks, farmland water conservancy, and cultural landscape the chapter elucidates how Suzhou's site selection, spatial structure, and urban form were grounded in a regional landscape system reshaped through sustained human intervention. The study demonstrates that the western suburban foothill zone functioned as a critical mediator linking Lake Tai's water system with the ancient city, integrating flood control, irrigation, transport, and landscape formation, and thereby consolidating a distinctive and resilient shanshui urban structure.

Suzhou City is one of China's most important historic and cultural cities, and it is also the birthplace of Wu culture. Its classical gardens are recognized worldwide as World Cultural Heritage. Situated in the heart of the Yangtze River Delta, Suzhou City occupies a strategic geographical location described as "bordered by the great sea in the east, controlling Zhenze in the west, blessed with fertile mountains and rivers, and converging by both land and water." The ancient city was established at the transitional zone between the lacustrine plains to the east and the foothill areas of the western suburbs, where natural Shanshui environment resources for tourism abound. The harmonious spatial arrangement of hills, rivers, lakes, farmlands, and the ancient city, combined with the temperate modulation of distinct seasonal climates, enabled a reciprocal integration of the city and nature, thereby shaping a regional landscape system rooted in, and integrated with, the natural order. Benefiting from such advantageous shanshui environment, Suzhou City developed a distinctive settlement pattern and profound cultural foundation. This natural and cultural endowment led to prosperity in both material and intellectual life, earning Suzhou City its enduring reputation as "heaven on earth." It may thus be said that Suzhou City's epithet as "paradise" is not only the result of human endeavor, but also the gift of heaven.

X. Zheng, *Terrapoetic Urbanism*, https://doi.org/10.1007/978-981-95-8676-9_8

The relationship between Suzhou City's city and its regional landscape system exemplifies the mountain–water–city traditional urban construction model. The unique endowment of shanshui provided the natural foundation for the birth of the city, while long-term human interventions—such as flood control, farmland irrigation, and canal transport—reshaped the suburban shanshui system. The combination of foothill landscapes and the hydrological network structured the regional landscape system, generating a suitable settlement environment and facilitating urban development. Within this process, the western suburban foothill area of Suzhou City integrated both in landscape and cultural dimensions with the intra-urban landscape system. The foothill area thus emerged as the critical linkage in shaping the urban–suburban regional landscape system.

Taking the regional landscape system of ancient Suzhou City as the research object, this study selects the ancient city together with the urban fringe of natural mountains and waters as the principal study area. Focusing on the development of four elements—natural substratum, river–lake water network, farmland water conservancy, and cultural landscape—it explores the construction of the regional landscape system, interprets its spatial pattern, and analyzes the relationship between the evolution of the regional landscape system and urban development.

Suzhou is situated in the hinterland of the Yangtze River Delta, with its ancient city established at the junction between the eastern lake–pond plain and the western low mountains and hills. To the west, the continuous, undulating low mountains and hills of the Lake Tai region, together with the abundant water resources of the Lake Tai Plain, constitute a mountain–water system of the urban fringe. This system has exerted a comprehensive and profound influence on the urban construction and urban development of Suzhou City.

For Suzhou City, the basis of its site selection lay in the surrounding grand mountains and waters; following the sorting and regulation of the water system, the urban construction proceeded on a regional landscape system shaped by artificial intervention. From this process emerged the city's distinctive dual chessboard pattern of land and water, together with a structure of urban gardens and an urban–suburban scenic system; in combination with the Shanshui system of the urban fringe, these elements formed a complete regional landscape system. In economic, cultural, and landscape dimensions alike, they constituted an urban–suburban complex of co-development, making Suzhou City a paradigmatic model of the Shanshui city (terrapoetic city) within the broader mountain–water–city framework of traditional practice.

Research on the city of Suzhou City begins with the study of the Suzhou City classical gardens. The courtyard gardens, celebrated for their refined aesthetics and regarded as the epitome of garden art, are by no means isolated artifacts enclosed within high walls and deep alleys; rather, they necessarily depend on and exist in relation to the city itself. The growth and development of the city are grounded in the land on which it stands—a natural environment that, once subjected to artificial intervention, gives rise to a regional landscape system, fundamentally distinct from architecture, which is entirely the product of human construction. From the scale of the garden, to that of the city, and further to the broader land on which the city is situated, the inquiry into the regional landscape system necessarily involves examining its constituent elements, its processes of formation, and its reciprocal relationships and interconnections with the city and with the gardens.

Taking the historical evolution of Suzhou City's mountain–water–city pattern as the blueprint, this study explores, from a regional perspective, the process by which the natural environment (i.e., mountains and waters) and the artificial environment (i.e., the city) adapt to and integrate with one another in the construction of a Shanshui city, thereby forming a regional landscape system [1]. From the vantage point of this system, the analysis further examines the development of the city and investigates the formation of the human settlement model embedded in the mountain–water–city paradigm.

The Suzhou City region underwent successive development stages from the Spring and Autumn and Warring States Period (770–221 BCE), through the Qin, Han, and Six Dynasties Period (221 BCE to 589 CE) characterized by stagnation and adjustment, the Sui, Tang, and Five Dynasties Period (581–960 CE) marked by restoration, the Song and Yuan Period (960–1368 CE) of steady growth, and finally reaching the zenith in the Ming and Qing Period (1368–1911 CE). Despite suffering devastating destruction during repeated wars and periods of political fragmentation, the city was consistently reconstructed and rehabilitated upon its original site. Overall, the development trend remained progressive, steadily advancing toward greater urban resilience and cultural flourishing.

1 The Exceptionally Endowed Regional Natural Pattern

Suzhou is situated in the Lake Taihu Basin at the lower reaches of the Yangtze River, bordered by Lake Taihu to the west, the Yangtze River to the north, and the sea to the east. In the western suburbs, a series of foothills encircle the city, while in the eastern suburbs a cluster of shallow lakes—including Yangcheng Lake, Jinji Lake, and Dushu Lake—are arrayed in sequence. The region is characterized by a dense hydrological network and an expansive hinterland. The surrounding mountain–water landform provided a solid natural foundation and ecological foundation for the emergence of the city (Fig. 1).

1.1 The Imprint of Primal Nature's "Water"

The Taihu region, though fertile and abundant in resources in later times, was in remote antiquity merely a low-lying, damp area with poor soil and low agricultural productivity. This condition corresponds precisely to the description in the *Yu Gong* chapter of the *Shangshu*[1]: "Their land is nothing but mud, and their fields are of the very lowest grade." In the process of struggling with this waterlogged natural environment, the early inhabitants of Wu engaged in water management and land reclamation, developing fishery and rice cultivation, and gradually forming primitive

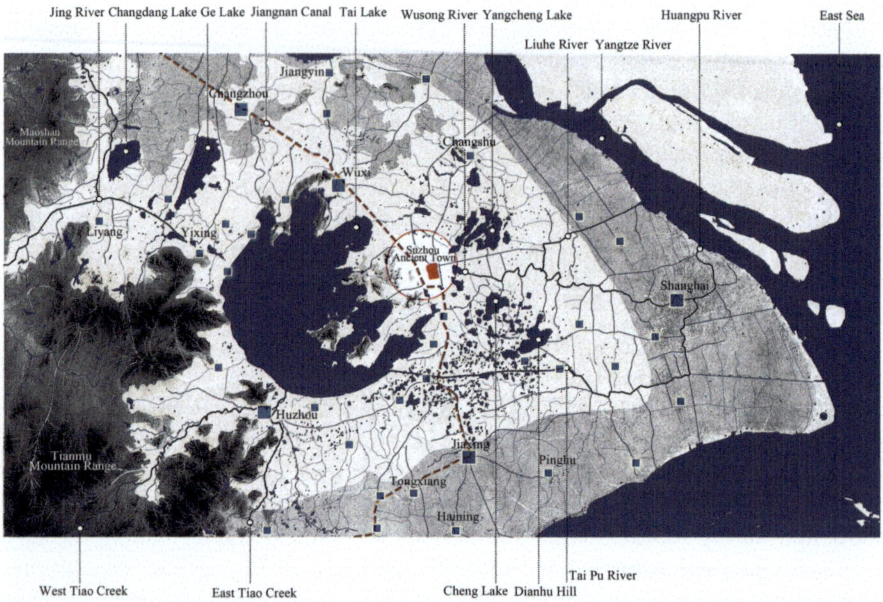

Fig. 1 Urban location of Suzhou City and the surrounding mountain–water pattern. Base map adapted from: Zhang Xiugui, Historical Process of Taihu Evolution [J], Collection of Chinese Historical Geography Studies, 2009, 24(1): 5–12, Fig. 1, Topographic Map of the Taihu Basin

techniques of irrigation, diversion, and drainage in agricultural production [2]. The urban site of Suzhou City, situated at the crucial point where upstream waters from the west converge and flow eastward into the rivers and seas, became from ancient times a place where the people of Wu both contended with water and depended upon it for survival. Supported by the lake-marsh system scattered across the landscape and the dense hydrological network, the abundant water resources have indelibly shaped the modes of production and daily life of Suzhou City's inhabitants over successive generations.

1.2 Lake Tai Hydrological System and the Western Mountain Mass of the City

Suzhou has since antiquity been characterized as a watery region, where its hydrological system developed in the course of the formation and reclamation of the Lake Tai plain. This system is constituted by a combination of natural lakes, natural river channels, and artificial river channels. The principal bodies of water include Lake Tai, Lake Shi, Lake Yangcheng, Lake Jinji, Lake Dushu, and Lake Cheng. The Beijing–Hangzhou Grand Canal skirts the western edge of the ancient city, with its Suzhou City section running to the north of the city. More than 20,000 rivers and streams, with a total length of 1457 km, converge to form a complete river–lake network system. This abundant provision of water resources laid the essential ecological and material foundation for the emergence and flourishing of the city.

Lake Tai has long served as the hydrological center of Suzhou City's hydrological network system, receiving inflows from the upstream and subsequently discharging eastward through the urban site of Suzhou City, which is situated between mountain–water landscapes. The waters then spread across the lake–marsh plain, ultimately flowing into the rivers and seas, thereby constituting a regional lake–hydrological network system. As recorded in the *Yu Gong* chapter of the *Shangshu*: "When the Three Rivers have entered, Zhenze is firmly established." The "Three Rivers" are generally regarded as the three outlets through which Lake Tai discharges eastward into the sea, while "Zhenze" refers to the ancient name of Lake Tai. This indicates that as early as the Xia dynasty, floods of Lake Tai followed natural waterways such as the Wusong River, moving from west to east, passing through the urban site of Suzhou City, and eventually entering the sea.

The surface natural morphology of Suzhou City is the outcome of evolutionary processes spanning long geological periods. In general terms, the region of Suzhou City belongs to the Lake Tai Plain, characterized by a topography that is higher in the northwest and lower in the southeast, with elevated terrain along the Yangtze River and comparatively depressed areas in the hinterland. Within this large dish-shaped basin, numerous smaller dish-shaped basins are also distributed.

The site of the ancient city was predominantly situated on the plain, with more than one hundred low hills located only in the western part, which represent the remaining spurs of the Tianmu Mountain range of western Zhejiang extending northeastward. In total, there are over one hundred mountain masses, covering an area of approximately 221 km^2. The major mountains are aligned along Lake Tai in a northeast–southwest orientation, forming four distinct groups of hills and islands:

(1) Qizi Mountain–East Dongting Mountain;
(2) Qionglong Mountain–Yuyang Mountain–Changsha Island–West Dongting Mountain;
(3) Dengwei Mountain–Tan Mountain–Manshan Island;
(4) The area of Dongzhu–Zhenhu, comprising residual hills such as Geshe Mountain, Xingshe Mountain, and Xiyang Mountain.

Between Qionglong Mountain, Yang Mountain, and the Qizi Mountains, there are granite hills including Lingyan, Tianping, and Tianchi. Along the Yangtze River, low hills such as Xiang Mountain are distributed, while isolated hills like Yushan and Yushan rise prominently on the plain between rivers and lakes.

The low mountains in the Suzhou City region generally have an elevation ranging from 10 to 200 m, with a few peaks rising above 300 m. Among them, Qionglong Hill, at 341.7 m, holds the highest elevation. Other notable peaks include Piao Miao Peak of West Dongting Hill (336 m), Mo Li Peak of East Dongting Hill (293 m), Yang Hill (338 m), Qizi Hill (294 m), Tianping Hill (201 m), Shangfang Hill (92.6 m), Lingyan Hill (182 m), Yu Hill (261 m), Tan Hill (252 m), and Yuyang Hill (170 m).

The historic city of Suzhou City itself has an elevation ranging between 4.0 and 4.5 m, whereas the plains to the north and east generally lie below 4.0 m. This indicates that, although the region as a whole is relatively flat, the terrain of the historic city is slightly higher than that of its surrounding areas.

From the perspective of topography, the series of low hills stretching from the western suburbs to the shores of Lake Taihu separates the lake from the historic city of Suzhou City. This configuration enabled the city not only to rely on the abundant resources of Lake Taihu but also to avoid effectively the threats of flooding, while simultaneously forming a natural hill defense barrier. At the same time, the historic city of Suzhou City was characterized by a terrain "as level as the palm of a hand," yet slightly elevated compared with the northern and eastern plains. Such an advantageous terrain not only facilitated flood discharge but also provided favorable conditions for urban construction, given the flat ground and the rich forest resources of the western hills [3].

By relying on these hill ranges as protective barriers and taking the Lake Taihu Plain as its foundational base, the urban site of Suzhou City already possessed the latent conditions for urban construction. Subsequently, through the implementation of hydraulic engineering that reorganized the natural mountain-and-water system, the advantages of agricultural irrigation and flood control were further strengthened. With continuous human intervention in nature, the natural environment gradually evolved into a regional landscape system, upon which the city was established. Over time, Suzhou City rose and developed into the beautiful and prosperous Paradise City.

2 Shaping a Suitable Human Settlement Environment Through Artificial Intervention in Natural Mountains and Waters

The surrounding natural resources of the ancient city site of Suzhou City possessed a latent degree of habitability. However, under its primordial natural state, the terrain of the site was characterized by low-lying and waterlogged conditions, which rendered it unsuitable for permanent human settlement.

Li Daoyuan's *Shuijing zhu (Commentary on the Water Classic)*[2] records a statement attributed to Wu Zixu, the city founder: "The region of Wu and Yue is ringed by the Three Rivers, leaving its inhabitants with little possibility of removal. Because the southeast lies low and gathers all waters, overflows from the lakes spread into sheet floods that become rivers; tributaries crisscross through man-made and natural channels, scattering lineages into separated settlements; hence the locations of old river courses and ancient waterways are difficult to determine." This description vividly highlights a distinctive topographical characteristic of the eastern Taihu region, where the terrain in the southeast is markedly lower. Waters from the west naturally converge eastward, causing lake breaches and extensive inundation. With channels and waterways intersecting in all directions, the river network spread across the plain, creating recurrent flooding.

During the Spring–Autumn Period, Wu Zixu,[3] the Grand Master of the State of Wu who presided over city construction, thoroughly assessed both the advantages and disadvantages of the foundational site. Benefiting from the exceptional natural mountains and waters that formed protective barriers together with the abundant natural resources of the surrounding region, the area still presented itself as an optimal location for urban site selection. Consequently, Wu Zixu examined the soil and surveyed the water, aligning his planning with the patterns of heaven and the principles of the earth, and applied context-sensitive design/planning. Through the systematic ordering and transformation of the natural mountains and waters, he established a grand city upon the marshlands to the east of Lake Tai.

2.1 *Early Settlement Civilization in the Foothill Region*

The Suzhou City region possesses a long-standing history of ancient civilization. The earliest traces of human activity were discovered in the foothill region to the west of the old city, on Sanshan Island of Lake Tai, where an archeological site dating back ten thousand years to the Paleolithic period was unearthed. At that time, Lake Tai had not yet formed; Sanshan Island and the surrounding foothill region were characterized by a coastal open woodland-grassland dotted with hills. The low hills were densely forested, while the caves and woods on the ridges provided natural shelter for early humans [3]. The favorable conditions in the Sanshan area enabled human survival, laying the groundwork for the creation of an early settlement civilization, and marking the initial steps in the historical process of habitation and reproduction on this land.

2.2 *Reshaping of the Regional River–Lake Hydrological Network System*

Located in the eastern Taihu area, Suzhou City is characterized by a west-high and east-low terrain, with its lowest points situated in the eastern urban districts and the neighboring regions of Kunshan and Wujiang. The eastward flow of collected waters from the western highlands frequently caused the hydrological network to overflow and rivers and lakes to breach, necessitating artificial diversion of regional waterways to mitigate urban flood hazards. First, both the external waterways surrounding the city and the internal water system had to ensure that the waters flowing from the west could smoothly discharge toward the lower-lying southeastern areas. Second, while maintaining flood control, the regulation of the water system was also integrated with suburban agricultural irrigation, urban water transportation, and the provision of domestic water use within the city, thereby ensuring the system's multifunctional role in supporting urban and rural development [4].

(1) The Water–Land Dual Checkerboard Pattern Within the City

When Wu Zixu first constructed the Great City of Helü (today's Suzhou City, Ancient City), each side of the city wall was equipped with two gates, totaling eight land gates and eight water gates. By utilizing the west-high-east-low topography, eight rivers were introduced to traverse the city, thereby initiating the construction of the hydrological network within the inner city. *Records of Yue—"Treatise on the Land of Wu"*[4] describes: "The Great City of Wu measured forty-seven li, two hundred and ten bu, and two chi in circumference. It featured eight land gates—two of which were equipped with towers—and eight water gates." By the 54th year of King Nan of Zhou (261 BCE), when Chunshen Jun of Chu governed the Wu region, an internal hydrological network system of "four vertical and five horizontal" was formed. Over successive transformations, by the Tang and Song dynasties, the city evolved into a dual checkerboard pattern of "juxtaposition of land and water, with parallel river routes." The structural framework of the hydrological network was further consolidated, culminating in a hydraulic system of "three horizontal, four vertical, and one ring." This configuration manifested a distinctive Jiangnan water town landscape, epitomized by the poetic image of "Homes all are built along the riverbank, and along the water lanes small bridges abound [5]." (Fig. 2).

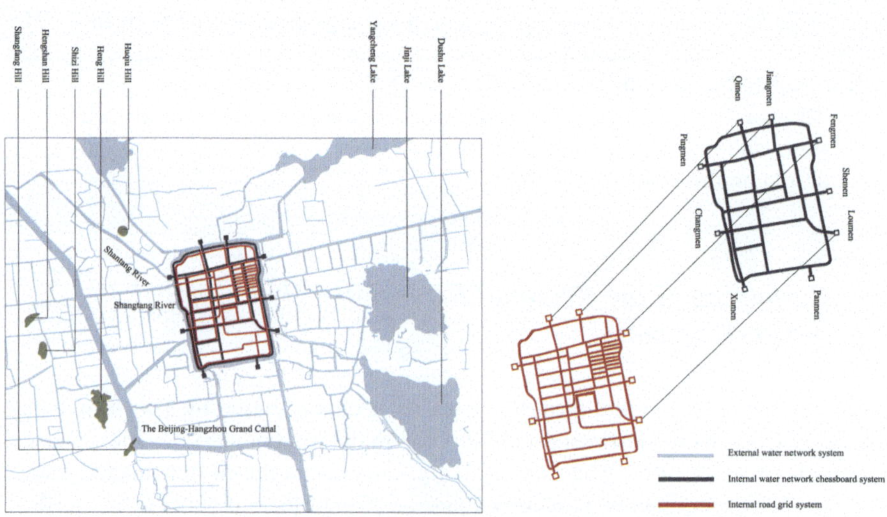

Fig. 2 Dual land–water grid pattern within Suzhou City and its connection with external water systems

(2) Development of Waterways and Grain Transport Outside the City

At the same time as the development of the inner city water system, the city's abundant surrounding water resources and the crisscrossing tributary waterways provided favorable conditions for the development of water transport. The growth of water transport was therefore an inevitable trend. During the Spring and Autumn Period, in order to contend for supremacy in the Central Plains, the State of Wu excavated such famous navigation channels as the Han Gou, which connected the Yangtze River with the Huai River, and the He Canal, which linked the Yellow River with the Huai River. Through these projects, Wu's waterway transportation extended to cover nearly half of China. As it was said at the time: "They could not dispense with boats and oars for even a single day [6]."

By the time of Emperor Yang of the Sui Dynasty, the Grand Canal had been fully connected from north to south, and Suzhou City emerged as the central hub of water transport in the Jiangnan region. The Grand Canal flowed past the western part of the ancient city of Suzhou City. By the Ming and Qing dynasties, following the dredging and reopening of the Grand Canal, the city's water transport reached its peak. The course of the Grand Canal through Suzhou City constituted a crucial element in the city's urban development, exerting a decisive influence on the prosperity and advancement of Suzhou City across multiple dimensions.

The development of grain transport in particular spurred the growth of the inland port and wharf district near Changmen in the western part of the city. Stimulated by water transport, the city broke through the limits of the city wall and expanded westward; outside Changmen, at Nanhao Wharf, merchants from across the realm moored in great numbers, and the area even became a new commercial center of Suzhou City, celebrated as the "Changmen Flourishing Age." From Changmen, the Shantang River and the Shangtang River led toward Tiger Hill and Fengqiao in the western suburbs, where guild halls clustered and shops stood in close ranks. These waterways not only served navigation and brought canal water into the city, but also formed a landscape axis extending from the western city to the foothill region of Tiger Hill and Fengqiao, thereby promoting, to a considerable extent, the development of tourism resources in the western suburban area.

These hydrological networks, formed through the utilization of existing resources and reshaped by human intervention, constituted a complete river network and lake–pond system that bore significance not only for water conservancy but also for transportation. Through its waterways, the Suzhou City Ancient City was linked eastward and westward to the Yangtze River, and north-southward to the Grand Canal. In doing so, it assumed the strategic position described as the "throat of Fujian and Zhejiang, and the key passage of Jianghuai," emerging as a crucial transportation hub of the southeastern region. At the same time, it became one of the most developed centers of commodity economy nationwide, thereby laying a profound material foundation for the enduring prosperity of the city [4].

2.3 Development of the Regional Farmland Water Conservancy System

The rice cultivation culture of the Wu region has a long-standing history. Since the Spring and Autumn period, the State of Wu actively promoted the principle of "filling the granaries,"[5] encouraging the reclamation of wastelands and placing great emphasis on water conservancy construction. At the same time, owing to the low-lying and flood-prone geographical conditions of Suzhou City, water conservancy became an essential measure to safeguard both agricultural production and the daily lives of the people in the Wu region. Through the excavation of river channels and the construction of revetments, a systematic connection was established with farmland, thereby transforming the water system from a natural state into one under human control. This enabled the full utilization of functions such as irrigation, drainage, and shipping, which in turn significantly advanced the agricultural production of the region [5]. Subsequently, with the southward migration of northern populations, new technologies and labor were introduced. By the Tang and Five Dynasties Period (618–979 CE), land development in the Wu region had achieved overwhelming dominance, and the economic center of the country shifted southward to the Yangtze River Basin.

During the Song Dynasty, agricultural development within the territory of Suzhou City experienced significant acceleration, primarily owing to the remarkable progress in water conservancy construction. For example, Fan Zhongyan[6] carried out large-scale management of flooding in Lake Tai, directing the dredging of the Wusong River as well as five rivers—Qianjing, Xiazhang, Qiya, Baimao, and Xupu—situated between Changshu and Kunshan. In addition, sluice gates were installed at river inlets along the Yangtze River, serving to block silt, resist tidal incursions, and discharge accumulated floodwaters, thereby bringing benefits to multiple prefectures [7]. After centuries of development, by the Song Dynasty most plains within Suzhou City had been transformed into fertile farmland. Against this backdrop, the people of Suzhou City redirected their efforts in land development toward expansive water areas and the surrounding hilly terrain, contending with mountains for land and with waters for fields.

The initial stage of development lay in the creation of polder fields and the associated water conservancy works. This land use method, involving the construction of embankments to block water and protect arable land, was already documented during the Spring and Autumn period. During the mid-Tang Dynasty (circa eighth–ninth century CE) through the Five Dynasties Period, the Kingdom of Wuyue dredged the outlets of Lake Tai leading to the sea, thereby establishing a systematic river network of embanked polders, organized into a pattern of longitudinal canals at seven-li intervals and transverse ponds at ten-li intervals. By the Song Dynasty, polder fields had already become fertile farmland capable of stable and high yields, representing one of the major achievements in the farmland water conservancy construction of the Wu region. According to the investigation of Jia Dan[7] in the Song Dynasty, there were more than 260 sites of pond-and-polder water conservancy systems in Suzhou City alone.

Secondly, the system of fengtian (floating raft fields) constituted a distinctive land use practice. As recorded by Cai Kuanfu in the *Poetic Remarks*[8]: "In the marshes and lakes of Wuzhong, masses of bulrush and cattail accumulate. Over the years, their roots are washed loose by the water and no longer cling to the soil; they then float to the surface, often extending for tens of zhang in length and measuring several chi in thickness. Upon them one can spread earth and carry out planting and tilling; people stand on them as on wooden rafts and can pole themselves back and forth. This is what is called fengtian (floating raft fields)." The southeastern city gate of Suzhou City, known as Fengmen Gate, derived its name precisely from the extensive stretch of such fengtian (floating raft fields) located beyond the gate.

Thirdly, the development of terraced fields, namely the practice of constructing successive embankments and leveling the soil on gentle slopes within the hills, represented a strategic mode of land use and the rational exploitation of mountain resources. Concurrently, with the improvement of production tools and the advancement of rice cultivation techniques, a relatively sophisticated agricultural technology system gradually emerged. Within this historical context, the Yangtze River Delta region, where Suzhou City is located, progressively became the most significant agricultural region and a vital grain supply base for the entire nation (Fig. 3) [7].

Fig. 3 Comprehensive water conservancy map of Suzhou City prefecture. *Source* (Ming Dynasty1368–1644) Zhang Guowei, Wu Zhong Water Conservancy Compendium, Volume 1, History Section, Geography Category, Rivers and Canals

Overall, within the farmland water conservancy system formed through the transformation of the natural environment of the region, water conservancy construction functioned as a critical measure to safeguard agricultural development as well as the sphere of production and livelihood. At the same time, the prosperity of agriculture stimulated and sustained the flourishing of the entire region.

3 Integration of Natural Mountains and Waters and the Urban Landscape System: The Construction of a Regional Landscape System

Through context-sensitive artificial interventions in natural mountains and waters, the city of Suzhou City and its surrounding areas developed a well-structured regional agricultural system and an interconnected water network. Ultimately, the low hills of the suburban periphery and the urban landscape within the city permeated one another in the process of regional landscape evolution, urban development, and cultural shaping. They were then further integrated through a shared water network system and historical context, thereby forming a trans-urban regional landscape complex composed of low hills, water networks, vegetation, classical gardens, temples and monasteries, pagodas, and cultural history (Fig. 4).

3.1 Landscape and Cultural Management of the Foothill Region in Suburban Areas

For Suzhou City, the foothill and hill areas of its western suburbs—such as Hengshan Hill, Shangfang Hill, He Hill, Lingyan Hill, Shizi Hill, and Huqiu Hill—form an enclosing "arch-like" configuration on the city's western side. Serving as the city's immediate visual setting, these sites, together with artificial development, have evolved into renowned suburban scenic sites, imbued with profound Suzhou City cultural characteristics and constituting a suburban landscape system where landscape and culture are deeply integrated. Meanwhile, Dongting East Hill, Dongting West Hill, Dengwei Hill, and Qionglong Hill create a vast verdant backdrop of background mountains for the city and its frontal mountain views [8]. Within this hierarchical and layered depth relationship, the residual ridges of the foothills serve both as extensions of the urban landscape and landscape culture, and as external background mountains for the city, offering visual appreciation and evoking reverence. As mediating resources, these landscapes establish a two-way connection between mountain-and-water landscapes and the city, incorporating the surrounding natural environment into the integrated urban landscape and cultural system (Fig. 5).

The foothill region stretching westward from Suzhou City's city limits to Lake Taihu is endowed with abundant natural resources and well-structured geographical

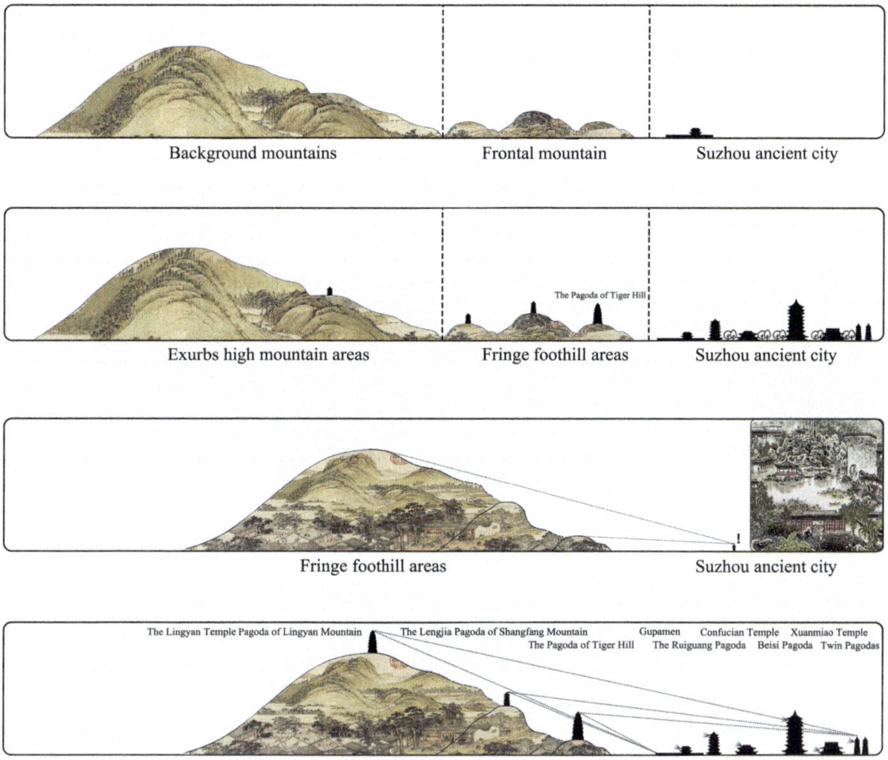

Fig. 4 Sightline analysis diagram of the low hills in suburban Suzhou City and urban landscape landmarks

Fig. 5 Distant view of the Dongting two hills from Xumen. *Source* Photograph by Wang Ruilong

assemblages. Diverse geological conditions have given rise to a variety of distinctive mountain landforms. Yet, the allure of the foothill landscapes lies not merely in their dramatic peaks, fantastic rock formations, or imposing ridges; rather, it emerges from the dynamic interweaving of natural resources with the currents of time and the enduring legacy of historical and cultural traditions, producing distinctive cultural

Fig. 6 Westward view from Changmen toward Tiger Hill and Tianping Mountain

landscapes. The most exemplary case is the Tiger Hill Scenic Area. As recorded in Records of the *Record of the Land of Wu*[9] *(Wudi Ji)*: "Tiger Hill's sheer cliffs and deep ravines, dense forests and thick bamboo groves stand as the premier expression of the valleys and ravines east of the Yangtze River." Its towering cliffs and plunging ravines present a spectacle of grandeur, while its rich ensemble of cultural features is epitomized by the celebrated "Three Wonders, Nine Prospects, and Eighteen Views." Among these, the Yunyan Temple Pagoda, rising atop Tiger Hill, stands with an unadorned yet graceful presence. As an enduring landmark building, it has long functioned as a defining symbol of Suzhou City's ancient city (Fig. 6).

Alongside the development of suburban scenic attractions, the westward expansion of the ancient city of Suzhou City toward Lake Taihu has given rise to the emergence and distribution of settlements such as Dongshan, Jinting, Mudu, Fengqiao, and Guangfu within the foothill and hilly region. These villages and towns are seamlessly integrated into the natural landscape, presenting an organic unity between human habitation and the surrounding environment. Their architectural styles, spatial layout patterns, and landscape design styles embody a profound accumulation of historical and cultural heritage (Fig. 7).

3.2 The Development of the Urban Landscape System and the Practice of Garden-Making

During the Six Dynasties Period, Suzhou City began to see the emergence of many temples and monasteries, such as the Bao'en Temple and the Western Jin Zhenyuan Daoist Temple (today's Xuanmiao Temple). At the same time, the private classical

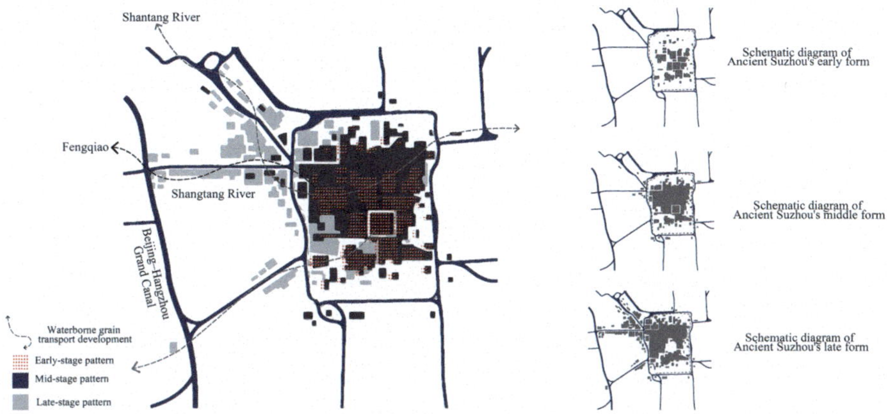

Fig. 7 Diagram of urban development distribution in Suzhou City across different periods

garden also gradually began to develop, with the Bijian Garden recorded as the first instance of a private classical garden [5].

By the Song and Yuan dynasties, more than fifty temples and monasteries existed throughout the city, together with pavilions, city walls, government offices, classical gardens, palaces, extensive residential quarters, and numerous stone bridges, as well as a crisscrossing waterway system, all of which together formed a rich and poetic urban landscape system. During this period, garden-making reached a climax, with the Canglang Pavilion and the Lion Grove Garden standing as precious historical legacies of the Song and Yuan Dynasties. Subsequently, as the ancient city gradually underwent gardenization, and with the prosperity and affluence of Suzhou City during the Ming and Qing Dynasties, officials and gentry purchased land to create gardens, imitating nature through rockery construction and water management. Suzhou City thus came to be described as a "half-city of pavilions and gardens," with its system reaching its peak; the majority of extant Suzhou City traditional classical gardens are also products of this period [5].

Suzhou's emergence as a celebrated center of Jiangnan private classical gardens, distinguished by natural mountains and waters gardens, was made possible by the rich Shanshui environment in its suburban areas. This environment provided abundant natural resources and creative inspiration for garden-making activities, thereby establishing both the essential material foundation and the prototype of garden-making.

First, in the case of Suzhou City gardens, "No garden can exist without water—it is water that brings a garden to life." Water served as the principal element in garden-making: pavilions, terraces, and towers were constructed in proximity to water, and most scenic compositions unfolded around it. The hydrological network of rivers and lakes provided abundant resources for garden-making, and the garden system was established upon the framework of the urban hydrological network. Ultimately, the water environment constituted the core factor underpinning the prosperity and flourishing of Suzhou City's gardens.

Secondly, the foothill ranges surrounding the city provide an excellent background for garden landscapes. While each garden is an independent entity in spatial composition, in terms of sightlines it merges seamlessly with the external mountain forms. For instance, from the Pavilion of Surging Waves (Canglangting) one may ascend to the Mountain-Viewing Tower; leaning against the balustrade, one's gaze extends outward to take in the distant vistas of Shangfang Hill, Tianping Hill, and Lingyan Hill, where the natural topography and the constructed garden scenery converge in a unified spatial and cultural continuum.

Finally, the surrounding grand mountains and waters provided the creative source for garden-making. The literati aspired to bring mountains and waters before their eyes; thus, through the construction of rockery and the arrangement of waterscape, and by employing various scenic composition techniques, they transformed the grand mountains and waters into small-scale mountains and waters immediately at hand. In doing so, they created literati mountain-and-water gardens imbued with profound poetic sentiment and pictorial charm.

3.3 Integration of Landscape Making in the Foothill Area with the Urban Landscape

The foothill area of western Suzhou City belongs, on the one hand, to the same natural system as the surrounding grand mountains and waters; on the other hand, it is located in the urban fringe, lying outside the city proper but closely integrated with the urban interior. This integration was realized through the development of suburban scenic resources, the construction of water and land transportation, the permeation of the historical context, and the linkage of iconic pagodas within a network of landscape architecture. Together, these elements constituted the scenic system of the western suburbs (Fig. 8).

Although the foothill extension area of the western suburbs of Suzhou City is not high in elevation, it is well-covered with vegetation and often situated in close proximity to water bodies, thereby producing a remarkable landscape effect. Consequently, these hilly areas historically became renowned suburban excursion sites or the locations of imperial retreats and private villas [9]. In the urban periphery zone, destinations such as Tiger Hill, Maple Bridge, Stone Lake, Mount He, and Lion Mountain have long traditions of suburban tourism, embodying both a profound history and substantial tourism value.

Based on the demand for the scenic resources of these foothill extension areas, artificial intervention opened both waterways and land routes leading from the city to the western suburbs toward Tiger Hill and Maple Bridge. The urban–rural fringe and the urban interior were connected through the hydrological network. Maple Bridge, as the entrance of the Grand Canal into the city, functioned as a grain checkpoint

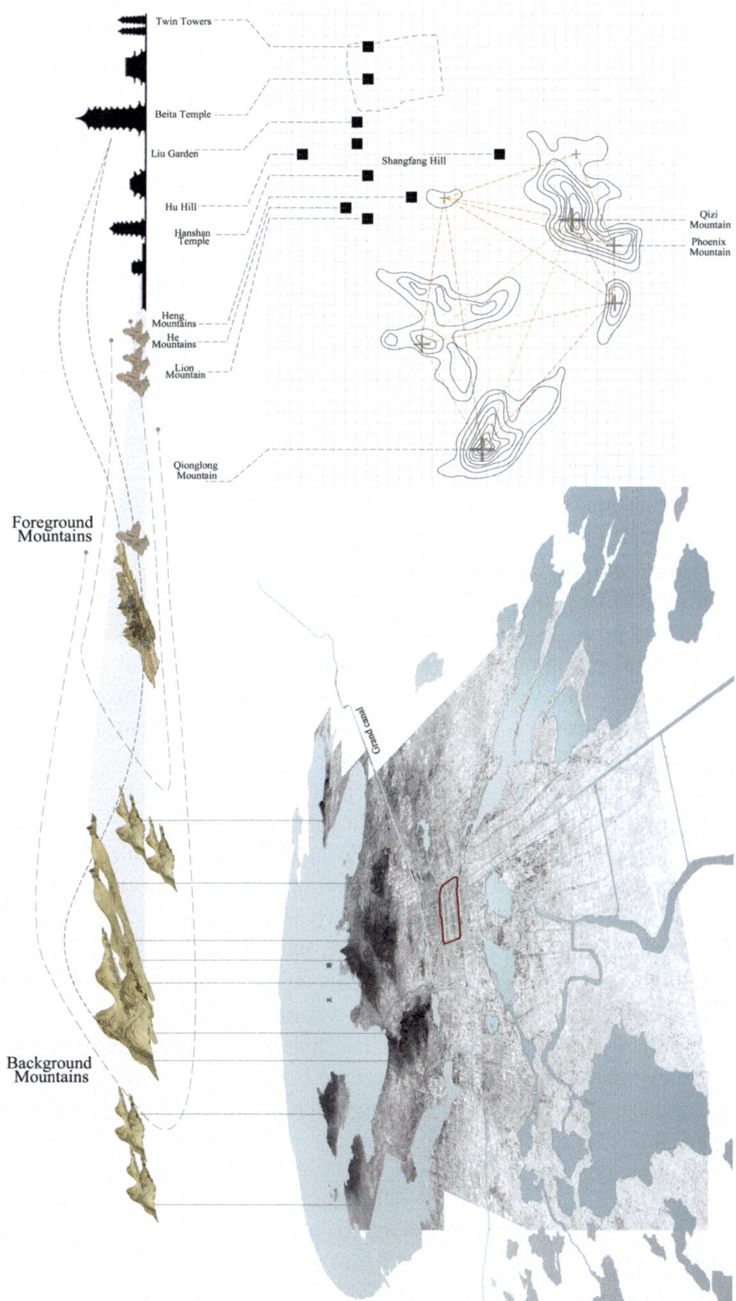

Fig. 8 Iconic urban landscapes in Suzhou City linking the city with its near and distant suburban environments

where grain boats traveled between the canal and Suzhou City. It became nationally renowned due to the poems of successive literati. Tiger Hill is a historical landscape model in the foothill extension area around the city. With the development of canal transport and commerce in the Changmen area, as well as the westward expansion of the urban spatial pattern, the connection between the foothill extension area and the city grew closer. The natural mountains-and-waters system of the suburbs and the garden system within the city both served as a carrier of historical and cultural heritage, gradually merging into one.

At the same time, among the various extant urban landscapes of Suzhou City, temples and gardens account for a significant proportion. The wooden or brick pagodas within them often served as commanding heights and became iconic features of the city. The network system formed by these tall pagodas, scattered both inside and outside the city, also functioned as an important medium linking the city with its surrounding areas—for instance, the Beisi Pagoda, the Ruiguang Pagoda, and the Luohan Monastery Twin Pagodas within the ancient city; the Yunyan Temple Pagoda at Tiger Hill outside the city; as well as the Lingyan Temple Pagoda of Lingyan Mountain and the Lengjia Pagoda of Shangfang Mountain. As the essence of ancient architecture and Buddhist culture, these pagodas constitute the very core of the urban character of ancient Suzhou City. They connected the city with its surrounding environment, together forming a complete Shanshui system.

The configuration of the natural mountains and waters surrounding Suzhou City constituted the foundation for the city's site selection. The city arose in reliance on the external natural mountains and waters, and through artificial interventions that reshaped nature, it achieved continuous development. After the evolution of thousands of years, it ultimately merged with the Shanshui system of the urban periphery, forming a highly integrated cultural and landscape complex. This constitutes a regional landscape system under the "mountains–waters–city" model and became the medium for the construction of a Shanshui city (Fig. 9).

The construction of a city must fully preserve the configuration of natural mountains and waters and the historical context formed and sedimented over the course of time. At the same time, in the processes of urban development and renewal, it is necessary to organically integrate the city with its surrounding natural environment from the perspective of the regional landscape system, thereby elevating urban planning to the scope of the regional landscape system and allowing the city to be immersed in natural mountains and waters. The historical process through which Suzhou City's Shanshui system in the urban fringe became integrated into the city–fringe regional landscape system provides important references for the contemporary construction of a Shanshui city.

Fig. 9 Illustration of the regional landscape system and urban landscape construction in Suzhou City. The near-suburban mountain–water environment provided background and resources for the construction of urban gardens. The areas outside the city and the urban core were integrated through the development of suburban scenic resources, the excavation of land–water transport routes, the infiltration of historical context, and the network of iconic landscape buildings such as pagodas, forming a coherent regional landscape system. *Source* (Ming) Zhang Guowei, *Wu Zhong Water Conservancy Compendium*, Volume 1, History Section, Geography Category, Rivers and Canals—General Map of Suzhou City Prefecture Waterways, Northeast Corner Map of Changzhou County Waterways in Suzhou City, Northwest Corner Map of Wu County Waterways in Suzhou City; (Qing) Xu Yang, *Prosperous Suzhou*; (Ming) Tang Yin, *Agricultural Scenes of Jiangnan*; (Five Dynasties) Dong Yuan, *Summer Mountains*

Notes

1. The *Yu Gong* chapter of the *Shangshu*, compiled during the Warring States period (fourth–third century BCE) and incorporated into the Book of Documents, presents an early geographic vision of China organized around river systems, terrain, and regional resources. Rather than a mythic narrative alone, it functions as a proto-spatial framework linking land, water, circulation, and governance.
2. The Commentary on the Water Classic, compiled by Li Daoyuan in the early sixth century (Northern Wei), is a comprehensive geographic work organized around river systems. By linking waterways with cities, landscapes, infrastructure, and historical sites, it offers a river-based framework for understanding regional space and human–environment relationships in premodern China.
3. Wu Zixu (c. 559–484 BCE) was a statesman of the Spring and Autumn period who served the state of Wu. He is associated with early urban planning, fortification, and water management, demonstrating how terrain and river systems were integrated into city defense and regional governance.
4. The "Treatise on the Land of Wu" in the Records of Yue, compiled during the Eastern Han dynasty (first–second century CE), documents the geography, waterways, cities, and historical traditions of the Wu region. Blending historical narrative with regional description, it provides early insight into how river networks, terrain, and settlements were understood as an integrated cultural landscape.
5. The principle of "full granaries" (shi canglin), articulated in the Warring States period (fourth–third century BCE), links food security and grain storage to social order and state governance.
6. Fan Zhongyan (989–1052) was a Northern Song statesman whose reforms and local administration emphasized water management, public works, and social welfare. His career illustrates how governance shaped urban infrastructure and regional landscapes in Song China.
7. Jia Dan was a Northern Song official active in the eleventh century. Like many Song dynasty administrators, his work involved land administration and local governance, illustrating how bureaucratic practice shaped settlement patterns and managed landscapes at the regional scale.
8. Cai Kuanfu's Poetic Remarks (twelfth century, Southern Song) is a collection of literary criticism that discusses poetic imagery and aesthetic experience, including perceptions of mountains, rivers, and travel. It offers insight into how landscapes were conceptually framed and appreciated in Song intellectual culture.
9. The Record of the Land of Wu, compiled by Gu Yewang in the sixth century during the Liang dynasty, documents the geography, waterways, settlements, and cultural traditions of the Wu region. It provides an early regional framework for understanding how river networks, terrain, and cities together shaped the Jiangnan cultural landscape.

References

1. Wu, Y. (2010). An analysis of the concept of "landscape city." *Chinese Landscape Architecture, 2*, 3–8.
2. Construction and Management of Water Conservancy in the Taihu Lake Region. http://www.360doc.com/content/12/1214/11/8282618_253961527.shtml. Accessed December 14, 2012.
3. Kuai, Y. (2010). Research on the development of popular science resources related to Suzhou landscape scenic sites. http://www.szst.cn/toupiao/021.htm. Accessed January 08, 2010.
4. Lu, X. (2015). A brief discussion on the water-city characteristics of ancient Suzhou. http://www.zx.suzhou.gov.cn/szzx/InfoDetail/?InfoID=ba124c72-53a6-488a-b722-8b9743ffaf98&CategoryNum=014008001. Accessed January 12, 2015.
5. Xu, S., Lei, Q., & Zhu, J. (2010). *Suzhou geography* (p. 11). Guwuxuan Press.

6. Liu, M. (2000). The historical-geographical foundations for the rise and development of Suzhou City. *Journal of Chinese Historical Geography, 1*, 173–184.
7. The past and present of Suzhou. http://blog.sina.com.cn/s/blog_538961f3010008ka.html. Accessed March 09, 2007.
8. Wang, D. (2002). *Chinese landscape culture and urban planning*. Southeast University Press.
9. Zhu, Z. (2009). *Research on the narrative urban space of Suzhou's old city*. Central South University.

Chapter 9
City: Regional Landscape System and Urban Zoning—The Yangzhou City on Shugang Hill and the Water Network

Abstract This Chapter investigates the urban structure of Yangzhou City during the Sui-Tang Period (581–907 CE) from the perspective of a regional landscape system, focusing on how the combined framework of the Shugang Hill and the engineered water network structured urban zoning and spatial organization. Located at the junction of hills, alluvial plains, the Yangtze River, and the Grand Canal system, Yangzhou City developed a city form in which natural landforms and hydraulic infrastructures together constituted a stable landscape structure for urban construction. From a landscape architectural and urban morphological perspective, this chapter analyzes how mountain barriers, canal systems, reservoirs, and inner-city waterways collectively shaped the distribution and evolution of major urban zones, including the government precinct, market and commercial areas, ward-based residential districts, gardens, and suburban extensions. Methodologically, the study integrates historical landscape interpretation with spatial analysis to examine the interaction between natural terrain, water management, and urban zoning processes. The analysis demonstrates that Yangzhou's urban development was not driven solely by political or economic factors, but by the long-term coordination between urban construction and a persistent regional landscape system. The city exemplifies a landscape-led model of urban zoning, in which water networks and topography functioned as formative mechanisms in shaping spatial order, offering an important reference for understanding landscape-based urban form in traditional Chinese cities.

Urban construction cannot overlook the close interconnection between the artificial environment and the natural environment. The zoning of a city is closely related to the regional landscape system, and the city together with its surrounding landscape pattern has always constituted an indivisible whole. Chapters Nine, Ten, and Eleven present three urban case studies Yangzhou City, Shaoxing City, and Nanjing City to elucidate the interaction processes between urban zoning and regional landscape systems.

X. Zheng, *Terrapoetic Urbanism*, https://doi.org/10.1007/978-981-95-8676-9_9

This chapter focuses on Yangzhou City during the Sui-Tang Period (581–907 CE). Beginning with the natural environment shaped by the Shugang Hill and the alluvial plain, it analyzes the engineered water network system, delineates the components of Yangzhou's regional landscape system, and examines the characteristics of the ancient city's construction based on this system. From this perspective, it explores the relationship between urban functional zones—such as the government precinct (yashu), the market zone (shisi), and the ward zone (lifang)—and the regional landscape system.

Yangzhou City is located in the northern part of the Yangtze River Delta, with its ancient city situated at the junction of the Shugang Hill and the alluvial plain [1]. Historically, the city experienced two major periods of development: first, the urban development during the Sui-Tang Period, and second, the renowned flourishing of garden construction in the Ming-Qing period (1368–1912 CE). The period of Sui–Tang Yangzhou City displayed two primary characteristics. First, the originally muddy and flood-prone alluvial plain, through long-term engineering interventions, was transformed into a complete water network system, which together with the Shugang Hill in the northwest constituted a regional landscape system. Second, the expansion of the ancient city from the Sui Dynasty's Jiangdu Palace to the Tang dynasty's Zicheng (Yacheng) and Luocheng[1] was consistently shaped by this landscape system and developed in close integration with it [1]. The city's expansion process vividly demonstrates the relationship between regional landscape systems and urban development, serving as a typical case of the close interdependence and joint development of ancient urban construction and regional landscape systems.

Important urban zones within the process of city construction—such as the government precinct, market zone, and ward zone—together with suburban derivatives such as temples and Daoist monasteries, town settlements, villa estates, and agricultural water-conservancy systems that complemented urban functions, were all built on the regional landscape system. Each functional area of the city continuously adjusted to and grew within the natural landscape, wherein the artificial and natural environment were deeply interwoven.

The siting, zoning, and construction of ancient cities involved multiple dimensions, including politics, economy, culture, military affairs, and geography. The focus of this section, however, is to articulate the spatial interrelationship between urban functional zoning, urban construction, and the regional landscape system from a regional perspective.

1 The Composition of the Regional Landscape System of Yangzhou City During the Sui-Tang Period (581–907 CE)

Yangzhou City is situated at the confluence of the Grand Canal (the Han Canal section of the Sui–Tang Grand Canal) and the Yangtze River. The local terrain slopes from west to east, with the highest elevation in the northwestern hilly area, known as the Shugang Hill region. From west to east the land gradually tilts in a fan-shaped pattern. Beneath the Shugang Hill lies the alluvial plain formed by the Yangtze River, while the river itself flows along the southern edge of the plain, together forming a configuration characterized by higher terrain to the west/northwest and the river bordering the south.

Built upon this landscape configuration, through long-term engineering works such as canal excavation (outside the city: Han Canal segment, Shanyang Ditch, Yilou River; inside the city: Shi River, Guan River) and reservoir construction (Chengong Pond, Goucheng Pond, Upper Lei Pond, Lower Lei Pond, Xiaoxin Pond) [2], a complete regional landscape system was gradually developed. This system provided a favorable natural foundation for the construction and subsequent development of the city (Fig. 1).

1.1 The Natural Barrier of the Mountainous Landform

The Jiang–Huai Plain, where Yangzhou City is located, was primarily formed by the alluvial deposits of the Yangtze River and the Huai River. The terrain is generally low-lying, with an elevation typically below 10 m. Influenced by geological structures and tectonic uplift, numerous low hills, and ridges are distributed across the plains along the Yangtze River. Among these, the most prominent in Yangzhou City is the Shugang Hill, situated to the northwest of the city. According to the *Dushi Fangyu Jiyao, Volume 23, Nan Zhi Wu*[2]: “The Shugang Hill lies four li (1 li $\approx$ 0.5 km) northwest of the prefectural city. It stretches for more than forty li, connecting westward to the boundaries of Yizheng and Liuhe counties, extending northeastward to Zhuyu Bay, and facing Jinling (Nanjing) City across the Yangtze River [4].”

From the perspective of urban resources, the Shugang Hill region contains numerous low hills, with abundant vegetation that contributes significantly to soil and water conservation. The undulating terrain creates valleys that offered natural sites for constructing artificial reservoirs (such as Chengong Pond, Goucheng Pond, Upper Lei Pond, Lower Lei Pond, and Xiaoxin Pond), which helped regulate and store regional stormwater. From the perspective of urban defense, the Shugang Hill functioned as a natural defense barrier. As recorded in the *Tu Jing*[3]: “The prefectural city stands to the southeast of the Shugang Hill. Its eastern, southern, and northern sides are all flatlands intersected by waterways; only the mountains of the Shugang Hill connect westward to Lu and Chu. Whenever northern troops invaded southward

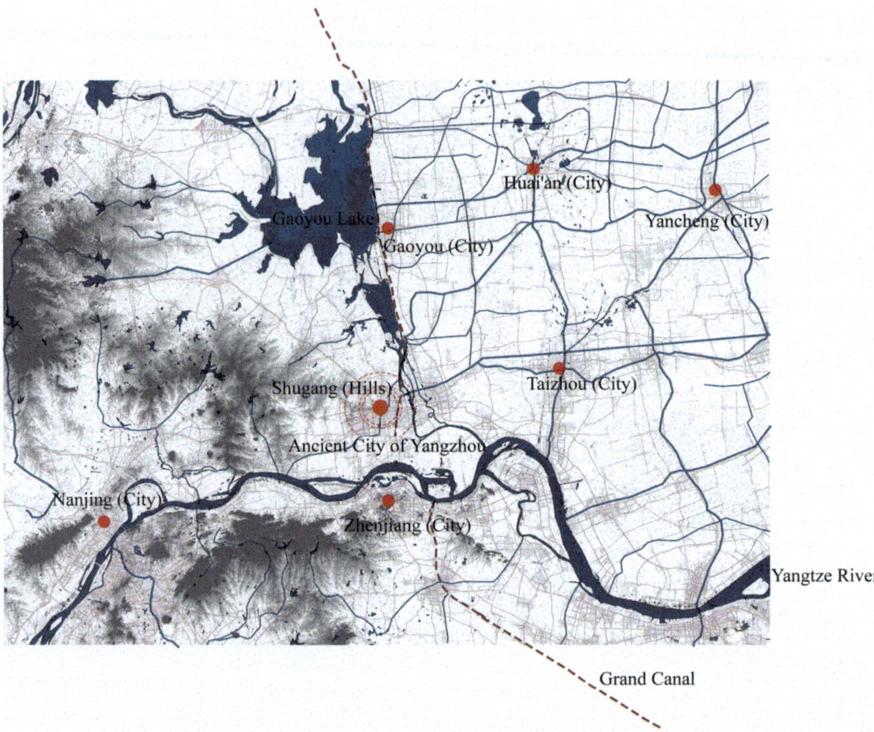

Fig. 1 Location of the ancient urban area of Yangzhou City and its surrounding natural landscape. *Image source* Redrawn by the author based on Notes 2 and 3

into Yangzhou City, they invariably followed the mountains, established encampments upon the heights, and thereby overlooked the city [5]." By relying on the natural landform of the Shugang Hill as a protective barrier, Yangzhou City had the essential physical conditions for the establishment of the city.

1.2 Artificial Interventions and the Reshaping of the Water Network System

The hydrological conditions of the Yangzhou City region in antiquity differed significantly from hydrological conditions seen today. Prior to the unification under the Sui Dynasty (581–618 CE), the Yangtze River carried sediments that accumulated along its northern bank, resulting in the silting and expansion of point bars. This process gradually pushed the river channel southward. By the time of Sui unification, the alluvial plain had broadened from its original twenty li to more than forty li. At this stage, the riverbank of the Yangtze lay roughly along the line of today's Sancha River, Yangzi Bridge, and Shiqiao (Fig. 2) [6, 7].

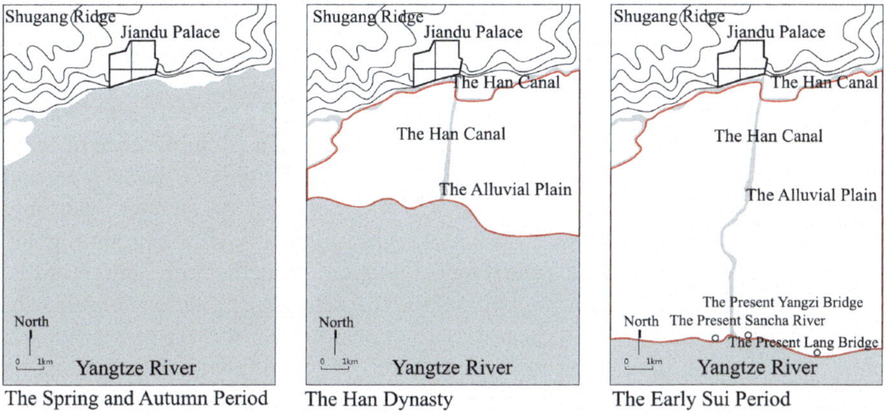

Fig. 2 Diagram of the changes in the Yangtze River shoreline in the Yangzhou City region from the Spring and Autumn Period (770–476 BCE) to the Early Sui Dynasty (581–618 CE)

Although the southward migration of the Yangtze River widened the alluvial plain, producing what contemporary sources described as "exceedingly flat and open land [7]." Thereby making room for later urban development, the plain itself remained a flood-prone floodplain, vulnerable to inundation from the Yangtze as well as from floodwaters descending from the Shugang Hill region in the west. This environment was not suitable for habitation. It thus required systematic engineered measures to channel and redirect the watercourses within the region—both to prevent the Yangtze from overflowing and to ensure that runoff from the Shugang Hill could drain quickly toward the slightly lower terrain in the southeast. Such hydraulic interventions reduced the threats of flooding to the city, while simultaneously facilitating agricultural production and sustaining economic activities.

After long-term artificial regulation, two north–south waterways (the Market River and the Official River) traversed the interior of Yangzhou City, running through the Luocheng area. In addition, two east–west waterways (the Zhuo River and the Han Canal) were formed. The Zhuo River ran along the northern part of the Luocheng, adjoining the Zicheng, and generally flowed eastward along the southern edge of the Shugang Hill. The Han Canal, excavated under Emperor Yang of the Sui, constituted the other east–west artery [6]. Together, these waterways formed a "two north–south, two east–west" water network pattern within the city.

The artificial hydraulic facilities of suburban Yangzhou City can broadly be divided into two categories. The first comprised canals excavated over successive dynasties for military, economic, and political purposes, such as the Han Canal of Spring and Autumn Period (770–476 BCE), Warring States Period (475–221 BCE), the Zhuyu Canal of the Han Dynasty (202 BCE–220 CE), the Shanyang Ditch excavated under Emperor Wen of the Sui, the reconstruction of the Han Canal under Emperor Yang of the Sui, and the Yilou River of the Tang [8]. These canals, located in the eastern part of Yangzhou City, served as navigational passages linking the lower Yangtze River with the lower Huai River. The second category consisted of agricultural water-conservancy facilities developed to meet productive needs. In the hilly region at the junction of the Shugang Hill, Yizheng, and Hanjiang, five interconnected reservoirs known collectively as the "Five Ponds" (wutang) were constructed: Chengong Pond, Goucheng Pond, Upper Lei Pond, Lower Lei Pond, and Xiaoxin Pond. Among them, Chengong Pond was the largest, with a circumference of more than ninety li; Goucheng Pond measured over ten li in circumference; both Upper and Lower Lei Ponds were six to seven li each; and Xiaoxin Pond, the smallest, was only two to three li around [2, 9]. The "Five Ponds" played an active role in promoting the agricultural development of the Yangzhou City region. From the mid-Tang onwards, they also functioned as reservoirs supplying water for the transportation of grain tributes and as supplementary water sources. The construction of these reservoirs and ponds helped to conserve water resources, thereby benefiting waterborne transport as well as agricultural production. The reservoirs were distributed across the hilly terrain of the Shugang Hill. Together with the west–east excavation and dredging of canals, this configuration made the Yangzhou City region a key node linking north and south. This network not only facilitated waterborne transportation but also provided significant convenience for Yangzhou's waterborne transport (Fig. 3).

2 The Ancient City Shaped by the Regional Landscape System

The construction of ancient Yangzhou City was built based on the natural landscape configuration and the water network system that had been reshaped through artificial regulation. From the siting of the ancient city, to the organic arrangement of its internal functional areas—including the government precinct, market zone, ward zone, Buddhist temples and Daoist temples, and academies—and further to the extension of urban functions into the suburbs, all were dependent upon a regional landscape system integrating the city and its suburbs. This system was composed of elements such as the Shugang Hill, the Yangtze River, canals (including the Shanyang Ditch, Han Canal, and Yilou River), the alluvial plain, gardens, and palace complexes (Fig. 4).

2.1 The Siting of the Ancient City

In ancient China, the locations of most cities were chosen in regions possessing a favorable landscape configuration. The natural environment exerted a decisive influence on the emergence and development of urban functions, while the placement of urban functional zones also required consideration of preexisting natural conditions. The two aspects interacted with and organically integrated into one another. The planning and construction of cities arose under a rudimentary worldview of nature,

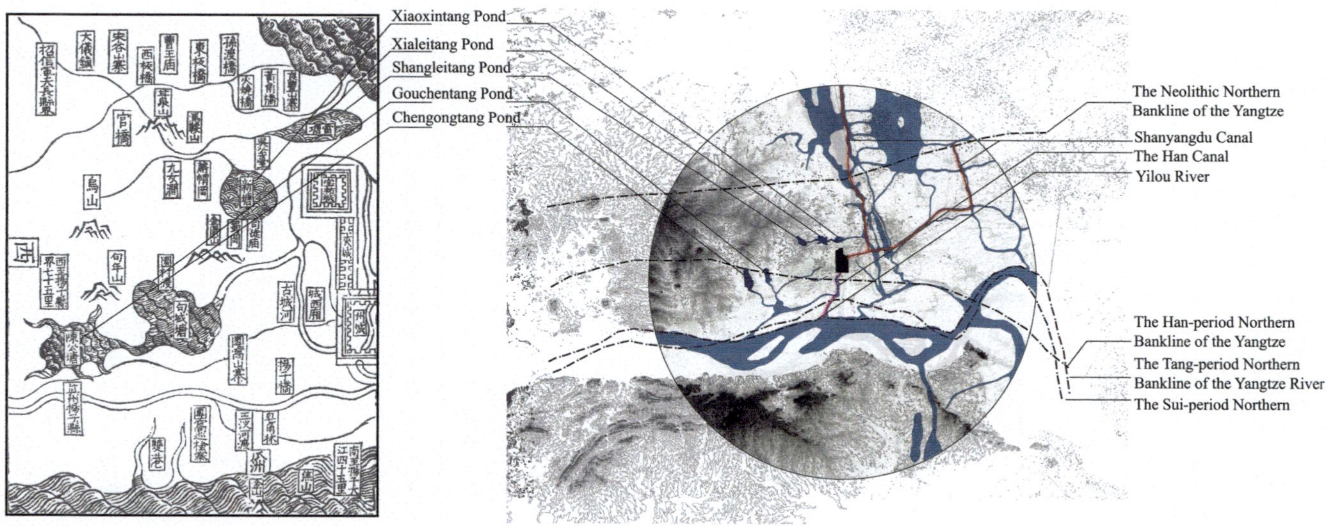

Fig. 3 Distribution map of major water-conservancy facilities in the Yangzhou City Area. Redrawn by the author based on Notes 2, 3, 6, and 9. *Source Jiajing Weiyang Gazetteer*④ (Ming Dynasty (1368–1644 CE)), Volume 1, Map of Jiangdu County (Song Dynasty (960–1279 CE)

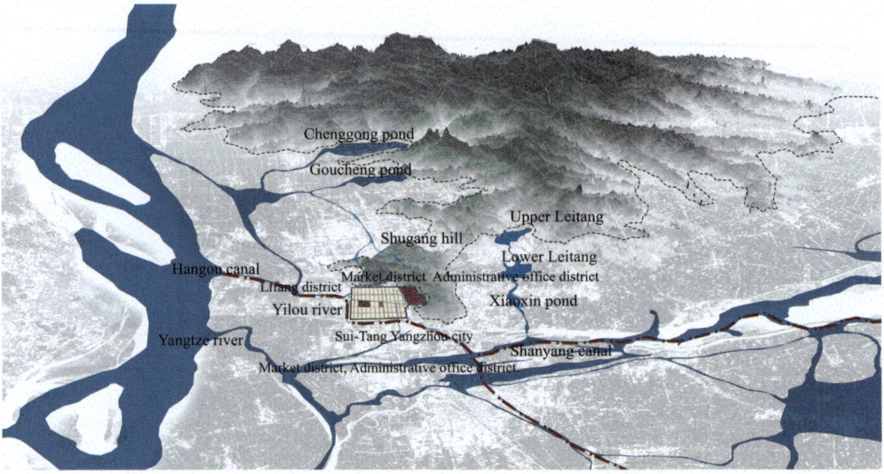

Fig. 4 Schematic diagram of the relationship between ancient Yangzhou City and the regional landscape system

an early ecological consciousness, and a simple form of human-centered thought, while simultaneously shaped by feudal ritual systems, class divisions, and religious culture [10].

Yangzhou City was no exception. Its site was located between the Shugang Hill and the alluvial plain: backed to the north by the Shugang Hill, which provided natural protection through its low hills, while the city proper extended across the alluvial plain. Facing the Yangtze River to the south, it benefited from fertile soils and level terrain as essential conditions for its development. Within the city, the ground level was slightly higher than its surroundings, while the eastern and southern areas possessed abundant water resources, which reduced the risks of flooding and ensuring adequate water supply. More importantly, Yangzhou City at this time occupied a strategic position at the intersection of north–south water and land transportation networks, which constituted the essential precondition for the city's rise.

2.2 *Coordination Between Ancient Urban Zoning and the Regional Landscape System*

During the Sui-Tang Period, the ancient city of Yangzhou City experienced three developmental stages: from the Jiangdu Palace of the Sui Dynasty, to the Zicheng (Yacheng) in Tang Dynasty (618–907 CE), and finally to the dual-city configuration of the Tang Zicheng and Luocheng.[4] Within the city were the government precinct, the market and commercial zone, and the ward-based residential quarters.

The government precinct was situated at the southeastern edge of the Shugang Hill, overlooking the foothills below, and its northwestern side benefited from scenic resources. The waterborne grain transport[5] and bazaar precinct[6] developed along the waterways; in the wake of commercial prosperity and scenic enhancement, these areas attracted increasing population flows and elevated the quality of urban life. The ward-based residential quarters made use of the terrain of the alluvial plain and the backdrop of the Shugang Hill to create a pleasant living environment.

Outside the city lay township settlements, villa estates, farmland, water-conservancy facilities, and the broader natural landscape environment. Each zone closely depended upon the natural setting as well as the artificially regulated water network system, while influencing and adapting to one another. Both within and beyond the walls, the built environment appeared to grow organically within the mountain–water configuration, coexisting in harmony with it. Through the organic integration of all functional zones with the regional landscape system, the city was able to sustain long-term prosperity (Fig. 5).

(1) Scenic Resources of the Shugang Hill and the Construction of the Government Precinct

From the construction of the Jiangdu Palace under the future Emperor Yang of the Sui Dynasty to the establishment of the Tang Dynasty Governing Office of Yangzhou City (Dudufu), the Military Governor's Office of Huainan, and various prefectural and county offices, Yangzhou's government precinct was consistently situated on the Shugang Hill. The formation of the Yangzhou City Zicheng began in the tenth year of the Kaihuang reign (590 CE), when Yang Guang—later Emperor Yang of the Sui Dynasty—served as Governor-General of Yangzhou City, established the Jiangdu Palace (official yamen complex) atop the Shugang Hill and installed the government precinct within it. At the beginning of the Tang Dynasty, Yangzhou's administrative seat continued to be located on the site of the Jiangdu Palace, thereby laying the foundation for the subsequent development of the city [11].

The northwestern suburbs of Yangzhou City, within the Shugang Hill region, abounded with scenic resources, including Ganquan Mountain, Pangu Mountain, Jinkui Mountain, Pushu Bay, Jiunü Stream, and the "Five Ponds." In addition, an artificial environment was created by drawing upon the rich forest resources and ecological conditions of the Shugang Hill. Among the most representative constructions were the Changfu Garden and the Sui Garden, both established by Emperor Yang of the Sui outside the Jiangdu Palace, upon the Shugang Hill, as palatial complexes with towers, terraces, and gardens. The *Taiping huanyu ji*[7] records: "The Ten Palaces are situated five li north of Jiangdu County, within the Changfu Garden. They were arranged along forests and beside streams, rising high over the ridges, and were positioned according to the terrain. They were called Guiyan, Huiliu, Jiuli, Songlin, Fenglin, Dalei, Xiaolei, Chuncao, Jiuhua, and Guangfen [12]." Regarding the Sui Garden, Du Mu[8] once wrote: "At Tianjie, the colors of autumn waters meet the Haimen Gate; in twilight mist, the sound of bells rises from the Sui Garden [13]."

The Changfu Garden, the Sui Garden, together with the landscape of Shugang Hill, spread around the northern side of the government precinct, serving as a transitional

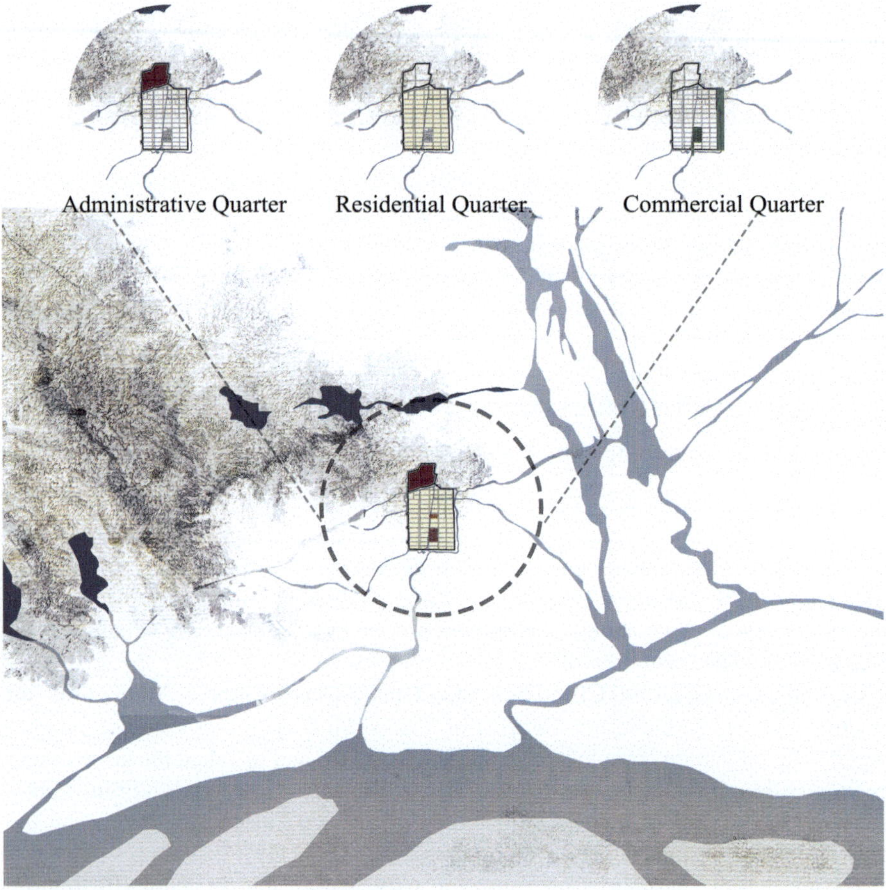

Fig. 5 Diagram of the relationship between the suburbs of the Ancient City of Yangzhou and the regional landscape system

zone from the natural environment to the artificial environment. The architectural ensembles within the Changfu Garden and the Sui Garden, nestled amidst the landscapes to the north of the city, created captivating scenery that invited prolonged appreciation. Their outstanding vistas acted as a medium through which the government precinct was visually and spatially connected to the suburban mountain–water environment, incorporating the natural scenery into the spatial experience of the government precinct. Supported by nearby palatial gardens and the natural landscapes of the suburban hills and waters, the government precinct stood prominently upon the elevated terrain of Shugang Hill, with the ridge as its backdrop, overlooking the entire Zicheng, and facing southward toward the Yangtze River, thereby forming a distinctive governmental landscape.

(2) The Development of the Canal System, and the Prosperity of Grain Transport and Markets

During the Sui-Tang Period, the development of Yangzhou City was inseparable from commercial activities, and these, in turn, were inseparable from grain transport. The expansion of grain transport drove the growth of the city. One of the principal reasons for constructing the Luocheng was the increasing prosperity of Yangzhou's commercial activity. The city's trade was organized into two categories: one comprised the officially planned markets established when the wards were laid out prior to the construction of the Luocheng; the other arose spontaneously in the course of urban development [14].

Within the Luocheng there were two official marketplaces, known as the Great Market and the Small Market. Their functions were analogous to the Western and Eastern Markets of Chang'an, but their spatial arrangements differed. In Yangzhou City, the Great and Small Markets were located in the central area of the Luocheng, aligned along the Market River. Separated by two wards, the Small Market was situated to the north of the Great Market. This official arrangement was closely related to Yangzhou's canal system. The Market River had existed prior to the construction of the Luocheng (ca. early Tang Dynasty [7]), and its water flowed from south to north; most freight along it also moved from south to north. To facilitate grain transport and market trade, the authorities adopted this configuration when delineating the Luocheng.

The reshaping of the water network system further stimulated the emergence of a second form of commerce. Transactions were no longer confined to official markets; rather, people began to establish trading places informally along streets and near canal banks—such as street markets along streets and near bridges, and grass markets outside the city along the canals. These marketplaces provided the physical basis for the saying, "Guangling stands at the great junction of north and south, where all commodities converge [15]." Coupled with the local custom that "the Jiang–Huai region places merchants above agriculture [16]," the commercial districts expanded further. Zhang Ku[9] described: "For ten li the streets are lined with markets; upon Mingyue Bridge one may behold immortals." Wei Yingwu[10] likewise wrote: "Along the river, trees flourish in green, with splendid halls stretching for ten li." Such depictions vividly illustrate how the street markets of Yangzhou City created an urban landscape distinct from that of the capital during the same period. As the closed ward system gradually transformed into an open street-based layout, the canal system and surrounding landscapes of Yangzhou City also underwent gradual changes under the influence of commerce. In the second year of the Guangde reign (764), Liu Yan[11] improved Pei Yaoqing's segmented grain transport method, resulting in the situation described as "River ships no longer entered Bian; all the shipments of Jiangnan accumulated at Yangzhou City [17]." Later, in the second year of the Baoli reign (826), when Wang Bo[12] served as Salt and Iron Transport Commissioner, it was recorded that "since the Official River within Yangzhou City became shallow, during droughts the grain transport vessels were impeded. Thus, from the city's southern Chang Gate, seven li westward to Gang Port, a new canal was excavated eastward, bending toward Chanzhi Temple Bridge and connecting to the old Official River. The excavation was deepened to facilitate navigation, making navigation easier [8]." From

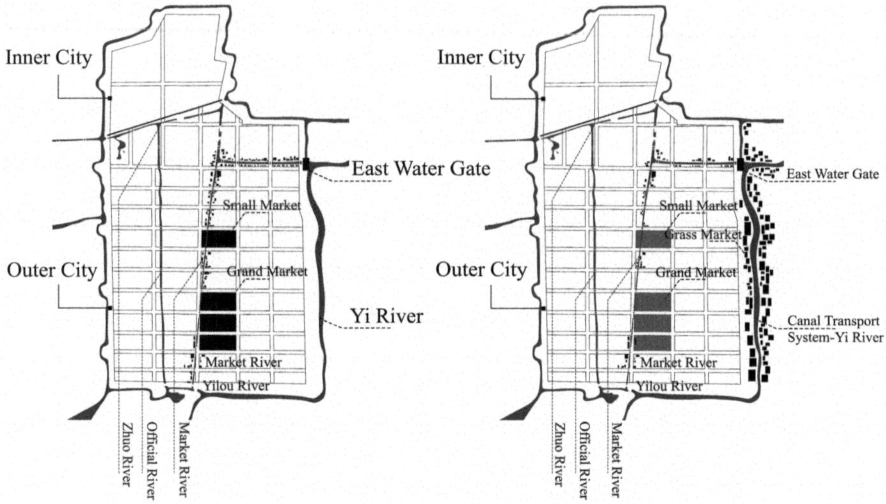

Fig. 6 Distribution and evolution of urban markets in Yangzhou City during the Sui-Tang Period (581–907 CE). Left: The "markets" established during the planning of wards and streets before the construction of the Outer City Wall (Luocheng). Right: As the city developed, the official canal within the city gradually lost its navigational function; the canal running along the southeast and east sides of the Outer City Wall became the main waterway, and an informal "grass market" gradually formed along the canal east of the city

this time, the Official River within the city gradually lost its navigational function, while the canals along the southeast and eastern sides of the Luocheng became the principal shipping channels, and a grass market gradually formed along the eastern canal (Fig. 6).

The prosperity of grain transport and markets greatly stimulated the development of surrounding landscapes. Initially, trees—especially willows—were planted along the canal-side imperial roads, providing both land transportation routes and roadside greening. Later, even privately built yet publicly accessible gardens appeared, such as teahouse and tavern gardens established for commercial purposes [18]. These improvements in canal-side landscaping, particularly along the southeastern waterways of the city, enhanced the habitability of the surroundings and strengthened the integration of the canal system with the city's functional districts.

(3) The Regulation of the Water System and the Planning of the Ward-Based Residential District

After the southward migration of the Yangtze River channel, the excavation of canals created a spatial configuration in which the alluvial plain was bounded by the Shugang Hill to the north, the Yangtze River to the south, and intersected by canals through its center. This spatial arrangement aligned with traditional principles of spatial arrangement and urban configuration [19]. Furthermore, the fertile alluvial plain, extending over forty li, offered naturally favorable conditions for settlement. Yet the abundance of rainfall posed great challenges for urban and residential construction. Through artificial regulation, however, a "two vertical, two horizontal" water network system was formed within the city, effectively controlling the groundwater level, ensuring water supply for daily use, and reducing the difficulties in residential construction and layout.

The residential morphology of Yangzhou's residential districts developed in three phases. The first phase extended from the establishment of the Jiangdu Palace to the construction of the Luocheng, when residences were scattered outside the palace grounds. The second phase began with the completion of the Luocheng and lasted until the dredging of the canals outside the city, when residences were more uniformly arranged within the city. The third phase occurred after the opening of the outer canals and continued until the fall of the Tang, when residences were primarily concentrated in the eastern part of the Luocheng [7]. The evolution of the ward-based residential districts was thus closely linked to the governance and restructuring of the water system.

During the Sui, the urban core consisted primarily of Jiangdu Palace, which contained no civilian residences. After the canals became navigable, artisans and merchants who gathered in Yangzhou City for trade and production established residences for ease of access and daily livelihood either on the flatlands at the foot of the Shugang Hill or along the canal banks. This led to an early spatial separation between the government precinct and the residential wards in the Tang Dynasty. In the Tang Dynasty, the Zicheng did not change in character and continued to serve as the government precinct. As the northern central regime increasingly relied on resources from the south, and as Yangzhou City became the key hub of grain transport from the south to the Central Plains, population influx grew substantially, prompting the construction of the Luocheng. When the Luocheng was built, its design largely followed the Tang urban principle of "within the walls of the prefectural city, land is divided into wards; outside the walls lie villages [20]." Inside the Luocheng, aside from the Great and Small Markets, most of the remaining space was allocated to residential quarters. The city was divided into thirteen wards from north to south and five wards from east to west, each measuring 450–600 m in length from east to west and about 300 m from north to south. Unlike the contemporaneous closed ward system in the two Tang capitals, however, Yangzhou's residences were loosely distributed, shaped by the intersecting watercourses and cross-axial roadways. The closer the location to waterways, the denser the housing, which was typically occupied by artisans and merchants.

By the late Tang Dynasty, the opening of the eastern canal stimulated further development of grain transport routes. Consequently, residences, warehouses, and hostels, attracted by commercial activities, gradually concentrated in the eastern part of the city, while the western wards gradually declined, with decreasing residential occupation.

2.3 The Extension of Urban Functions into the Suburbs

As the urban area continuously expanded, the city's functions also became increasingly diversified. The original urban planning could no longer accommodate the demands of increasing socio-economic activities; consequently, urban functions gradually extend beyond the boundaries of the city walls and extended outward into the surrounding natural landscape. Among these extensions were Buddhist and Daoist temples, villa estates, and town settlements—forms of the built environment that were distributed across the suburban landscape. These suburban developments supplemented the functions of the inner city while simultaneously strengthening the connection between the city and its encompassing regional landscape system.

(1) Buddhist and Daoist Temples as Shapers of Localized Cultural Landscapes

Records of Daoist and Buddhist activities in the Yangzhou City region date back to the late Eastern Han [17, 21], although their true development and flourishing occurred during the Sui-Tang Period. In this era, numerous Buddhist and Daoist Temples were established, mainly concentrated in the suburban Shugang Hill region near the city. Representative examples include Qiling Temple (later renamed Daming Temple), Guanyin Chan Temple, and Chanzhi Temple. Within the city itself, certain temples also existed, often situated adjacent to the waterways, such as Kaiyuan Temple and Zhengsheng Temple (later renamed Tianning Temple).

These temples often occupied pivotal positions within the broader region, serving as visual focal points of the landscape. Their siting took advantage of favorable scenic conditions that afforded commanding views. For example, Qiling Temple was celebrated in Li Bai's[13] verse:

The pagoda piercing azure sky;

Looks o'er the earth low and high.

Its top is blended with vital air;

Its crest rises over sea of clouds fair.

The world's seen divided into a dual sphere;

Its roof to the palace in heaven is near.

The water reflects its gold cone in a dream;

The sun is moving its sunlike flame.

Chanzhi Temple, located at the end of the Shugang Hill and on the edge of the high ridge, commanded superb vistas—eastward toward the canals, westward toward the ancient city, and southward across the plains below. Because of these temples, the people of Yangzhou City found places of both recreation and worship, where they could pray, make offerings, and enjoy the natural scenery.

As cultural landscapes, the temples were an extension of the city's spiritual and cultural life into the suburbs. They also served as regional landmarks, whose construction promoted the development of public garden landscapes and strengthened the interaction between the city and its surrounding landscape. In doing so, they enriched the cultural significance of both the urban center and its suburban environs, fostering a more diverse and multifaceted urban landscape.

The natural landscape pattern, with Shugang Hill to the north and the Yangtze River to the east, provided both protection and space for the development of the Ancient City of Yangzhou. The canal to the east of the city promoted the growth of waterborne transport. Cultural landscapes dominated by temples were mainly distributed in the Shugang Hill area near the city's outskirts. The four suburban towns of Yizheng, Guazhou, Liuxu, and Wantou all developed in close connection with the canal.

(2) Settlements and Towns Enriching the Regional Transportation System

With the expansion of Yangzhou City, a number of small towns gradually developed along the course of the Yangzhou canals, namely Yizheng, Guazhou, Liuxu, and Wantou, collectively known as the "Four Towns." Each of these towns was closely integrated with the canal system. The presence of ferry crossings and facilities for the loading and unloading of grain shipments stimulated their growth, enabling them to develop into secondary transshipment centers for the grain transport along the canal as well as hubs of land transportation [11].

Together, Yangzhou City, the "Four Towns," the roads that connected them, and the great waterways of the Yangtze River and the Grand Canal, which traversed the entire Yangzhou City region, constituted a comprehensive transportation network. This network linked the natural mountain–water environment, the urban landscape, and the settlements along the routes into an integrated spatial network of development (Fig. 7).

2.4 Gardens and Scenic Construction Within the Region

The prosperity of Yangzhou City during the Sui-Tang Period provided the economic foundation for the construction of gardens, while the region's natural landscape supplied the material basis. During this time, the number of gardens in Yangzhou City increased steadily, and two distinct types emerged: public gardens and private gardens. The building of gardens did not occur in isolation but was instead closely linked with the landscape environment and with urban functions. Different types of gardens appeared within different urban functional zones, each with unique stylistic features and serving different groups. In the process of interaction between the regional landscape system and urban zoning, the construction of gardens was subtly but profoundly shaped.

Within the government precinct, there were official gardens associated with administrative offices; near canal sections used for grain transport and commercial

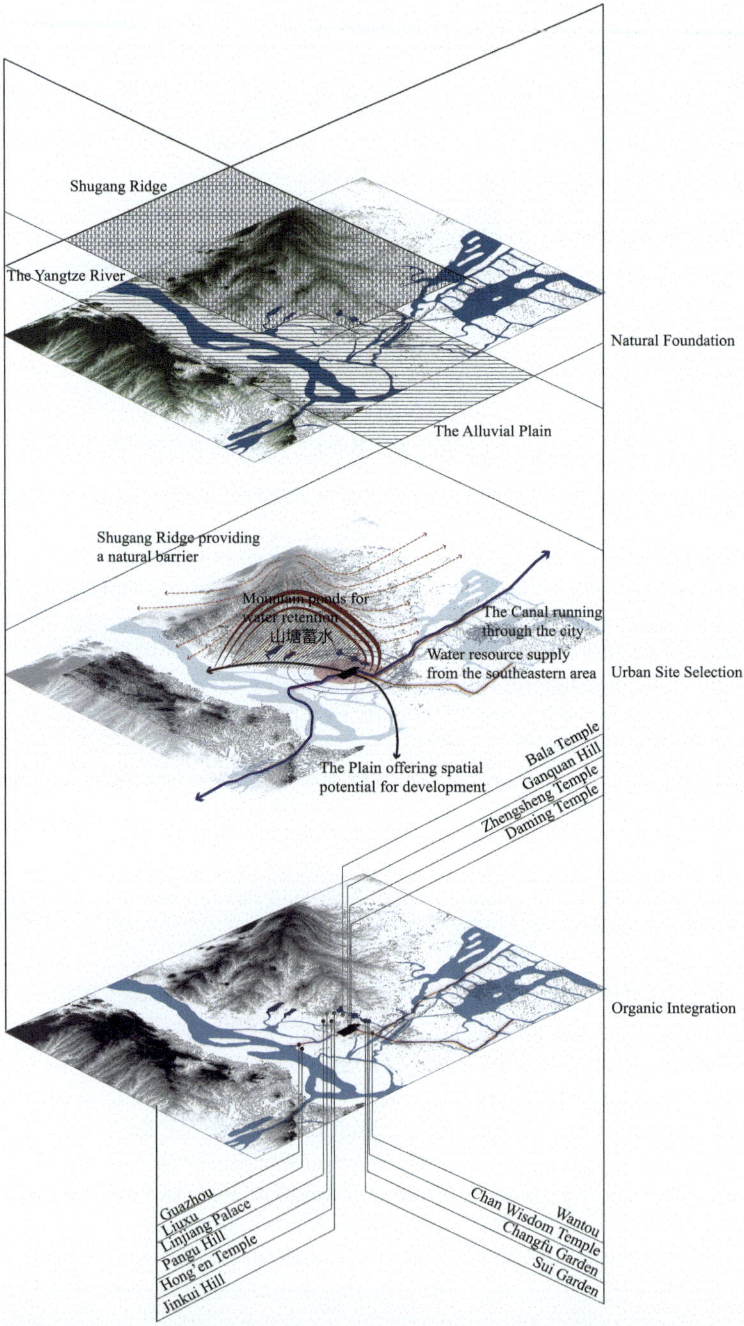

Fig. 7 Diagram of the relationship between the Ancient City of Yangzhou and the regional landscape system

marketplaces, there were teahouse and tavern gardens; within ward-based residential districts and villa estates, there were private gardens; and within temples and Daoist monasteries, there were temple gardens [18]. The relationship between these types of gardens and urban functions can be summarized as follows. First, within the city, official gardens and private gardens predominated. Official gardens were typically attached to governmental institutions such as prefectural offices or academies, and were concentrated within the government precinct. For example, *the Revised Gazetteer of Yangzhou City Prefecture*[14] records the "Zhengchun Pavilion" within the prefect's garden. Both official and private gardens were intimately related to the water network system. Fang Qian's *verse in Dwelling in the Lin Pavilion of the Hao Family in Yangzhou City*[15]: "The chilly moon pours in at my window, While, tired, I watch it from my slanting pillow; The winding stream that surrounds the islet, Then our wine-cups float slow on its billow". Akedang'a [9] suggests that private gardens at that time had already begun incorporating water into their layouts. Second, the Shugang Hill to the north of the city, as part of the external natural environment, not only contained imperial gardens such as Changfu Garden and the Sui Garen—with lofty halls, tiered corridors, pavilions over water, winding paths, and fragrant groves—but also provided a magnificent scenic backdrop for the city's internal gardens. Thus, the genuine mountains and waters beyond the city corresponded with the miniature mountains and waters within the gardens. Finally, in the southern suburbs near Guazhou, palatial gardens such as the Linjiang Palace were constructed to offer views of the Guangling Tide that still existed at the time, as well as landscapes transformed by hydraulic engineering.

The construction of gardens in Yangzhou City was thus inseparably connected with both the natural environment and the landscapes transformed through artificial regulation. Among the region's terrains, only the Shugang Hill in the north could be relied upon for landscape making. As Wang Yinggeng remarked in the Comprehensive Record of Mount Ping's Scenic Views and Ping Shan Hall Gazetteer[16]: "Here in Guangling, situated between the Yangtze and the Huai, the plain stretches endlessly without high mountains, deep valleys, or rushing streams to delight the eye; only the Shugang Hill to the north of the city stands as the preeminent site of the prefecture, commanding views of the verdant mountains of Run and Zhen, shimmering green against the horizon." Accordingly, the making of Yangzhou's gardens drew primarily on the Shugang Hill as their principal landscape setting, while the water systems at its base infused vitality into the gardens. Ultimately, a composite garden system emerged, consisting of multiple types of gardens, set against the backdrop of the Shugang Hill, and interconnected by the watercourses flowing below (Fig. 8).

After the flourishing of the Sui-Tang dynasties, Yangzhou City, by virtue of the terrain of the Shugang Hill, attained an overall urban appearance celebrated in verse as "streets shaded by willows for a thousand paces, rosy clouds reflected upon the double city walls." [13]. Within the city, the dense water network system created a cultural atmosphere evoked in the line "a moonlit night at the Twenty-Four Bridges, where does the jade-like beauty teach the flute?" [13]. Beyond the walls, in the suburban areas, the excavation and navigation of canals brought a scene of painted boats plying the waters, merchants gathering in abundance, and temples and pagodas

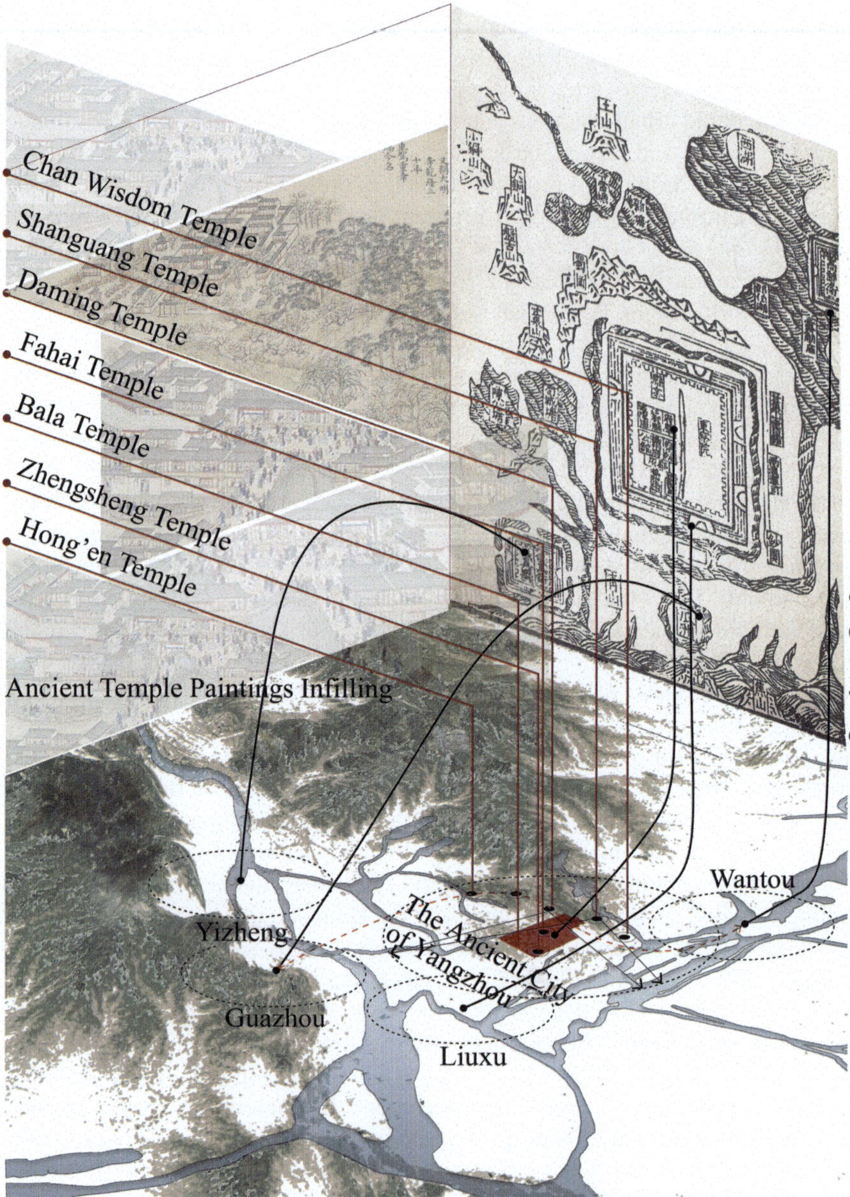

Fig. 8 Distribution of temples and garden complexes in the Shugang Hill area of the suburbs of Yangzhou City. *Source* Jiajing Weiyang Gazetteer (Ming Dynasty), Volume 1, General Map of the Prefectures and Counties under Yangzhou City Prefecture; Qian Weicheng, Paintings of the Scenery and Imperial Palaces along the Route of the Qianlong Emperor's Southern Inspection Tour (Qing Dynasty)—Fajing Temple and Ancient Qiling Temple of Yangzhou City (also known as Daming Temple)

nestled among the wooded hills, evoking the bustling spectacle of "bridge lights stretching to the Milky Way, sails and masts crowding by the riverbanks beneath the constellations [22]." The reason Yangzhou City could present such a landscape was that measured artificial construction was closely dependent upon the natural mountain–water configuration, producing a regional landscape system with distinctive local characteristics. At the same time, well-planned spatial development within both the city and its suburbs ensured integration of the urban fabric with the surrounding natural environment, ultimately forming an organic and pleasant living environment.

An analysis of Yangzhou's development during the Sui-Tang demonstrates that ancient urban functional zoning was not composed of isolated segments, but rather operated as interdependent parts of an integrated whole. The regional landscape system, grounded in the mountain–water environment and extending across both city and suburb, functioned as the essential link connecting urban functions. In the course of urban development, the regional landscape system facilitated the well-structured distribution of urban zones, while those zones, in turn, drove further refinement of the system. Together, they enabled a more livable and sustainable form of urban growth. At the core of the relationship between urban development and the regional landscape system lay the integration of the natural mountain–water environment with the built environment. Through more than four centuries of deliberate transformation of nature across the Sui-Tang, the Yangzhou City region established a well-defined landscape configuration, ensuring that the city and its natural setting remained in a state of dynamic equilibrium. By the Ming and Qing, Yangzhou City had earned the reputation that "Yangzhou City excels through its gardens and pavilions [23]," and today it endures as a city renowned for its cultural depth, prosperity, and refinement (Fig. 9).

The construction of any city is inseparable from the natural landscape environment upon which it depends. Therefore, urban development must take into account the relationship between urban functional zoning and the regional landscape system. Only by establishing well-structured zoning that enables the city to operate efficiently while remaining in harmony with nature can an urban settlement achieve resilience and sustainable development across millennia.

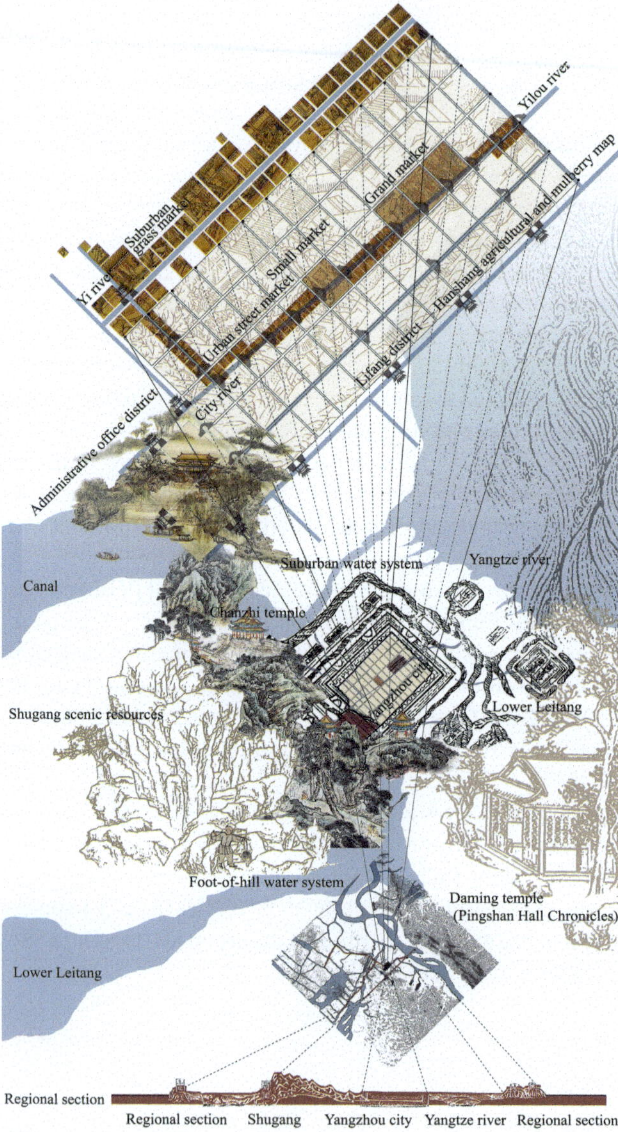

Fig. 9 Diagram of the progressive relationship among the regional landscape system, urban site selection, and internal functional zoning of Yangzhou City. *Source* Jiajing Weiyang Gazetteer (Ming Dynasty), Volume 1, General Map of the Prefectures and Counties under Yangzhou City Prefecture; Wang Yun (Qing Dynasty), Xiuyuan Painting; Yuan Yao (Qing Dynasty), Four Views of Yangzhou City: Spring Terrace under the Bright Moon; Zhao Zhibi (Qing Dynasty), Pingshantang Illustrated Records, 10 Volumes—Agriculture and Sericulture in Hanshang; Yuan Yao (Qing Dynasty), Immortal Realm of Penglai; Ma Yuan (Southern Song Period), Water Album: Layered Waves and Surging Billows

Notes

1. "Zicheng" and "Luocheng" were the core and expanded areas of Sui-Tang Period (581–907 CE) Yangzhou City, showing early Chinese integration of natural landscape, waterways, and urban zoning.
2. *Dushi Fangyu Jiyao, Volume 23, Nan Zhi Wu*, compiled by Gu Zuyu in the early Qing Dynasty (1368–1912 CE), is a foundational work of Chinese historical geography that interprets regional landscapes through the integrated roles of mountains, water systems, administrative space, and urban form.
3. *Tujing* were early Chinese geographical treatises, fully developed during the Tang Dynasty (618–907 CE), that documented regional landscapes through integrated descriptions of mountains, water systems, settlements, and urban boundaries, serving as precursors to later landscape and urban surveys.
4. Compiled during the Jiajing reign of the Ming Dynasty (1522–1566 CE), the *Jiajing Weiyang Gazetteer* is a local gazetteer of Weiyang—historical Yangzhou, a major commercial and cultural city along the Grand Canal. Like other Ming local gazetteers, it systematically records administrative boundaries, waterways, urban wards, transportation networks, public buildings, gardens, temples, and local customs.
5. Waterborne grain transport (Caoyun) in the Sui–Tang Period (581–907 CE) was an official, state-managed system for transporting grain, salt, and other essential commodities along rivers and canals.
6. Bazaar precinct (Shisi) in the Sui–Tang period (581–907 CE) were officially designated commercial areas within or adjacent to urban centers. These precincts were typically organized along major streets or near city gates, incorporating permanent shops, stalls, warehouses, and occasionally temples or guild halls.
7. *Taiping Huanyu Ji* Northern Song Dynasty (980 CE) is a geographic gazetteer documenting cities, landscapes, and infrastructure, useful for understanding medieval Chinese urban-environment interactions.
8. Du Mu Tang Dynasty (803–852 CE) was a poet and official whose works vividly describe urban landscapes and gardens, offering insight into Tang city–environment interactions.
9. Zhang Ku, Tang Dynasty (618–907 CE), was a poet whose work documents Yangzhou's streets, markets, and waterways, highlighting urban–landscape interactions.
10. Wei Yingwu, Tang Dynasty (737–792 CE), poet documenting Yangzhou's urban riverscapes, streets, and gardens.
11. Liu Yan, Tang Dynasty (747–820 CE), Tang official who reformed Yangzhou's Grand Canal and shaped canals and markets.
12. Wang Bo, Tang Dynasty (826 CE), served as Salt and Iron Transport Commissioner and proposed digging a canal to improve Yangzhou's river transport.
13. Li Bai Tang Dynasty (701–762 CE) was a renowned poet and scholar whose works often depict natural landscapes, rivers, gardens, and urban settings.
14. Revised Gazetteer of Yangzhou City Prefectur, Qing Dynasty (1736 CE) is a gazetteer of Yangzhou compiled under official supervision. It documents the city's administrative divisions, waterways, streets, bridges, gardens, and urban infrastructure, providing detailed historical information on urban planning, river management, and landscape features during the Qing period.
15. Fang Qian, *Dwelling in the Lin Pavilion of the Hao Family in Yangzhou City,* Qing Dynasty (1750 CE) is a poetic account of a temporary residence at the Hao family's garden pavilion in Yangzhou. The work describes the garden layout, plantings, water features, and surrounding landscape.
16. Wang Yinggeng, Mount Ping's Scenic Views and Ping Shan Hall Gazetteer, Qing Dynasty (1783 CE) is a local gazetteer and illustrated record of the Pingshan Hall area. The work documents topography, watercourses, garden layouts, pavilions, and scenic sites.

References

1. Wang, W., & Zhang, Y. (1993). *Collected essays on Chinese archaeology* (p. 5). Science Press.
2. Wang, H. (2016). *World heritage of the Yangzhou Canal* (p. 1). Nanjing Normal University Press.
3. Tan, Q. (1996). *The Historical Atlas of China, Vol. 5: Sui, Tang, Five Dynasties and Ten Kingdoms Periods* (p. 6). China Cartographic Publishing House.
4. Gu, Z., He, C., & Shi, H. (Qing). (2005). *Records of historical geography* (p. 3). Zhonghua Book Company.
5. *Gazetteer of Ganquan County, Guangxu Revised* (Vol. 2). Mountains and Rivers.
6. Luo, Z. (1980). The discovery of Tang-Dynasty Ancient River Channels in Yangzhou and related issues. *Cultural Relics* 21–27.
7. Li, X. (2015). *The historical space of Chinese Cities* (p. 3). Peking University Press.
8. Liu, X. (Later Jin) (1975). *Old book of Tang* (p. 5). Zhonghua Book Company.
9. Akedang'a (Qing), Yao, W. (Qing) (2006). *Gazetteer of Yangzhou Prefecture, Jiaqing Revised* (p. 12). Guangling Book Company.
10. Zheng, X. (2006). *Research on landscape planning approaches for new urban districts*. Beijing Forestry University.
11. Li, T. (2002). *A historical study of Tang-Dynasty Yangzhou* (p. 10). Jiangsu Ancient Books Press.
12. Le, S. (2007). *Taiping Huanyu Ji (Universal Geography of the Taiping Era)* (p. 11). Zhonghua Book Company.
13. Du, M. (Tang), Annotated by Feng, J. (Qing) (1962). *Collected Poems of Fanchuan with annotations* (p. 9). Shanghai Ancient Books Publishing House.
14. Jiang, Z. (1994). Tang-Dynasty Yangzhou waterways and the twenty-four bridges. *Studies on Han-Tang and Frontier Archaeology*, 162–168.
15. Wang, P. (2012). *Tang Huiyao (Institutional History of the Tang)* (p. 4). Shanghai Ancient Books Publishing House.
16. Liu, S. (Tang). (2013). *New Anecdotes of the Great Tang* (p. 6). Zhejiang Publishing Group Digital Media Co., Ltd.
17. Li, F. (Song) (1961). *Extensive records of the Taiping Era* (p. 9). Zhonghua Book Company.
18. Du, M. (2014). *The transformation of Yangzhou Gardens: People and scenery* (p. 6). Tongji University Press.
19. Wu, L. (2014). *A history of the human habitat* (p. 10). China Architecture & Building Press.
20. Li, L. (Tang) (2014). *Tang Liudian (Six Statutes of the Tang)* (p. 7). Zhonghua Book Company.
21. Chen, S. (Jin), Annotated by Pei, S. (Song) (2011). *Records of the Three Kingdoms* (p. 1). Zhonghua Book Company.
22. Li, S., Annotated by Lu, Y. (2009). *Collected works of Li Shen with commentary* (p. 11). Zhonghua Book Company.
23. Li, D. (Qing). (1960). *Record of painted boats in Yangzhou* (p. 4). Zhonghua Book Company.

Chapter 10
Urban Subarea: Regional Landscape System and Zonal Construction—"Nine Hills and Five Rivers" and the Ancient City of Shaoxing

Abstract This Chapter examines the urban subarea structure of the ancient city of Shaoxing from the perspective of a regional landscape system, focusing on how the natural framework of the Nine Hills and Five Rivers shaped long-term urban zoning and spatial organization. Situated within the mountain–plain–water landscape of the Shanhui Plain, Shaoxing City developed a distinctive urban form in which topography and hydrology functioned as a stable landscape substrate for urban construction. From a landscape architectural and urban morphological perspective, this chapter elucidates how hills, rivers, lakes, and water networks collectively structured urban functional zoning, including administrative precincts, residential quarters, commercial districts, storage and dock areas, gardens, and religious sites. Methodologically, the study integrates historical landscape analysis with spatial interpretation, using the Southern Song period (1127–1279) as a focal moment to examine the interaction between natural landscape systems and artificial hydraulic intervention in shaping subarea patterns. The analysis demonstrates that Shaoxing's urban development was not the result of incremental urban expansion alone, but rather a process of continuous adaptation to a persistent landscape framework. The enduring role of the Nine Hills and Five Rivers reveals a landscape-driven logic of subarea construction, offering a representative model for understanding regional landscape systems as formative forces in traditional Chinese urban morphology.

Shaoxing City is an ancient city with a history of 2500 years. Since its founding during the reign of King Goujian of Yue (496–465 BCE), the Shanhui plain's hydrological network has undergone several major changes; yet the city has continued to develop on its original site and has formed a distinctive urban character—an outcome inseparable from traditional principles of human settlement-making. Therefore, Shaoxing City during the Southern Song Period (1127–1279) is selected as a representative case: taking Nine Hills and Five Rivers as the natural foundation for urban construction and, on the basis of an artificially reorganized hydrological network, the discussion examines the city's the construction of urban functional areas—including the formation of functional spaces such as government offices (yamen), residential quarters, marketplaces, granaries, docks, market towns, gardens, and temples

X. Zheng, *Terrapoetic Urbanism*, https://doi.org/10.1007/978-981-95-8676-9_10

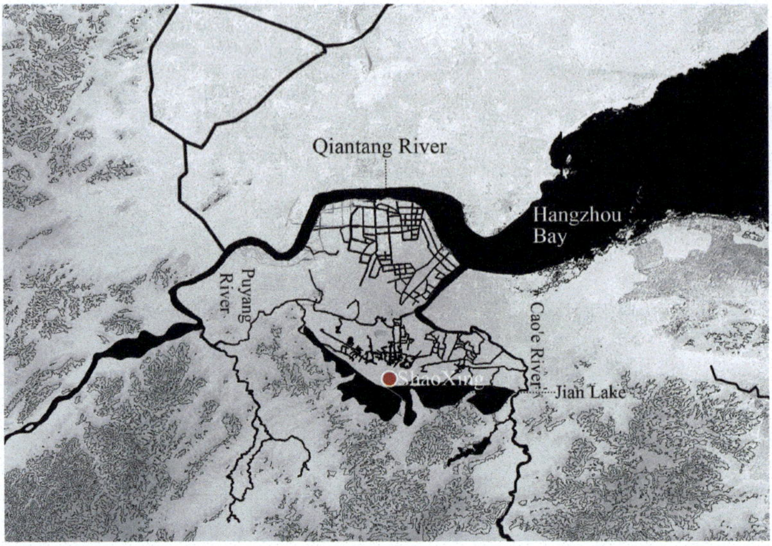

Fig. 1 Location of Shaoxing City

and monasteries—and explores how these spaces adapted to the regional landscape system.

Shaoxing City (formerly called Dayue, abbreviated as Yue, and so named because Emperor Zhao Gou of the Southern Song Period proclaimed "Shaozuo Zhongxing" [1]) lies on the southern shore of Hangzhou Bay. The ancient city was established upon the Jibei Hills of the Kuaiji Mountain Range and the Shanhui Plain. To the south, continuous hill ranges extend into the plain, forming multiple low hills, river valleys, and basins which, together with the abundant waters of Hangzhou Bay, constitute an urban fringe landscape system integrating mountains, plains, and water. In addition, mountain runoff and a relatively high groundwater table produced a reticulated lake–river network across the Shanhui Plain, exerting a profound influence on the city's site selection, urban functional development, and landscape construction (Fig. 1).

Shaoxing City has long been renowned as a watertown and land of marshes. As Wang Anshi[1] remarked, "The Yue mountains remain ever green, the waters ever clear; the Yue people have always dwelt in a mountain–water country [2]." This statement is a tribute to the city's landscape-based urban formation and its refined aesthetics.

As early as the Hemudu Culture period (ca. 5000–3000 BCE), traces of the Yue people's activities can be found in the Kuaiji Mountains and along the shores of Hangzhou Bay. The urban site selection of the city, the transformations of Jian Lake, and the spatial evolution within the urban area all hold great research value for the study of hydraulic works and urban planning. Scholars from various disciplines have already produced a considerable body of studies. For instance, in 2001, Che Yueqiao and Chen Qiaoyi co-authored The Historical Geography of Shaoxing

City, which provides a detailed account of historical geographical changes in the Shaoxing City region. In 2004, Li Lei published research on the historical evolution of Shaoxing's urban spatial structure, analyzing its development from the perspective of urban planning. In 2008, Qiu Zhirong published Gardens and Water in Shaoxing City, discussing the unique characteristics of Shaoxing's gardens. In 2010, Tu Jianhong and Chen Guocan each published studies on urban construction in Shaoxing City during the Southern Song Period. In 2014, Chen Qiaoyi, Qiu Zhirong, and other experts founded the Shaoxing City Jian Lake Research Association, which in 2015 published multiple volumes on Shaoxing's mountain–water environment, hydraulic works, ancient city, ancient bridges, historical atlases, poetry collections, and anthologies of scholarly papers. In 2015, Mao Lei published research on the ecological adaptability of traditional watertown vernacular dwellings in Shaoxing City. These studies have largely concentrated on hydraulic works, urban planning, gardens, architecture, and heritage conservation, with each focusing primarily on its specialized field. As a result, mountains, waters, and the city have often been treated as relatively independent components in scholarly studies.

Mountains, waterbodies, hills, and lakes are regarded as the natural foundation of urban development, forming the natural elements of the regional landscape system. The city is constructed upon this system, and the patterns of functional zoning emerge in relation to the constraining factors embedded within the regional landscape system. Moreover, the construction of urban subareas integrated with landscape planning further incorporates the built environment into nature, thereby enabling the city to sustain its development within a dynamic natural setting.

During the peak of marine transgression about 6000 years ago, the Shanhuai Plain was still an expanse of open water. The Yue people chiefly relied on the forests and natural resources of Kuaiji Mountains: "They cultivated the land along the hills and ridges or pursued deer and other game for subsistence [3]." In this migratory form of agriculture, they lived what could be described as a mountain-dwelling mode of life. After the sea receded and plains emerged—where "the waters were turbid and stagnant" [4]—the environment remained harsh. Building on adaptation to nature, continuous human intervention progressively steered the region toward a more habitable condition. Around Shaoxing's ancient city, the landscape pattern underwent four major phases of transformation: during the Spring and Autumn Period (770–476 BCE), King Goujian of Yue, "Woxin changdan" (lying on firewood and tasting gall),[2] founded the capital at the lower reaches of the Ruoye Creek, and the Yue people moved down to settle on the plains. To remedy the adverse water regime—"All streams converge here; when the great waters of the lakes overflow, they spread and break forth, forming rivers upon reaching the land, while tributaries and branches interweave into a network of channels" [5]—the Yue people initiated hydraulic works for flood control and irrigation (e.g., ancient waterways and Fuzhong Pond). In the Eastern Han Dynasty (25–220 CE), with the ancient city as the center, Jian Lake was constructed to collect thirty-six upstream sources; in the Western Jin Dynasty (265–316 CE), a canal was excavated from the west of Shaoxing City to Guling on the bank of the Qiantang River; by the Southern Song Period, Jian Lake was gradually reclaimed, the upstream system largely unchanged, while the

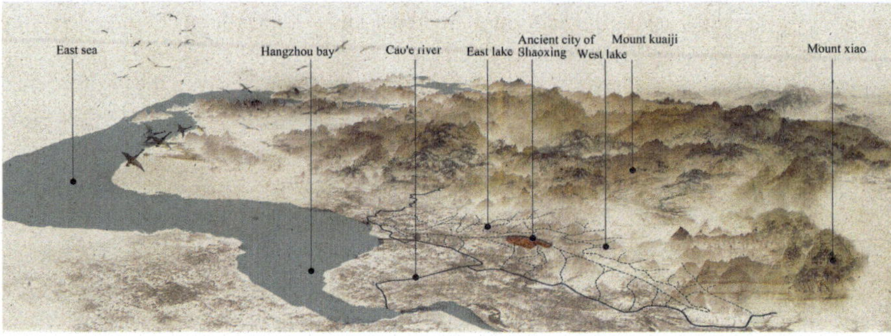

Fig. 2 Urban site selection and regional landscape pattern of Shaoxing City

plain's hydrological network as a whole migrated northward. Reviewing Shaoxing's urban history—from urban site selection and capital establishment in the Spring and Autumn Period, through the first expansion in the Sui-Tang, to the comprehensive street-and-canal planning completed in the Southern Song—the city's form continued into the present. By the Southern Song, urban construction had reached a peak, the city's functional zoning was refined, the surrounding hydrological network stabilized, and the relationship with the natural environment approached a condition of harmonious development; hence this study focuses on the Southern Song period. The urban image in the *Record of the Xiaoyao Pavilion*[3]—"Around the tower are windows; beyond the windows lies the city; encircling the city is the lake; and beyond the lake rise the mountains"—epitomizes the regional landscape system in which mountains, waters, and urban area are interwoven in Shaoxing City (Fig. 2).

1 The Natural Pattern Within the Regional Landscape System

The ancient city of Shaoxing is situated in the transitional zone between the Kuaiji Mountains and Hangzhou Bay, characterized by an overall topography sloping from the higher elevations in the south to the lower terrains in the north. The southwestern part, defined by the Kuaiji Mountain Range, marks the highest elevation, from which the mountainous terrain extends outward and gradually declines northeastward. To the north of the mountains lies the Shanhuai Plain, a geomorphic feature formed through marine regression, which further extends northward and protrudes into Hangzhou Bay. This mountain–plain–water sequence creates a terraced landform that constitutes the natural foundation for the development of the ancient city of Shaoxing.

1.1 Four Mountains, Two Rivers, and Numerous Streams—The Landscape Foundation of the Shanhuai Plain

On the Shanhuai Plain, the four major mountains are Kuaiji Mountain, Siming Mountain, Tiantai Mountain, and Longmen Mountain, with Kuaiji Mountain serving as the most prominent. Its mountain mass extends broadly, exhibiting a rugged and intricate terrain that projects outward, thereby generating multiple intermontane basins and alluvial fans at the mountain foothills. Abundant in forest and natural resources, this area constituted a crucial material resource base for the ancient city of Shaoxing.

The Shanhuai Plain is endowed with abundant water resources. In its early stages, the two principal river systems traversing the region were the West Xiaojiang (Puyang River) and the East Xiaojiang (Cao'e River). Extending northward, the Kuaiji Mountain Range gives rise to the Jibei Hills, within which numerous rivers flow along a north–south orientation, interspersed with catchment areas of varying scales.

Before the creation of Jian Lake in ancient times, the waters of the Jibei Hills were all received and conveyed by the downstream sections of the Cao'e River and the Puyang River, before ultimately discharging into Hangzhou Bay [6]. Consequently, the distribution of the Thirty-Six Sources of Jian Lake across the Shanhuai Plain laid the essential foundation for Shaoxing's transformation into a water-network city.

2 The "Nine Mountains and Five Rivers" Together Fostered the Construction of the Ancient City

The Yue people gradually ended their lifestyle of mountain dwelling and migrated north of the mountains, where most communities established settlements upon relatively elevated and well-drained isolated hills. The ancient city of Shaoxing developed in the same manner. Since its foundation in 490 BCE, the ancient city has continuously evolved on its original site. Beyond the overarching influence of the surrounding landscape environment, the site selection of the city—structured around the "Nine hills of Yue and the four-vertical-one-horizontal water network"—generated an internal shanshui pattern that profoundly shaped the city's urban development and construction.

The nine isolated hills surrounding the waterway hub of the Ruoye Creek included Mount Zhong (also known as Mount Wolong, now called Mount Fushan), Mount Ji, Mount Peng, Mount Guai (also referred to as Mount Feilai, Mount Gui, or Mount Baolin, now called Mount Tashan), Mount Baima, Mount Baolang, Mount Emei, Mount Huozhu, and Mount Huangzhuo. Among them, Mount Zhong stood as the tallest, resembling a coiled dragon resting on the plain, with steep northern slopes and a broad, gentle southern expanse, and was thus regarded as the most suitable site for city construction. Fan Li established a small city to the south of the mountain, where a palatial terrace was constructed on the gentle foothills. From then on, the

Wolong Mountain area became the administrative seat of commandery and prefecture throughout successive dynasties in Shaoxing City. In the same year, a larger city was constructed on its eastern side, incorporating the remaining eight isolated hills within its boundaries. This laid the foundation for the subsequent scale of Shaoxing City Prefecture and gave rise to the landscape pattern of "Eight Hills Enclosed within the City" (with Mount Baolang originally outside the small city but later included after urban expansion). From the Southern Song Period onward, the Jian Lake gradually disappeared, water bodies shifted northward, and the northern terrain rose, leading to the partial submergence of hills within the city. By the time of the *Map of Shaoxing City Prefecture in the Kangxi Reign of the Qing Dynasty*,[4] only Mount Wolong, Mount Ta, Mount Ji, Mount Baolang, and Mount Baima remained clearly visible. Today, only Mount Fushan, Mount Ji, and Mount Ta survive within the ancient city of Shaoxing, constituting distinctive urban features and forming its three major scenic spots and historic sites.

Apart from the mountain resources, the location of the small city and the large city encompassed five natural river courses. To the south of the city lay the Nan chi River, flowing northwards into the city, later known as the Fu River; the Potang River, also running north–south, entered the city through the Shuipian Gate and functioned as the main river course of the western half of the city; in the southeast, the Pingshuixi River entered through the Dongguo Gate and passed under the Bazhi Bridge, serving as the main river course for the eastern half of the city. On the western side, the Lougong River connected with the internal water system through the Shuipian Gate and the Ying'en Gate. The Jiuwuqiao River traversed the large city from east to west, linking with the small city through its water gate. Collectively, these rivers constituted a "four vertical and one horizontal" river course, which marked the earliest water-network pattern of the ancient Yuezhou city (Fig. 3) [7].

Through human intervention, the city derived a distinctive watertown character and landscape pattern from the Nine Hills and Five Rivers. Research on the historical evolution of the spatial form of the ancient Shaoxing city demonstrates that the relationship between mountain, water, and city has been one of constant transformation and adaptation. At its core, urban development has consistently relied upon the Nine Hills and Five Rivers. The internal landscape framework has delineated the spatial pattern for the city's expansion and growth (Fig. 4).

3 Artificial Intervention in Reshaping the Regional Water Network

The regional landscape pattern of "mountain–plain–water" not only provided abundant water resources but also resulted in frequent flood hazards. The first was the runoff from the Kuaiji Mountains, often causing floods; the second was the intrusion of seawater, rendering the water saline and unsuitable for irrigation. Consequently, the development of water conservancy became particularly crucial. Through artificial

Fig. 3 Diagram of the urban landscape pattern: "Nine Hills of Yuezhong" and "Intersecting Five Waterways"

intervention, the water resources of the Shanhuai Plain were effectively managed, forming a more habitable water-network system.

3.1 Evolution of the Water-Network Pattern Within the City

The water-network system of the ancient city of Shaoxing relied on the "four vertical and one horizontal" natural water system, which, through artificial intervention, evolved into a checkerboard water network pattern. Later generations referred to this as the "Seven-String Waters" (Qixian Shui), although the network encompassed far more than seven waterways. As Ren Guiquan vividly described in The Formation

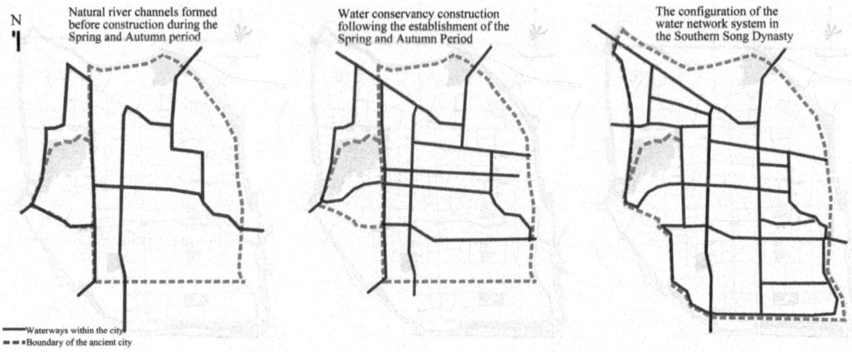

Fig. 4 Distribution of Shaoxing City urban water network in Pre-Spring and Autumn (c. 1046–770 BCE), Spring and Autumn (770–476 BCE), and Southern Song Dynasties (1127–1279)

and Characteristics of Shaoxing's Water Towns, the natural river channels within the city can be likened to the "backbone of a fish," while the transversely constructed canals resemble its "spines," a metaphor that effectively illustrates the formation of the water-network system. By the Southern Song Period, the street-and-canal layout of the city's water-network system was largely finalized. According to the *Map of the Main Roads in Shaoxing Prefecture*,[5] the urban area spanned 7.4 km^2, containing thirty-three waterways with a total length of approximately sixty kilometers [8]. This intricate network provided highly convenient waterborne transportation and, with its interlacing structure, became an important boundary framework for the subdivision of urban wards.

The Fu River, as a natural river channel flowing into the ancient city of Shaoxing, gradually evolved into a boundary river demarcating "one city and two counties" during the city's historical development. Its strategic importance and commercial value were considerable, as evidenced by records from the Southern Song Period that describe the prosperity of commercial markets and shops along the Fu River. According to the Map of Shaoxing City Prefecture from the Qing Dynasty (1644–1912), the Fu River served as the principal waterway within the city, into which multiple east–west natural river channels converged, forming an internal water-network pattern known as the "Seven-String Confluence" (Qixian Hui Shui). Here, the "seven strings" referred to the east–west waterways intersecting with the Fu River. Throughout its history, the transformation of the city's water-network pattern has remained fundamentally dependent on the natural waterway system, and the traces of these natural river channels within the city remain distinctly discernible to this day.

3.2 *Out-of-City Waterways and the Development of Waterborne Grain Transport*

The abundant water resources and the dense water network surrounding the ancient city of Shaoxing created highly favorable conditions for the development of shipping. During the Spring and Autumn Period, King Goujian of Yue initiated the excavation of the Shanyin Ancient Waterway, establishing an early route. In the Eastern Jin Dynasty (317–420 CE), Administrator He Xun[6] of Kuaiji utilized this ancient channel to construct the Xixing Canal, linking the Zhejiang River to the Western Xiao River and connecting with the canals east of Shangyu, as well as the Yao and Yong Rivers, ultimately reaching Mingzhou (modern Ningbo). This project marked the full completion of the Eastern Zhejiang Canal, which became the principal transport artery of the region. Throughout the Northern and Southern Song dynasties (960–1279 CE), the Eastern Zhejiang Canal supported intensive water transport, playing a decisive role in driving Shaoxing's prosperity across multiple dimensions—most notably in shaping its urban zoning and the evolution of its urban form [9].

3.3 *The Development of the Regional Farmland Irrigation System*

The agricultural development of ancient Shaoxing can be traced back to the era of migratory agriculture. During the Spring and Autumn Period, following King Goujian of Yue's return from Wu, his long-term policy of "Ten Years of Nurturing the People, Ten Years of Training the Army" placed particular emphasis on agriculture [1]. As rice was the staple grain crop of ancient Shaoxing, the construction of embankments and ponds to store fresh water and to keep out saline water for irrigation became essential, making farmland irrigation a critical undertaking. Beginning in the Spring and Autumn Period, organized efforts were made to build dike-and-pond engineering works such as the Wutang and Shitang embankments, thus exerting deliberate control over the natural water system to optimize its irrigation and drainage functions.

With the increase in population and the resulting pressure on agricultural production, the extensive construction of the dike-and-embankment system across the Shanhuai Plain led to the construction of the Jianhu Hydraulic Project during the Later Han Dynasty (25–220 CE). Owing to its substantial retention capacity, this monumental hydraulic project effectively controlled the incidence of mountain floods and significantly improved the water environment of the Shanhuai Plain.

In response to the saltwater intrusion from Hangzhou Bay, the construction of seawalls became particularly crucial following the development of Jianhu. Combined with the excavation of canals for grain transport, specifically the Xixing Canal, which organized the inland river system, the seawall system was largely completed and

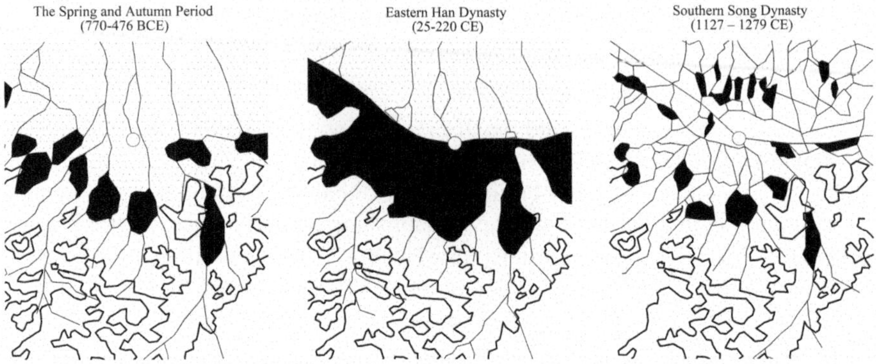

Fig. 5 Changes in the peripheral landscape pattern of ancient Shaoxing City during the Spring and Autumn, Eastern Han, and Southern Song dynasties

reinforced by the Song Dynasty (960–1279 CE). Consequently, a dense river–lake network was established across the Shankuai Plain (Fig. 5) [1].

4 Urban Development upon the Landscape Pattern

The natural landscape pattern, under moderate artificial intervention, fostered a well-functioning regional water-network system and favorable conditions for agricultural cultivation, within which the ancient city of Shaoxing developed and prospered. The city was strategically located upon the Nine Hills of Yue, and its urban planning was structured around the foothills and water network, enabling zoning and construction that was fully responsive to topography and hydrology. Within the city, government offices, residential quarters, and commercial district were arranged in situational accordance, creating a harmonious urban landscape. As the city expanded beyond its original boundaries, the surrounding hills and dense woodlands, river–lake network, and peri-urban zones were integrated into landscape construction endeavors. These included private gardens and residences, the construction of Buddhism and Taoist temples, thereby linking the intra-urban spatial structure with the suburban environment. Over time, Kuaiji Mountains and Jian Lake, together with the water-network system of the plains, the foothill government offices, the waterborne commercial district, waterside dwellings, private estates, and temple-pagoda complexes, became interwoven into a unified landscape city that transcended the conventional boundary between city and countryside. This integration embodied the evolution of the regional landscape system and the flourishing of an urban morphology characterized by a symbiotic integration of landscape and city.

4.1 Site Selection and Capital Establishment

Fan Li[7] constructed the capital north of the Kuaiji Mountains and south of Hangzhou Bay. The city site was backed by mountains and faced the sea, with the Western Xiaojiang River and Eastern Xiaojiang River serving as natural barriers to the west and east. Firstly, the location occupied the junction of the Ruoye Valley and the principal east–west thoroughfare of the Shanhuai Plain, ensuring convenient transportation and facilitating access to the basins behind the Kuaiji Mountains for resources essential to production and daily life. Secondly, during that period, the Shankuai Plain was a marshland "exposed to tidal incursions and inundated by brackish waters," and elevated isolated hills provided the fundamental sites for urban construction. The nine isolated hills essentially defined the scale of the ancient city of Shaoxing. Shaoxing City was established upon the Nine Hills and Five Rivers as its urban foundation, relied on the Kuaiji Mountains as a natural resource base, and used the two rivers as natural barriers in the east–west direction, developing within a landscape encircled by mountains and interlaced with waterways.

4.2 Boundary Transformation

During the Spring and Autumn period, Fan Li constructed both the Minor and Major Cities with the Nine Isolated Hills of Yue as the natural foundation. The Minor City was strategically positioned with Wolong Hill as its natural barrier to the northwest, where no city wall was erected. From the western corner of Wolong Hill, a city wall was built extending to the Hanpian Gate and eastward to the Fengyi Bridge, forming the southern city wall, while the eastern city wall connected northwards to Wolong Hill. This configuration, described as "a circumference of one hundred and twenty-one *bu* (paces), round on one side and square on three," Yuejue Shu [10] defined the initial scale of the Minor City. The Minor City contained four land gates and one water gate. In the same year, the Major City of Shanyin was constructed to the east of the Minor City, with a perimeter of twenty *li* and seventy-two *bu* (paces), comprising three land gates and three water gates. Collectively, the Minor and Major Cities incorporated eight of the Nine Isolated Hills of Yue within the city's boundaries. The city's development was thus harmoniously integrated with the natural environment, resulting in the *'Bashan Zhongcang'* (eight hills enfolded within the city) landscape pattern.

The period from the Sui-Tang Dynasties (581–907 CE) to the Northern Song Dynasty (960–1127 CE) marks the most pivotal phase in the urban transformation of Shaoxing City. Yang Su constructed the Zicheng (Inner City) and the Luocheng (Outer City), expanding upon the foundations of the Minor City and Major City. The Zicheng (Inner City) was first established at the foot of Wolong Hill, incorporating four land gates and one water gate. The southwest continued to utilize the hills as the city wall, while the southeast was reinforced with city walls. The Luocheng

(Outer City), with three land gates and six water gates, roughly corresponding to today's ring road [11], effectively shaping the enduring urban outline of Shaoxing City. Apart from modifications such as the fortification works by Liang Zhongxian during the Xuanhe (1119–1125 CE) period to defend against pirates, the reduction of its southwestern corner, and the expansion in the 13th year of Zhizheng (1353 CE) that incorporated one *xiang* (township-level district) into the city, the Luocheng (Outer City), remained largely unchanged throughout the Song, Yuan (1271–1368 CE), and Qing dynasties (1644–1912 CE) [12].

The ancient city of Shaoxing ultimately adopted a rare dual-moat configuration in which a watercourse was formed between two concentric city walls. Historical records indicate that by the Baoli era (825–827) of the late Tang Dynasty, the moat had gradually become interconnected. The construction of Jian Lake and the presence of a natural river terrace to the north of the city provided the essential conditions for the formation of the moat. The poetic line "At dusk, bamboo shadows chill the window; withered willows line the moat of the ancient commandery." Yuan [13] suggests that by this time, the southwestern moat had also been fully encircled. Subsequently, the addition of the inner watercourse led, by the early Ming dynasty, Jiajing reign (1522–1566 CE), to the establishment of a fully developed urban boundary defined by both an inner and an outer watercourse, forming a distinctive dual-ring water system.

The city boundary, while constituting a defensive system, simultaneously established an enclosed, inward-facing space. By adopting mountain masses as natural urban boundaries and dispensing with a city wall, the city effectively integrated the natural environment into its urban form. This strategy further capitalized on landscape resources for the development of government office complexes and gardens, embodying a landscape-oriented spatial planning approach that followed natural forms and was guided by scenery. Over successive periods, the city wall of Shaoxing underwent multiple expansions and repairs; however, the spatial extent of the ancient city remained consistently defined by the surrounding nine isolated hills, underscoring the pivotal role of mountain masses in the city's evolution. Distinct from the regular orthogonal layouts of other historical cities, Shaoxing's city gates were arranged in an asymmetrical layout, reflecting a planning philosophy of adapting to available materials and local conditions, following the terrain and waterways [14].

4.3 Construction and Development of Urban Functional Zoning and Integration with the Regional Landscape System

Shaoxing's ancient city underwent political transitions—from a Yue capital to a county seat under a commandery, and then to a city where both commandery and county administrations coexisted—and shifted in urban form from a western city with an eastern suburb to a nested (inner–outer) configuration, while the ward-and-market system declined or was dismantled. By the Southern Song, the city pattern had largely stabilized: functional divisions comprised the government precinct, the commercial district, residential quarters, and peri-urban market towns. Evolving over time, these zones displayed consistent logic of spatial differentiation and integrated closely with the natural environment, together shaping a distinctive water-town cityscape. Palaces and the government precinct consistently occupied the Mount Zhong area, forming a transitional belt toward its natural scenery; commerce required convenient waterborne transport and accordingly organized riverside street-alley spaces that, tied into the hydrological network, extended beyond the walls and blurred the urban boundary; residences were dispersed within the outer ward, leveraging the intra-city hydrological network to create pleasant living environments; and public gardens were sited amid picturesque mountain-and-water settings, integrating city with its suburban fringe (Fig. 6).

(1) Landscape Resources of Mount Zhong and the Construction and Planning of the Yamen District

Since the Spring and Autumn Period, the inner city around Mount Zhong served as the location of palatial buildings, the government precinct, and the prefectural seat. The early Li City, as the capital of the State of Yue during the Spring and Autumn period, was developed by Fan Li, who utilized the landscape resources of Mount Zhong to construct palatial terraces and build enclosed gardens, making it the site of the Yue king's palace. During the Sui-Tang Dynasties, the inner city in the Mount Zhong area continued to serve as the seat of the prefecture and of the local counties [12]. Subsequently, the Shanyin County Yamen and Kuaiji County government were relocated to the foothills of Baolin Mountain and to the site of Kaiyuan Temple, respectively—both locations characterized by open terrain and scenic environments [15]. In the Southern Song Period, the city was upgraded to Shaoxing Prefecture, with the prefectural government seat rebuilt from the Tang Dynasty prefectural seat, situated on the southern slopes of Mount Zhong, and served for a time as the imperial palace of Emperor Gaozong.

It is conceivable that the area surrounding Mount Zhong at that time not only possessed abundant natural landscape resources—its encircling mountain ridges resembling a reclining dragon; its elevated terrain offering panoramic views of the entire city and distant vistas of the rear sea; its luxuriant forests, pristine springs, and rugged cliffs—but also featured human-made constructions that fully exploited the ecological and topographical advantages of Mount Zhong. These included the Yue King Terrace, the Wanghai Pavilion (originally named Feiyi Tower), the Zhendong Pavilion, and the Penglai Pavilion. As recorded by Song scholar Diao Yue in the "*Record of the Wanghai Pavilion*"[8]: "As the Yue lord's insignia gradually extended east of the river, the seat of the military governorship was acclaimed. Perched at Mount Zhong, this yamen occupied a naturally magnificent location. South of the mountain lay the expansive Jianhu Lake, while to the north the land connected directly with both river and sea. Spanning hundreds of li around, it coiled upon the lakes and rivers like a resting dragon: the governor's mansion lay in the dragon's belly; the eastern gate of the government precinct marked the dragon's mouth; the western garden corresponded to its tail; and the Wanghai Pavilion stood along its ridge." This

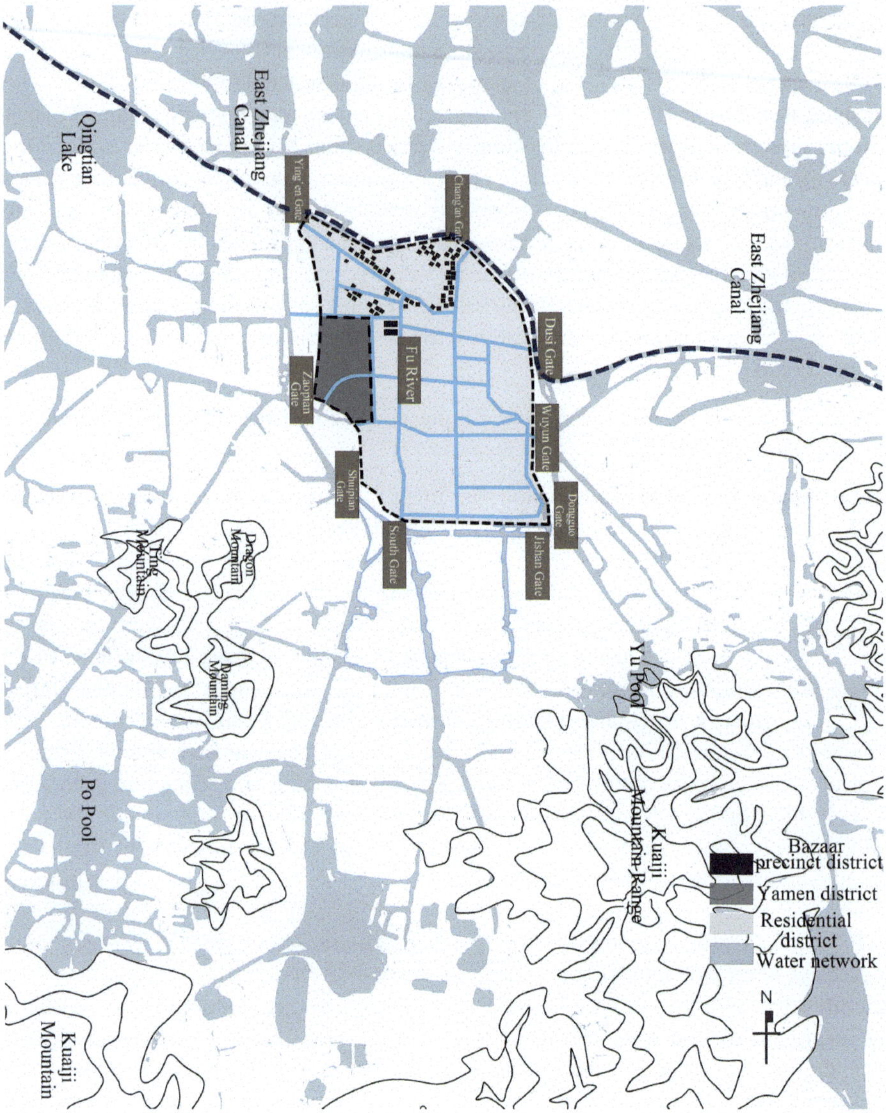

Fig. 6 Diagram of Shaoxing City urban districts and their relationship with the surrounding landscape

integration of natural environment and human intervention created the spectacular scenic district of Mount Zhong. The construction of the prefectural seat here represented an extension of nature into the urban fabric, serving as a transitional zone where the built environment of the city seamlessly merged with the surrounding natural landscape.

The natural environment of Mount Zhong and the artificial landscape construction atop it encircled the yamen precinct, extending along its northern flank. The yamen, with its orderly architecture, sat with composure at the southern foot of Mount Zhong, the two forming a mutual aesthetic dialogue: the human imprint of the yamen receded into the verdant forests of Mount Zhong, while the mountain's natural setting provided a verdant backdrop to the government precinct. The soaring hill and the disciplined architectural form together produced a distinctive landscape for the government precinct. As aptly rendered in the verse: "The county seat encircles the cloud-swept peaks, with mirrored waters of Kuaiji Mountains dominating the view. On all four sides it is naturally shielded; one house remains perched all day on its tower. The Milky Way seems to descend before the eaves, and drum horn echoes startledly rise from the earth. I, an official at the Jade Spring incense altar, though exiled, still reside in a place as serene as Penglai [16]." This poetic imagery epitomizes the harmonious integration of the natural environment and the constructed government precinct, exemplifying an landscape arrangement where the built environment dissolves into nature, and nature frames and dignifies governance.

(2) Development of Water Transport and the Construction of Marketplaces and Granary and Storage Facilities

During the Northern and Southern Song dynasties (960–1279), the flourishing canal network, combined with the disintegration of the Ward-Market System (Fang-Shi System),[9] spurred the commercial prosperity of Shaoxing's ancient city and drove its urban development. In addition, the convenience of water transport facilitated the circulation of goods and the concentration of population, leading to the establishment of granary and storage facilities as well as post stations and guesthouses in areas thriving with water transport.

(1) Prosperous Spatial Fabric of Water Alleys and Commercial Establishments

During the Southern Song Period, apart from the capital Lin'an, Shaoxing City Prefecture ranked foremost among the nation's forty major metropolises, reflecting the remarkable level of economic prosperity at the time. The city's economic development was intimately linked to its water transport system. Prior to the Song Dynasty, the ancient city of Shaoxing contained only a single market, located at the southeastern corner of the Outer City (Luocheng), known as the Old Market. However, the silting of Jian Lake, the reconfiguration of the water network, and the flourishing of the canal system—particularly the Eastern Zhejiang Canal [9]—allowed vessels to bypass the city center entirely, navigating instead along the city's northern reaches. This transformation in navigational routes catalyzed a northward shift of Shaoxing's commercial center, with the number of markets expanding from one to eight (Table 1). Among these, the South Market and North Market served as the primary urban hubs, strategically positioned along the Fu River, which functioned as the boundary river demarcating the two counties within the city. Originating from an existing natural watercourse within the city, the Fu River evolved into a bustling commercial street, its prosperity extending northward to connect with marketplaces situated along the artificial canal of the Eastern Zhejiang Canal, ultimately forming a comprehensive commercial network driven by both natural and engineered waterways.

The prosperity of commerce stimulated increased population mobility, leading to the construction of post stations and guesthouses along the canal to accommodate travelers. As recorded in the *Jiatai Gazetteer of Kuaiji*,[10] "The prefectural city's Penglai Pavilion lay to the left of Wolong Mountain; to the east stood the Wenjin Pavilion, and northward the Chuanting Pavilion. All were connected by the great eastern main river of the prefecture, providing convenient stopping points for arriving boats and carriages. On the west of the city lay the Ying'en Gate, and Wuyun Gate to the east—each also aligned with a pavilion [17]".

(2) Busy Granary, Warehouse, and Wharf Spaces

The convenience of water transport not only attracted an influx of merchants and the proliferation of commercial markets but also made the exchange of goods indispensable. To facilitate the transportation, handling, and storage of materials, the majority of granary depots, storage administrations, and wharves were concentrated along the canal (Table 2).

The development of water transport facilitated the distribution of commercial establishments and markets, post stations and guesthouses, and granary and wharves

Table 1 List of commercial establishments and markets within the city

<table>
<tr><th>Market name</th><th>Location</th><th>Remarks</th></tr>
<tr><td>Zhaoshui Fang Market</td><td>200 paces southeast of the city</td><td></td></tr>
<tr><td>South Market</td><td rowspan="2">1 li west of the city</td><td rowspan="2">Both the South and North Markets were located on either side of the Qingdao Bridge. Qingdao Bridge was situated south of the current Liberation North Road Xuan entrance. The "West" here seems to be a textual error for "South"</td></tr>
<tr><td>North Market</td></tr>
<tr><td>Dayunqiao East Market</td><td>2 li south of the city</td><td></td></tr>
<tr><td>Mei Market</td><td>45 li west of the city</td><td></td></tr>
<tr><td>Dayunqiao West Market</td><td>North of the city</td><td></td></tr>
<tr><td>Longxing Temple Front Market</td><td>2 li north of the city</td><td></td></tr>
<tr><td>Xudi Market</td><td>2 li north of the city</td><td></td></tr>
<tr><td>Jiangqiao Market</td><td>2 li north of the city</td><td></td></tr>
</table>

Source Historical Urban Spatial Structures of China
Note Present-day Renmin Road in Shaoxing City serves as the division between the northern and southern halves of the city. The majority of the aforementioned markets were located in the northern sector of the city

Table 2 List of warehouses and storage facilities within the city

Warehouse name	Location
Salt distribution warehouse	2 li east of the Prefectural Yamen, east of the Long Bridge along the canal
Jia-di warehouse	1 li northeast of the Prefectural Yamen
Glutinous rice receiving warehouse & yard	Outside the West Gate
Reward warehouse	2 li east of the Prefectural Yamen
Metropolitan tax office	1 li east of the Prefectural Yamen
Huiyi warehouse, bag-making office, bamboo processing yard	Northeast of the Prefectural Yamen

along the canal and along the water-network. The resulting commercial prosperity, in turn, drove the transformation of the canal's water network and its surrounding landscape. "Weirs control the rivers, and river crossings serve the grain transport; navigation stretches from Ou Bay to Fujian, reaching Ningbo and the State of Wu; oars and sails ride the waves [18]." vividly depicts the bustling scene of official and commercial traffic, with countless vessels converging on the canal.

(3) Water-Network and Urban Streets for Shaping a Pleasant Residential Environment

The ancient city of Shaoxing evolved upon a hydrological network, with its urban construction almost entirely organized around water. Although, at the initial stage of its establishment, the Shanhuai Plain shaped by marine regression and characterized by marshland—was not the optimal choice for urban construction, its hydrological network interwoven across the "low-lying flatlands," combined with the enclosure of foothills, suggested a residential suitability. Through hydraulic engineering, by the Southern Song Period, the city had essentially formed a water-oriented urban pattern, characterized by configurations such as "one river and one street," "one river and two streets," and "a river without streets." Within the city, abundant domestic water supply and convenient transportation were achieved.

The layout of residential areas in the ancient city of Shaoxing underwent several significant stages of transformation. Prior to the Tang Dynasty, "the Yue city was filled with old wards and winding lanes," with residential areas and commercial areas distinctly separated. During the Northern Song Dynasty, a new ward-and-lane residential system was implemented, in which the function of the wards (fang) evolved from purely residential areas to units of administrative division, incorporating certain commercial outlets. By the Southern Song Period, the adoption of the ward-and-neighborhood system completely dismantled the segregation of official and civilian residences, and the separation of residential wards from commercial markets [19]. Beyond the reforms of administrative systems, the transformation of the hydrological network exerted a profound influence on the spatial organization of residential areas. The intersecting and interlaced water-oriented urban pattern became the defining boundary for the division of wards and lanes. Wang Gang[11] subdivided Shaoxing

City Prefecture into five wards and ninety-six neighborhoods, with most names associated with water and bridges. Although no rigid zoning constraints existed between the residential compounds of officials and commoners, the spatial pattern reveals that the elevated, dry terrain in the southern part of the city was predominantly occupied by residences of officials and local gentry, while the residential areas of commoners were dispersed across other sections. Along the canal and major water transport routes, merchants primarily resided. The pervasive presence of the hydrological network created a unique lifestyle in which nearly every household was adjacent to water and possessed a boat—"wherever there are temples and hills, water reaches every household; no home is without access to it" [20]—becoming a defining feature of Shaoxing's residential dwellings. The iconic image of "small bridges, flowing water, and households" epitomizes the residential environment of Shaoxing City.

The building layout oriented toward and defined by water integrated the residential environment with the landscape system. The city's spatial organization was founded upon the hydrological network, which constituted the urban construction foundation. Within this framework, the built environment and the natural landscape were seamlessly integrated into the city's developmental trajectory, facilitating the formation of distinctive urban landscape characteristics and identity.

(4) Expansion of Market Towns into the Suburban Area and the Strengthening of Urban–Suburban Linkages

Following the Southern Song Period relocation of the imperial court,[12] the market towns of the Shaoxing region reached an unprecedented level of prosperity, with their forms becoming increasingly diversified. The more active types included peri-urban market towns, agricultural market towns, craft-based market towns, commodity transshipment market towns, and rural fairs [21]. These were predominantly distributed across the urban periphery, surrounding rural areas, grain-producing zones, regions with convenient transportation access, or areas endowed with distinctive geographical conditions. From the classification of these market towns, a differentiated commodity specialization had already emerged: there were grain-exchange markets such as the Rice Market, fruit-trading markets such as the Mei Market and the Xiangli Market, and a salt market that capitalized on the brine resources of Hangzhou Bay. The prosperity of these market towns not only signified the outward extension of urban activities into the countryside but also reinforced the connectivity between the city and its hinterland, representing an built environment extending beyond the urban boundary and integrating into the natural landscape. The division of labor among different markets and the regional specialization of product production epitomize a practice of "adapting to local conditions," where human intervention was exercised in a manner that respected natural processes while gradually transforming the landscape [22]. Expansive bayberry groves, shaded trading venues beneath green canopies, and scenes of "boats as numerous as pearls" laden with gorgon fruit and "crimson as ruby beads" [23] of bayberries evoke a vivid image of prosperity. Viewed from the Eastern Zhejiang Canal, the suburban landscape of Shaoxing City unfolds as a picturesque panorama of stone bridges, pavilions, fishermen's huts, bamboo weirs, water caltrop ponds, and lotus ponds, each scene a testament to a harmonious integration of human settlement and natural beauty [24].

(5) Kuaiji Mountain, Jian Lake, and the Creation of the Nine Hills Scenic Area with a Rich Cultural Landscape

Among the Nine Hills of Yue, Wolong Mountain, Tashan Mountain, and Jishan Mountain stand as the three tallest elevations and are the only ones preserved to this day. Situated respectively to the northwest, south, and west of the urban area, they form a tripod-like configuration, serving as the optimal foundation for landscape-making within the city's scenic planning and design.

The earliest landscape garden development in Shaoxing City can be traced back to the Spring and Autumn period, when King Goujian of Yue "established a royal garden at Leye [3]." Later, Fan Li selected the southern foot of Mount Zhong as the capital site, constructing the Feiyi Tower atop the mountain, rising to a height of fifteen *zhang* (3 m), from which one could gaze far across to the banks of the Qiantang River. This became a landmark structure within the ancient city of Shaoxing. After several reconstructions, it was renamed the Wanghai Pavilion during the Tang Dynasty (618–907 CE), its function gradually transforming from a military watchtower into a scenic architectural feature. The advantageous topography of Mount Zhong, with its gently sloping southern foot, provided an ideal site for the establishment of the palatial complex [25]. The early construction of imperial gardens laid the foundation for the city's garden tradition. Records note that Mount Zhong contained seventy-two towers, pavilions, and halls, including the Penglai Pavilion, lauded as "the foremost scenic jewels of landscape," and the Qingbai Hall. As documented in *The Moon over Yue's Inscriptions on Metal and Stone*[13]: "Encircled on all eight sides, the wondrous winding form is known as Wolong (reclining dragon). On the mountain grows a hundred stalks of bamboo, a thousand peach and plum trees; in autumn they harvest tea, in winter they convey pines, and in spring they plant flowers in abundance." Renowned for its large-scale hillside landscape garden, Mount Zhong became one of the most celebrated gardened landscapes in Jiangnan [26].

The earliest recorded construction on Ta Mountain dates to the Spring and Autumn period, when the King of Yue established an Astronomical Platform atop its summit. With the subsequent spread of religion, numerous Buddhist and Daoist Temples were erected on the mountain. By the late Eastern Jin Dynasty, the Yingtian Pagoda had been built at its peak. From the beginning of the Qing Dynasty, the site underwent comprehensive design interventions, gradually developing into a refined and picturesque scenic destination.

Ji Mountain (also known as Wangjia Mountain) derives its name from its association with Wang Xizhi, whose former residence was located at its foot. Historically, the mountain was adorned with numerous significant structures, including the Wangjia Pagoda, the Jishan Pavilion, the Shrine of Dong Changsheng, the Shrine of the Three Fans, the Northern Tianzhu Temple, and the Ji Mountain Academy. These attest to the flourishing landscape construction and garden making of Ji Mountain in antiquity.

Ancient city of Shaoxing developed a "three-mountain-and-three-pagoda" landscape pattern centered on Wolong Mountain, Ji Mountain, and Ta Mountain. The three pagodas emerged as the most prominent visual control points defining the skyline of ancient city of Shaoxing.

(6) Foothill Hydrological Network and Garden Construction

The natural landscape environment, distinguished by its elegance and charm, attracted numerous literati and poets who sought to establish estates and cultivate a life of refined retreat. Coupled with periods of social upheaval, significant numbers of Central Plains migrants, particularly members of the scholar-official class and prominent clans, settled in the region. Early residential compounds were predominantly situated within the Kuaiji Mountain-Jian Lake landscape, aspiring to a state of "artfully artificial, divinely natural." As society evolved, garden construction progressively expanded into the city, with sites frequently located along waterways and in proximity to low hills, fostering the development of urban private gardens as a defining element of the city's spatial and cultural landscape.

Among these were Shen Garden, Qingteng Study, and the former residence of Wang Xizhi, a collection of small-scale private gardens. Unlike typical Jiangnan gardens, which feature an abundance of pavilions and towers, the gardens of Shaoxing City placed greater emphasis on the natural landscape environment, pursuing an aesthetic of antiquity and simplicity rather than ornate display.

The ancient city of Shaoxing, situated amidst Kuaiji Mountains and Jian Lake, is endowed with a natural landscape environment of exceptional beauty, providing an ideal foundation and resource for landscape construction. Beyond the city lie the lofty ridges of the Kuaiji Mountain range and the Jian Lake hydrological network, while within the city the Nine Hills of Yue and an interlaced hydrological network form a spatial framework where hills beyond the city wall respond to those within, and waterways connect seamlessly. Through human intervention, the scenic potential of the Nine Hills of Yue was developed, waterways and urban streets were dredged, and emblematic structures such as temples and pagodas were erected. Around Jian Lake, landscape construction flourished, and within the embrace of Kuaiji Mountain and Jian Lake, scholars and literati established private gardens and retreats. Over centuries, the integration of the natural landscape environment with human-built features, infused with cultural heritage, gave rise to a regional landscape system of profound humanistic character (Fig. 7).

In conclusion, the spatial zoning, site selection, and construction of the ancient city of Shaoxing were fundamentally based on a regional landscape system shaped by an urban–suburban landscape framework. This reflects how the ancients, by fully respecting nature and applying artificial intervention with moderation, achieved a state of harmonious development with the natural environment—one that has endured with remarkable stability for over a millennium (Fig. 8).

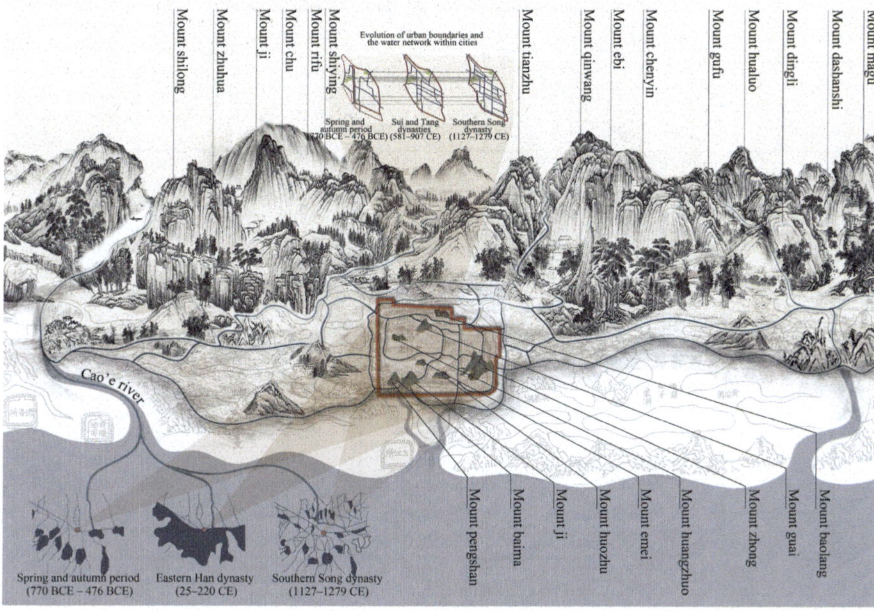

Fig. 7 Analysis of Shaoxing City urban and regional landscape system

The development of the ancient city of Shaoxing up to the present has been fundamentally shaped by the nine isolated hills that, since the period of capital site selection, have defined the framework for controlling the urban boundary. Their abundant natural resources provided the foundation for garden construction, serving as the prototype of the city's scenic system. Through artificial intervention, an urban characteristic landscape style emerged, anchored by the nine hills and the water network. The palace and the government precinct, making strategic use of the landscape resources and elevation of Mount Zhong (Wolong Mountain), established gardens that guided the evolution of the city's landscape pattern. Buddhist and Taoist temples erected pagodas on elevated sites, integrating temple environments with garden aesthetics; the combination of tall pagodas with mountainous terrain formed the city's iconic "Three-Mountains-and-Two-Pagodas" landscape pattern. The Dashan Pagoda and Yingtian Pagoda, positioned at opposite ends of the Prefectural River, emphasized the river's role as the city's central axis and visual corridor. Furthermore, scholars and literati built gardens around the landscape environment, embedding poetic sentiment into daily life and enriching the cityscape with a profound cultural landscape.

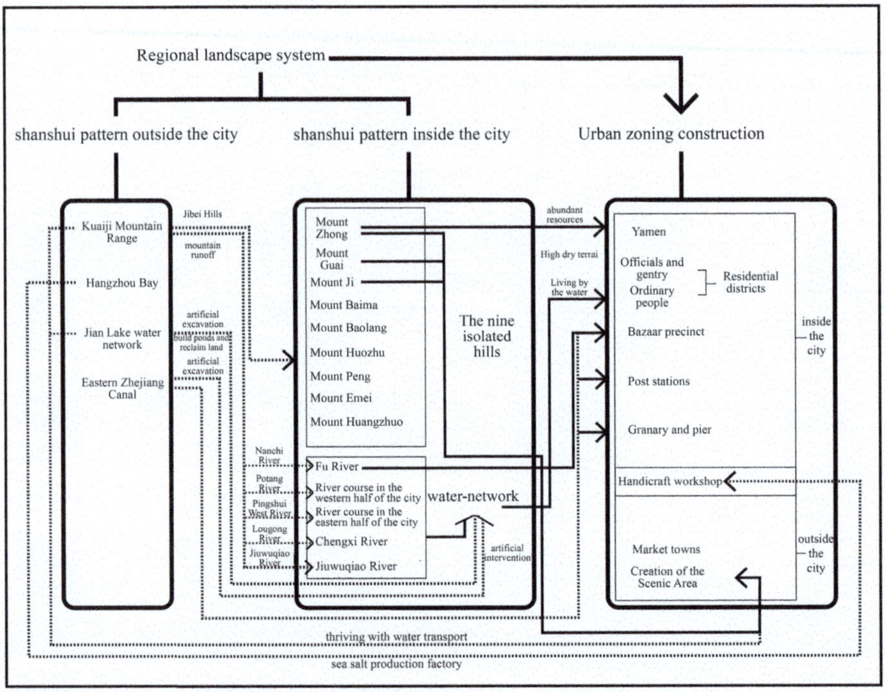

Fig. 8 Analysis of the formation process of Shaoxing City urban and regional landscape system

The development of the ancient city of Shaoxing diverges markedly from that of most other cities. The harsh conditions at its founding and the profound transformations of its water network environment during its evolution ultimately contributed to its rise as one of the foremost Jiangnan metropolises during the Southern Song Period. The integration of artificial intervention with natural conditions proved indispensable to this process. The synergy between zoning-based construction and the regional landscape system gave rise to a watertown urban characterized by distinct regional identity (Fig. 9).

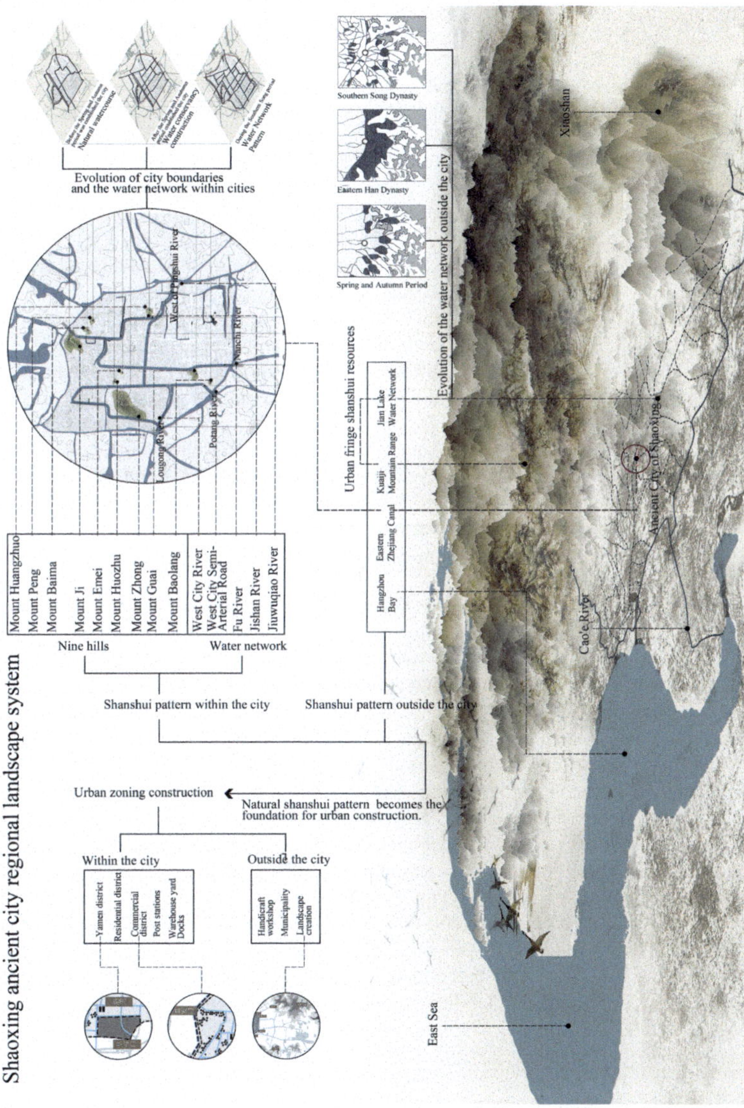

Fig. 9 Development of Ancient Shaoxing City urban districts and the regional landscape system

Notes

1. Wang Anshi, Chancellor of the Northern Song Dynasty (1021–1086 CE), was a major reformer whose New Policies reshaped land administration, water management, and infrastructural governance.
2. "woxin changdan" (lying on firewood and tasting gall) originates from *Shiji* and *Wu Yue Chunqiu*, describing King Goujian of Yue's self-imposed hardship as a strategy of long-term territorial and political restoration.
3. *Record of the Xiaoyao Pavilion*, written by Wang Anshi in the Northern Song Dynasty (1021–1086 CE), records the siting and cultural meaning of a pavilion as an interface between architecture, landscape, and scholarly life.
4. *Map of Shaoxing City Prefecture in the Kangxi Reign of the Qing Dynasty* (Qing dynasty, 1662–1722 CE) is a Qing-period urban map depicting city walls, waterways, streets, and surrounding landscape features, illustrating the water-oriented urban structure of Shaoxing.
5. *Map of the Main Roads in Shaoxing Prefecture* (Shaoxing Fucheng Qulu Tu) (Qing Dynasty, before 1912 CE) records the historic road–water network structure of Shaoxing, illustrating the spatial framework of a traditional water-based city.
6. He Xun (325–366 CE), Eastern Jin official, used existing waterways in Kuaiji to support transport and irrigation, illustrating early landscape and urban water management.
7. Fan Li (c. 536–448 BCE), Yue statesman and strategist, exemplified early planning and land–water management practices in territorial and urban development.
8. Diao Yue (Song Dynasty, 1047–1106 CE), *Record of the Wanghai Pavilion* (Wanghai Ting Ji). This work is a descriptive essay by Diao Yue, a literati of the Northern Song, recording the construction, layout, and scenic views of the Wanghai Pavilion and its surrounding gardens and waterways.
9. The Ward–Market System (Fang–Shi System), a traditional Chinese urban layout dividing cities into residential wards (fang) and commercial markets (shi), regulating streets, markets, and neighborhood structure.
10. Jiatai Gazetteer of Kuaiji, compiled during the Jiatai reign of the Southern Song Dynasty (1201–1204 CE), the gazetteer records the administrative divisions, waterways, roads, commercial districts, temples, gardens, and notable landscapes of Kuaiji (modern Shaoxing).
11. Wang Gang, during the Southern Song Dynasty (approx. 1190–1250 CE), Wang Gang, a local official and urban planner, reorganized Shaoxing Prefecture's urban layout by dividing the city into five xiang (wards) and ninety-six fang (residential/commercial blocks).
12. Southern Song court relocation, during the Southern Song Dynasty (1127–1279 CE), following the Jurchen conquest of northern China, the Song imperial court fled southward, establishing its capital at Lin'an (modern Hangzhou). This migration, known as the Nan Du (Southern Relocation), profoundly reshaped urban centers, waterway networks, and the administration of commerce in the south, influencing city layouts, riverine transport, and the development of garden and market spaces.
13. *The Moon over Yue's Inscriptions on Metal and Stone* (Yue Zhong Jin Shi Yue), compiled in the Ming Dynasty (1368–1644 CE) by an unknown author, this work collects inscriptions on bronze and stone from the historical Yue region (modern Zhejiang and surrounding areas). It documenting historical settlements, landmarks, and water-landscape patterns.

References

1. Che, Y., & Chen, Q. (2001). *Historical geography of Shaoxing*. Shanghai Bookstore Publishing House.
2. Wang, A. (Song Dynasty). (1974). *Ascending the City Tower of Yuezhou: Collected works of Wang Wengong* (Vol. 45, p. 7). Shanghai People's Publishing House.

3. *Wuyue Chunqiu, from twenty-five alternative histories* (Vol. 6). Qilu Press (2000).
4. Guanzi Shuidi (39). (2006). In *Guanzi: Newly annotated Edition* (p. 315). Qilu Press.
5. Commentary on the Water Classic. (2009). *Mian River Commentary* (Vol. 29, p. 10). Zhonghua Book Company.
6. Chen, Q. (1962). The rise and fall of Ancient Jianhu and agricultural water conservancy in the Shanhui Plain. *Acta Geographica Sinica, 3*, 187–202.
7. Ren, G. (2015). *The formation and characteristics of Shaoxing's water town: In China's Jianhu* (p. 7). China Literature and History Press.
8. Yan, G., & Li, L. (2007). The evolution and characteristics of water-town urban spatial structure—A study of structural change in Shaoxing's Yuecheng Area. *Journal of Liaoning University of Science and Technology, 30*(1), 53–57.
9. Li, X. (2015). *Historical urban space of China*. Peking University Press.
10. *Yuejue Shu* (Vol. 8). Qilu Press (2000).
11. Chen, Q. (1991). The formation and development of Shaoxing City in historical periods: Essays on Wu–Yue Culture (p. 12). Zhonghua Book Company.
12. Li, Y. (2012). *General history of Shaoxing* (p. 10). Zhejiang People's Publishing House.
13. Yuan, Z. (Tang Dynasty). (2002). *In Harmony with Governor Li Deyu of Zhexi: Forty-rhyme poem based on his dream account—Chronologically annotated collected works of Yuan Zhen* (p. 6). Sanqin Publishing House.
14. Wang, F. (1991). *The influence of the Jianhu water system on Shaoxing's water-city layout and river regulation: In "Jianhu and Shaoxing Water Conservancy"* (p. 7).
15. *Jiatai Gazetteer of Shanyin County* (Vol. 5). Reprinted by the Shaoxing County Gazetteer Editorial Committee (1992).
16. Yuan, Z. (Tang Dynasty). (2002). *Boasting of the prefectural residence to Liu Yuxi: Chronologically annotated collected works of Yuan Zhen* (p. 6). Sanqin Publishing House.
17. Shaoxing County Gazetteer Editing Office. *Jiatai Gazetteer of Kuaiji: "Penglai Pavilion" (with Baoqing Supplement).*
18. Wang, S. (Song Dynasty). *Fu on the Customs of Kuaiji, with Preface: The Three Fu of Kuaiji.*
19. Tu, J. (2012). *A brief discussion on the flourishing period in the construction of Ancient Shaoxing* (p. 3). Shaoxing Yue Culture Research Association.
20. Zhang, J. (Tang Dynasty). *Seeing Zhu Qingyu off after passing the imperial examination: Complete Tang Poems* (Vol. 384).
21. Yao, P., Chen, G., & Qiu, K. (2011). Market towns and rural economy in the Shaoxing Area during the Southern Song Dynasty. *Journal of Zhejiang Normal University (Social Sciences Edition), 36*(4), 88–94.
22. Lu, Y. (Song Dynasty). (2005). *Watching the picking of bayberries in Liufeng mountain and staying in the hills for days: Annotated Jian'nan poetry manuscript* (p. 4). Shanghai Ancient Books Press.
23. Lu, Y. (Song Dynasty). *Playfully chanting village food to show neighbors.*
24. Chen, W. (2015). *The former glory of the Eastern Zhejiang Canal* (p. 1). CPPCC Shaoxing Committee.
25. Yuan, Z. (Tang Dynasty). (2002). *Responding to clerical official Zheng during a Banquet at the Sea-viewing Pavilion in the Ninth Month of the Fourth Year: Chronologically annotated collected works of Yuan Zhen* (p. 6). Sanqin Publishing House.
26. Feng, G., & Xiu, S. (1990). The naming of Shaoxing Gardens. *Journal of Shaoxing University, 2*, 124–126.

Chapter 11
Urban Subarea: Rivers and Subarea Development—Evolution of the Landscape Character Along the Qinhuai River in Nanjing City

Abstract This Chapter investigates the landscape character of urban subareas shaped by the Qinhuai River in Nanjing, examining river-driven spatial development at the subarea scale. As a major river system integrated into the urban fabric through long-term hydraulic modification, the Qinhuai River provided a fundamental landscape framework for the formation and differentiation of water-oriented urban subareas. From the perspective of landscape architecture and urban morphology, this chapter elucidates the multiple functions of the river system, including water management, transportation, spatial organization, and environmental regulation, and their role in structuring subarea landscapes. Methodologically, the study combines historical analysis with spatial interpretation to examine the interaction between natural hydrological conditions, artificial water infrastructure, and urban form during key historical periods, particularly the Six Dynasties (222–589 CE) and the Ming-Qing Period (1368–1912 CE). The analysis demonstrates that the Qinhuai River functioned not merely as an infrastructural corridor, but as a persistent generator of landscape space, shaping subarea structure, landscape character, and patterns of urban life. The long-term continuity and transformation of river-based landscape structures reveal how water systems sustained urban development at the subarea scale, providing a critical foundation for understanding river-oriented landscape formation in traditional Chinese cities and complementing subsequent discussions at canal and basin scales.

The natural rivers—reshaped through artificial water networks—became closely intertwined with the city, fostering the growth of functional urban subareas along their banks and giving rise to landscape regions characterized by water-centered development. As the Ancient Capital of Six Dynasties (222–589 CE),[1] Nanjing City was profoundly shaped by the Qinhuai River, which played a pivotal role in its urban development. In particular, the evolution of subareas along the river included infrastructural elements such as ferries, bridges, imperial roads, wards, and markets, as well as cultural and spatial landscapes including temples, imperial gardens, and private gardens. Using Nanjing City as a case study, this chapter examines the composition,

X. Zheng, *Terrapoetic Urbanism*, https://doi.org/10.1007/978-981-95-8676-9_11

formation, and characteristics of these landscape subareas, thereby exploring the interactive relationship between urban rivers and the landscapes of riverside subareas.

Renowned as the "land of Six Dynasties' splendor, imperial capital of Jinling," Nanjing City underwent numerous historical appellations: during the Spring and Autumn (770–476 BCE) and Warring States Periods (475–221 BCE) it was established as Yuecheng and Jinling Settlement; under the Qin Dynasty (221–206 BCE) it was renamed Moling; in the Eastern Wu (222–280 CE) it became Jianye; in the Western Jin Dynasty (265–316 CE) it was called Jianye again; during the Eastern Jin Dynasty (317–420 CE) and the Southern Dynasties (420–589 CE) (Song, Qi, Liang, and Chen) it was known as Jiankang; subsequently it bore names such as Danyang, Jiangning, Shengzhou, and Jinling, before being designated Nanjing City as the capital of the Ming Dynasty (1368–1644 CE), from which its present name derives. To avoid ambiguity, the general term "Nanjing" is adopted throughout this study. Geographically, the city is situated in the Ning-Zhen-Yang Hills region of the lower Yangtze River, adjoining the Yangtze River Delta to the east, backed by the southern Anhui hills to the west, bordered by the Taihu Lake water network to the south, and linked to the Jianghuai Plain to the north, with the Yangtze River cutting across its northwest. The Qinhuai River—historically known as Longcangpu, renamed Huai River in the Han Dynasty, and later widely recognized as the Qinhuai from the Tang Dynasty onward—has since retained the appellation "Qinhuai River [1]." Flowing east to west across the southern part of the city, the Qinhuai River constitutes Nanjing's most significant urban river, whose evolving landscape character along both banks has been closely tied to the city's development. Accordingly, the spatial scope of this study encompasses both the capital city proper and its suburban areas, extending approximately 500 m outward from each side of the Qinhuai River system. Within this flexible boundary, landscape elements are most densely distributed and maintain the most intimate relationship with the river (Fig. 1).

Viewed in the longue durée of Nanjing's urban construction history, the primary phases of capital construction were concentrated in four major periods: the Six Dynasties, the Southern Tang (937–975 CE), the early Ming Dynasty, and the Republic of China. Among these, the periods of greatest flourishing in terms of riverside landscapes and cultural activities occurred during the Six Dynasties and the Ming–Qing Dynasties. Consequently, the temporal scope of this study focuses on two key stages: the Six Dynasties (222–589 CE) and the Ming Dynasty (1368–1644 CE). These periods were chosen because urban construction activities in Nanjing City reached their peak, the city's influence was profound, and—most importantly—the landscapes along the Qinhuai River system attained their richest forms. The Six Dynasties marked the beginning of Nanjing's history as a capital city, during which the evolution of its landscapes reveals the formative process of urban riverfront landscapes, from emergence to development. By contrast, the Ming Dynasty, characterized by unprecedented advances in social productivity and extraordinary cultural prosperity, represents the apex of landscape character [2]. This chapter thus seeks to trace the formation process of subarea landscapes along urban rivers and to examine the dynamic interaction between urban rivers, functional zoning, and the landscape spaces that derived from them.

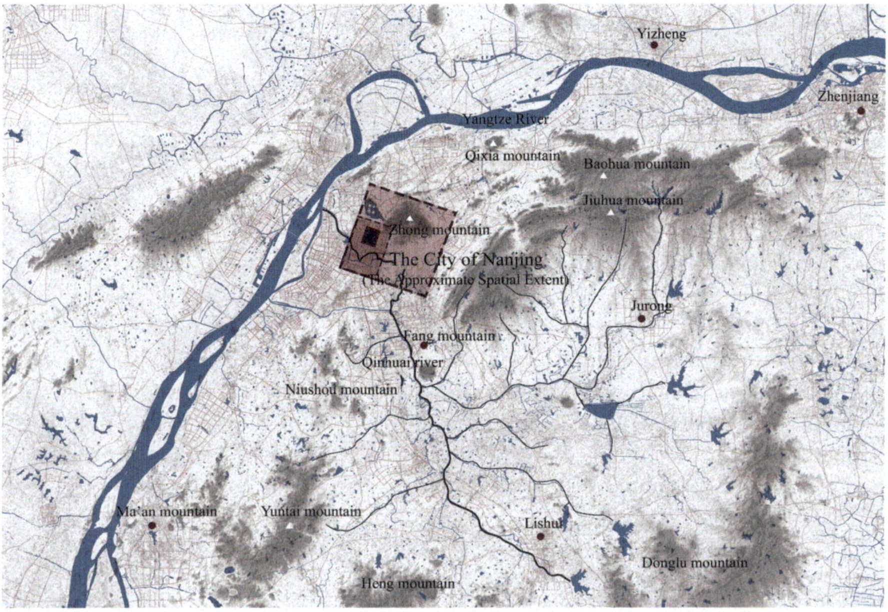

Fig. 1 Location map of the ancient city of Nanjing City. Base map adapted from Qinhuai River Basin Map (http://blog.sina.com.cn/s/blog_558ae5d0100v82v.html) and redrawn by the author based on Notes 15 and 16

1 The Natural Foundation of the Qinhuai River System Subarea Landscape

As the saying goes, "The waters of Jinling take the Huai as their axis." The Qinhuai River system comprises the natural river course of the Qinhuai in the southern part of the city, together with the artificial waterways encircling the capital city—including the Yundu Canal, Chaogou Canal, Qingxi Canal, and the Chengbei Canal—as well as the external canal Pogangdu, which connected the Qinhuai River to the San Wu[2] region (namely Wu Commandery, Wuxing, and Kuaiji). The Pogangdu served as a vital lifeline linking Nanjing City with the Taihu Basin, a principal grain-producing area, and it constituted the most important navigable waterway for the capital, fostering the prosperity described as "tens of thousands of merchant vessels." However, as it falls beyond the urban section under consideration, it will not be further discussed here.

The morphology of the Qinhuai River system played a decisive role in shaping the water-landscape complex, which was simultaneously influenced by the overall mountain-and-water environment of Nanjing City.

1.1 Overview of Nanjing's Mountain-and-Water Environment (Shanshui)

Nanjing City is renowned for its picturesque mountains and rivers. Geographically, its northeast consists of mountainous areas, the southwest of hilly areas, the central region of the Qinhuai Plain, and the west of the alluvial zone along the Yangtze River [3].

With mountains as its structural backbone, Nanjing's terrain belongs to the Ningzhen mountain range, which is divided into northern, central, and southern branches (Fig. 2).

Northern branch—stretching along the north bank of the Yangtze River, beginning with Shitou Mountain (modern Qingliang Mountain) and extending eastward to include Bazi Mountain (also called Siwang Mountain), Shizi Mountain, Mufu Mountain, Wulong Mountain, Qixia Mountain, and Baohua Mountain. These mountains were historically equipped with beacon towers and fortresses, serving strategic defensive functions.

Central branch—starting again from Shitou Mountain, this branch includes Wutai Mountain, Gulou Hill, Jilong Mountain, Fuzhou Mountain, and Zhong Mountain. As the northern support of the city, this range provided natural foundations for the construction of city walls, temples, manorial estates, and gardens, all of which were adapted to the mountain's topography.

Southern branch—beginning at Sanshanji on the east bank of the Yangtze River and extending northeastward, this branch comprises Niushou Mountain, Jiangjun Mountain, Yuhuatai, Fang Mountain, Qinglong Mountain, and Tang Mountain, forming the natural protective barrier for southern Nanjing City.

With water as its lifeline, Nanjing's principal water systems consist of the Yangtze River, the Qinhuai River, and the Jinchuan River. The Yangtze River is the most significant of these, flowing from the southwest, encircling the city, and then turning eastward. Both the Qinhuai River and the Jinchuan River are tributaries of the Yangtze, flanking the northern and southern sides of the central branch mountains, and connected to each other through Xuanwu Lake. In the Six Dynasties period, hydraulic construction in Nanjing City was concentrated on the Qinhuai River system, which—compared to the Jinchuan River system—exerted a far deeper and more lasting influence on the city's spatial form and landscape character.

1.2 The Formation of the Mainstream of the Qinhuai River

The Qinhuai River, according to historical records, "has two sources: one rising from Baohua Mountain in Jurong, flowing southwest; the other from Donglu Mountain in Lishui, flowing northwest. Entering the territory of Jiangning, the two merge at Fangshan Dyke and then flow westward into the Yangtze River [4]." The eastern source originates from Baohua Mountain in Jurong County, and the southern source

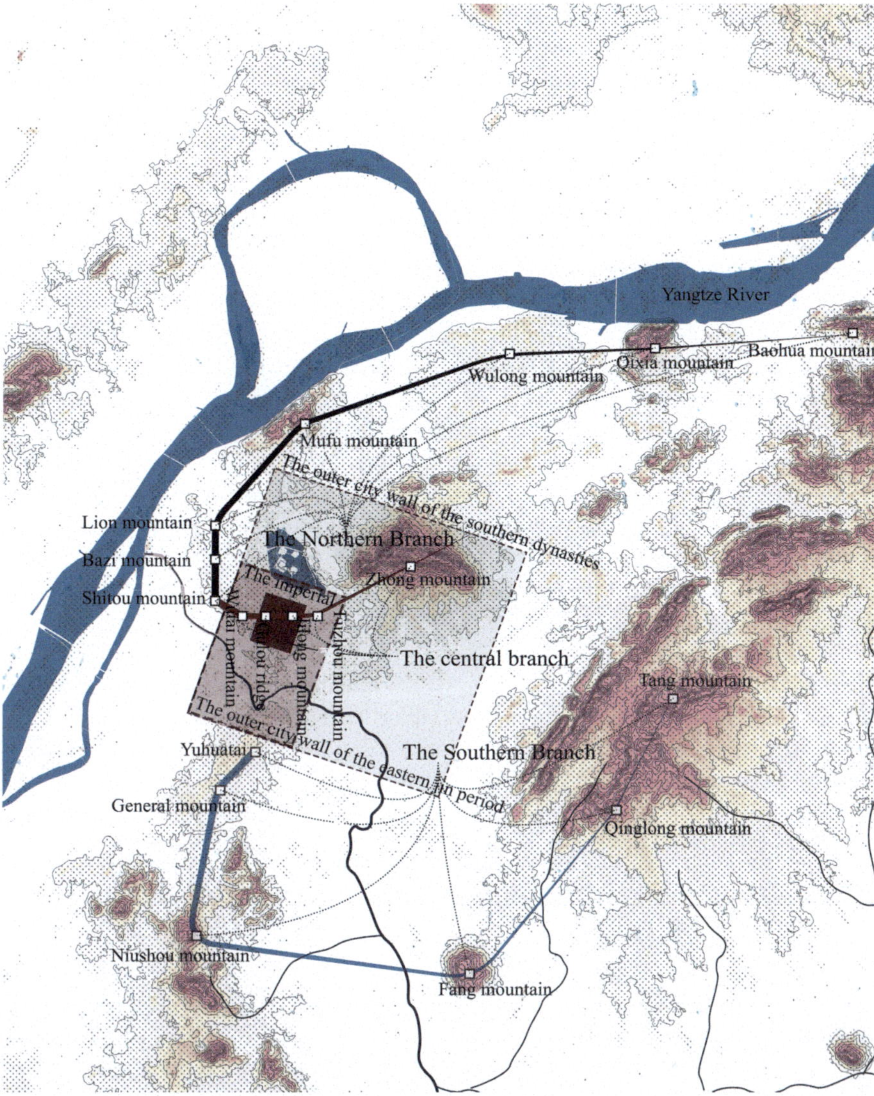

Fig. 2 Map of the mountain and river structure of Nanjing City. Base map adapted from Qinhuai River Basin Map (http://blog.sina.com.cn/s/blog_558a7e5d0100v82v.html) and redrawn by the author based on Notes 2 and 33

from Donglu Mountain in Lishui County. These two tributaries converge at Fangshan Dyke in Jiangning County, forming the mainstream, which flows northwestward through the capital city of Nanjing, entering the Yangtze during the Six Dynasties at the site of today's Shuiximen Gate. The river extends approximately 110 km from its sources to its confluence with the Yangtze, earning the name "Hundred-Li Qinhuai."

Its basin forms an alluvial-plain micro-basin, composed successively from the inside outward of riverside polder plains, extensive hilly uplands, and low mountains, with Nanjing City situated at the northern edge of the plain [5].

The precursor of the Qinhuai River was the "Great Qinhuai Lake" of the Neolithic Period (c. 10,000–2000 BCE), in the Chinese context. From the upper to the lower reaches, this consisted of three interconnected lake groups centered around present-day Chishan Lake, Baimiwei, and Dongshan. In dry seasons these remained as separate lakes, while in flood seasons they merged into a single expanse. The fertile plains surrounding the lakes, suited for cultivation, together with the hilly uplands conducive to habitation, fostered the initial settlements of the Hushu Culture.[3]

During the height of the agricultural civilization of the Xia, Shang, and Zhou dynasties, early inhabitants began to reclaim land from the lakes, "constructing earthen embankments to control water and cultivating within [6]," thereby creating a system of over 390 km of flood-diversion and irrigation channels, which constituted the embryonic form of the Qinhuai River. By the Qin Dynasty, legend holds that Emperor Qin Shihuang, while passing through Jinling, was told by a geomancer that "five hundred years hence, Jinling would bear the aura of an emperor [4]." To suppress this auspicious energy, "he ordered the cutting of Fangshan, severing the ridge, and channeling the waters into the Yangtze—hence the name Qinhuai [4]." In reality, this was a large-scale dredging and channel-cutting project. Following land reclamation, the Qinhuai River's course had become winding and sluggish, obstructed at the ridge between Fangshan and Shikuishan, which hindered flood discharge and water diversion. Upon completion of the project, the river channel took shape, achieving smooth flow for both drainage and irrigation. At that time, it was then referred to as Huai Water (Huai-shui) [7].

1.3 The Reshaping of the Water Network in the Construction of the Capital City

Large-scale hydraulic construction during the Six Dynasties was concentrated in the Eastern Wu, resulting in the formation of the Qinhuai River system, in which natural waterways and artificial canals were interconnected with the Qinhuai River as the core. On this foundation, the Eastern Jin and the Southern Dynasties carried out several smaller-scale hydraulic works (Fig. 3).

The river courses of this water system were divided into three categories: (1) fully utilized natural rivers, such as the Qinhuai River; (2) modified natural rivers, such as the Qingxi; and (3) excavated artificial canals, such as the Yundu, Chaogou, and the Chengbei Canal [3].

Yundu Canal: A north–south artificial canal located west of the city, connecting with the Qinhuai River in the south and reaching Cangcheng (the institution for storing

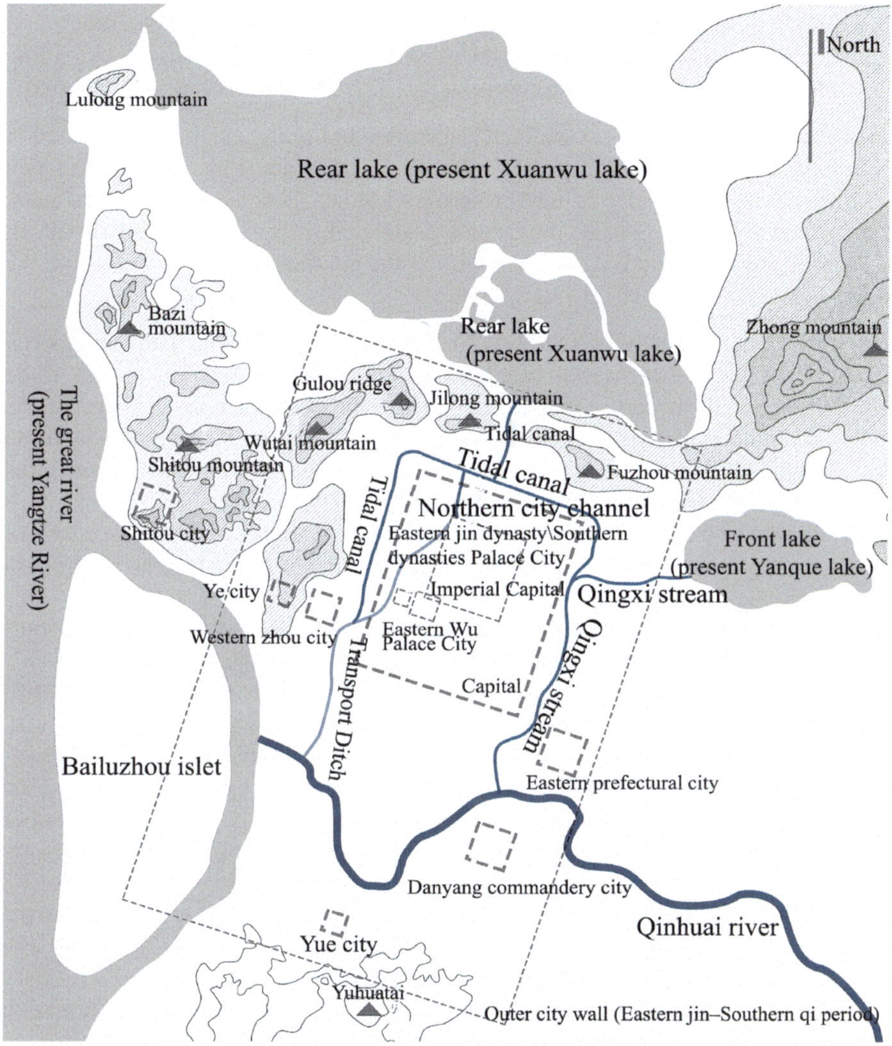

Fig. 3 Distribution map of the Qinhuai River water system in Nanjing City during the Six Dynasties (222–589 CE). *Source* Redrawn by the author based on Notes 2, 33, and 35

grain and supplies within the imperial palace city) in the north. It was the only grain-transport canal in Six Dynasties Nanjing City, responsible for conveying tribute grain and transporting goods to the capital [8].

Qingxi Canal: Also known as the East Canal, originally a natural river that gathered runoff from the western slopes of Zhong Mountain into Qianhu Lake (today's Yanque Lake). To discharge lake water and strengthen military defenses, a north–south artificial canal was excavated on the city's eastern side, connecting northward to the

Chaogou Canal and southward to the Qinhuai River. Ancient records describe Qingxi Canal as "five *zhang* wide, eight *chi* deep, with vast and surging waters extending for ten *li*." Because of its winding course, it was also called the "Nine-Bend Qingxi."

Chaogou Canal: An east–west artificial canal north of the city, named for its function of introducing tidal waters from the Yangtze River into Houhu Lake (today's Xuanwu Lake). It was also referred to as the Northern City Moat. "Introducing the tidal waters, it connected to Qingxi, reached the Qinhuai, extended westward to the Yundu Canal, and northward to Houhu Lake," thus ensuring the continuous flow of water into the Yundu Canal.

Chengbei Canal: An artificial waterway linking the Chaogou Canal with the imperial palace city. The *Jiankang Shilu* records: "Another canal was excavated north of the city, introducing the waters of Houhu Lake to flow into the palace, circling around the halls and chambers, exhaustively displaying craftsmanship at a cost ten-thousandfold [9]." Compared with the Chaogou Canal, the Chengbei Canal was primarily intended for garden design, aesthetic enjoyment, and imperial leisure.

In addition to these four main artificial waterways, moats surrounded sections of the imperial palace city not reached by rivers, and imperial ditches ran along both sides of the main thoroughfares.

The purposes of hydraulic construction in the Qinhuai River system were fourfold: Military defense, using watercourses as moats. Until the second year of Qi's Jianyuan reign (480 CE), the capital's fortifications remained bamboo palisades [10]; thus, defense relied heavily on the four waterways of the Qinhuai River, Qingxi Canal, Yundu Canal, and Chaogou Canal [11]. Political consolidation and economic development. For instance, the Qinhuai River conveyed grain and goods transported via Pogangdu Canal from the Sanwu region into the city, and these were then further carried into the imperial palace via the Yundu. Flood control and water management, including supply, drainage, and hydrological regulation between channels—for example, the Chaogou Canal, which introduced the waters of Houhu Lake into the Yundu Canal. Landscape construction, exemplified by the Chengbei Canal. Through these developments, the Qinhuai River system came to assume Nanjing's primary role in transportation, while simultaneously laying the foundation for the formation of the water-landscape complex (Fig. 4).

2 Subarea Landscapes Along the Qinhuai River System During the Six Dynasties

The construction of the ancient capital of Nanjing was closely interrelated with the Qinhuai River system. Even before the Eastern Wu established their capital, multiple urban settlements were already distributed along both banks of the river.

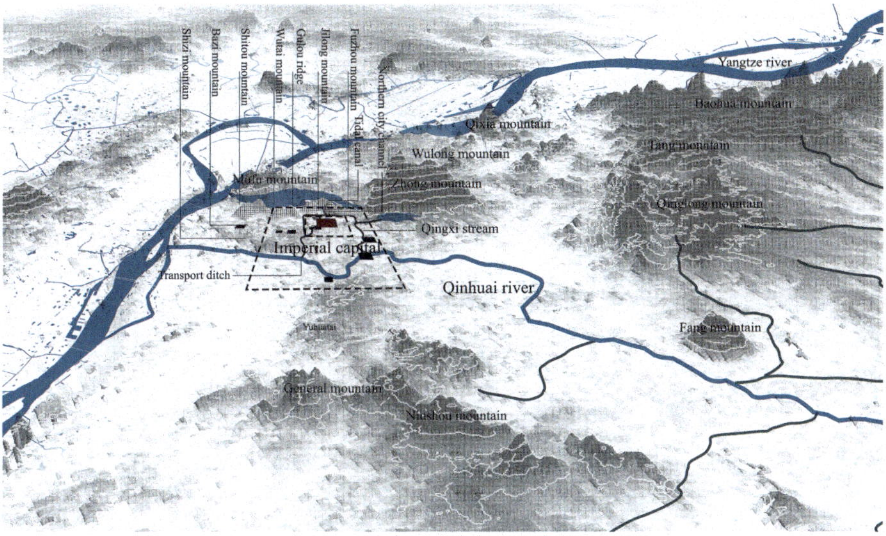

Fig. 4 Landscape environment of the Ancient City of Nanjing. Base map adapted from Qinhuai River Basin Map (http://blog.sina.com.cn/s/blog_558ae5d0100v82v.html) and redrawn by the author based on Notes 2 and 33

After the founding of the capital, the Qinhuai River functioned as the southern military defense line of the city. During the process of capital planning and construction, functional elements such as transportation, residential quarters, and commercial activities were organically integrated according to context-sensitive design along both banks. Consequently, the landscape space gradually became enriched, ultimately forming a regional landscape system composed of riverine water networks, ferry crossings, ward-based residential communities, waterborne marketplaces, and temples and pagodas.

2.1 Building Cities by Water and Integrating Rivers into Urban Form: The Evolution of the City–River Relations from the Spring and Autumn Period (770–476 BCE) to the Ming Dynasty (1368–1644 CE)

"As for the establishment of capitals, if not beneath great mountains, they must surely be above broad rivers [12]." In antiquity, the siting of a capital city was invariably determined by consideration of the landscape environment. In addition to the natural barriers of "Zhongshan coiling like a dragon and Shitou Mountain crouching like a tiger [13]," the Qinhuai River constituted a crucial natural advantage for city building and capital establishment. The *Spring and Autumn Annals of Emperor Xian*[4] records: "The Xiao River (that is, the Qinhuai River), extending for more than one hundred li, can accommodate large vessels." On the one hand, the Qinhuai River, together with the Yangtze River, served as a dual military barrier southwest of the ancient city; on the other, it provided immense convenience for transportation and commerce.

The relationship between the Qinhuai River and the city originated during the Spring and Autumn period and became largely fixed after the Southern Tang established its capital. While the boundaries of the capital city shifted during the Ming and Qing dynasties, these changes had little direct impact on its connection with the Qinhuai. In the Spring and Autumn and Warring States periods, settlements were only scattered along the Qinhuai, such as Yecheng, a fortified armory city established by King Fuchai of Wu,[5] and Yuecheng, built by Fan Li of the State of Yue in Changganli—"the building of the city began with Yuecheng [14]," which marked the beginning of Nanjing's urban construction history. Later, King Wei of Chu, Xiong Shang, established Jinlingyi (renamed Moling by Emperor Qin Shihuang to suppress imperial auspices). Yecheng, Yuecheng, and Jinlingyi were all founded as military strongholds on the ridges at the Qinhuai's confluence with the Yangtze River, enabling control over waterway traffic as well as providing vantage points for surveillance and defense.

After the Eastern Wu established its capital, the Qinhuai lay outside the imperial city but within the outer fortifications (whose precise location remains uncertain; the diagram is extrapolated from the *Six Dynasties Capital*,[6] reflecting the layout from the Eastern Jin to the Southern Qi; the Southern Liang expanded it further, but the extent was too great to be depicted). Together with the Yundu Canal, Qingxi Canal, Chaogou Canal, and Chengbei Canal, the Qinhuai became an organic part of the city. Its form and functions extended throughout the Six Dynasties, laying the fundamental pattern of the city–river relationship. Along both banks, prosperous commercial quarters and residential districts took shape, while the riverfront also became a renowned site for pleasure outings and social gatherings. At the same time, the sites of Yuecheng and Jinlingyi were preserved to build Shitou City as a military fortress, while to the south of the Qinhuai, the Danyang commandery city was established as the local administrative seat. In the Eastern Jin period, Dongfu City and Xizhou City served as fortified garrisons.

After the Sui Dynasty pacified the city, the Southern Tang expanded the capital southward, during which the Qinhuai was divided into the Inner and Outer Qinhuai: the Inner Qinhuai was incorporated into the city and transformed into a landscape river, while the Outer Qinhuai, excavated as a moat beyond the city walls, functioned both as defensive fortification and grain-transport canal. When the Ming Dynasty expanded the capital northward, its relationship with the Qinhuai remained unchanged. By the Ming–Qing period, the Inner Qinhuai reached the zenith of its landscape character, lined with waterside pavilions, bustling with merchants and visitors. As a famous verse depicted:

> The pear blossoms look like snow, willows green mist make;
> Spring reigns along both sides of Qinhuai Stream.

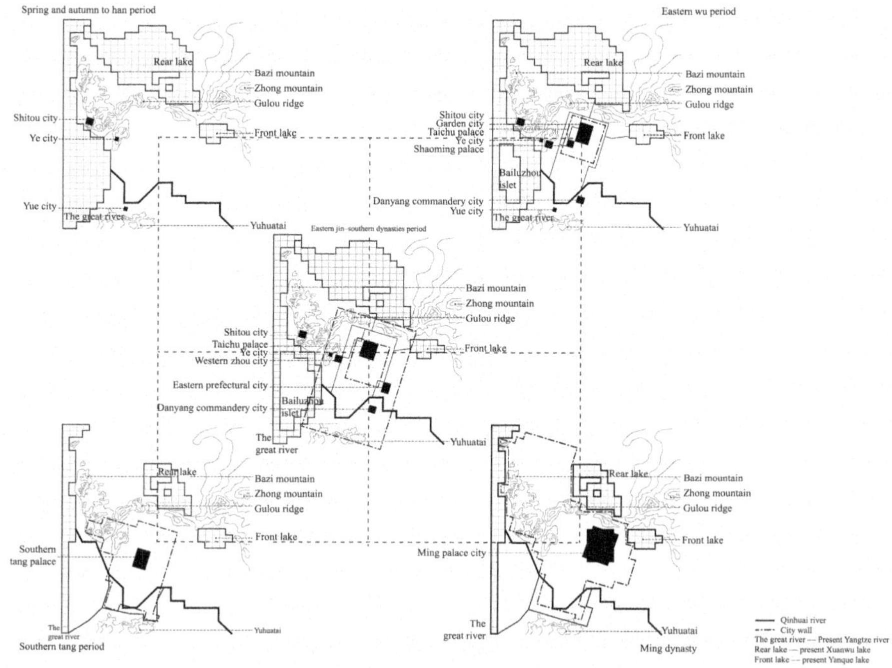

Fig. 5 Diagram of the evolution of the relationship between the city and rivers of Nanjing City in different dynasties. *Source* Redrawn by the author based on Notes 2, 3, 6, and 35

A row of bowers along the shore vies in splendor;
The reflection of beauty vies with beauty in splendor (Fig. 5) [15].

2.2 *Flourishing Through Water, Adapting to Local Conditions: Water Transport and Environment as Determinants of Urban Functional Distribution in the Six Dynasties*

The well-developed waterborne transportation conditions of the Qinhuai River system facilitated the rapid growth of urban functions. These functions extended along the river, and, through context-sensitive design/planning in accordance with environmental conditions, were predominantly distributed on both banks of the Qinhuai River as well as in the scenic areas of Qingxi.

(1) The Development of Waterborne Transport and the Emergence of Ferries, Bridges, and Imperial Roads

The prosperity of water transport not only stimulated the economic development of both banks of the Qinhuai River system but also gave rise to various transportation nodes: ferries formed at points of transfer between waterborne and land-based routes; bridges were built in density to facilitate river crossings; and the imperial road, as the principal thoroughfare of the capital, provided the essential link between the imperial palace city and the Qinhuai River (Fig. 6).

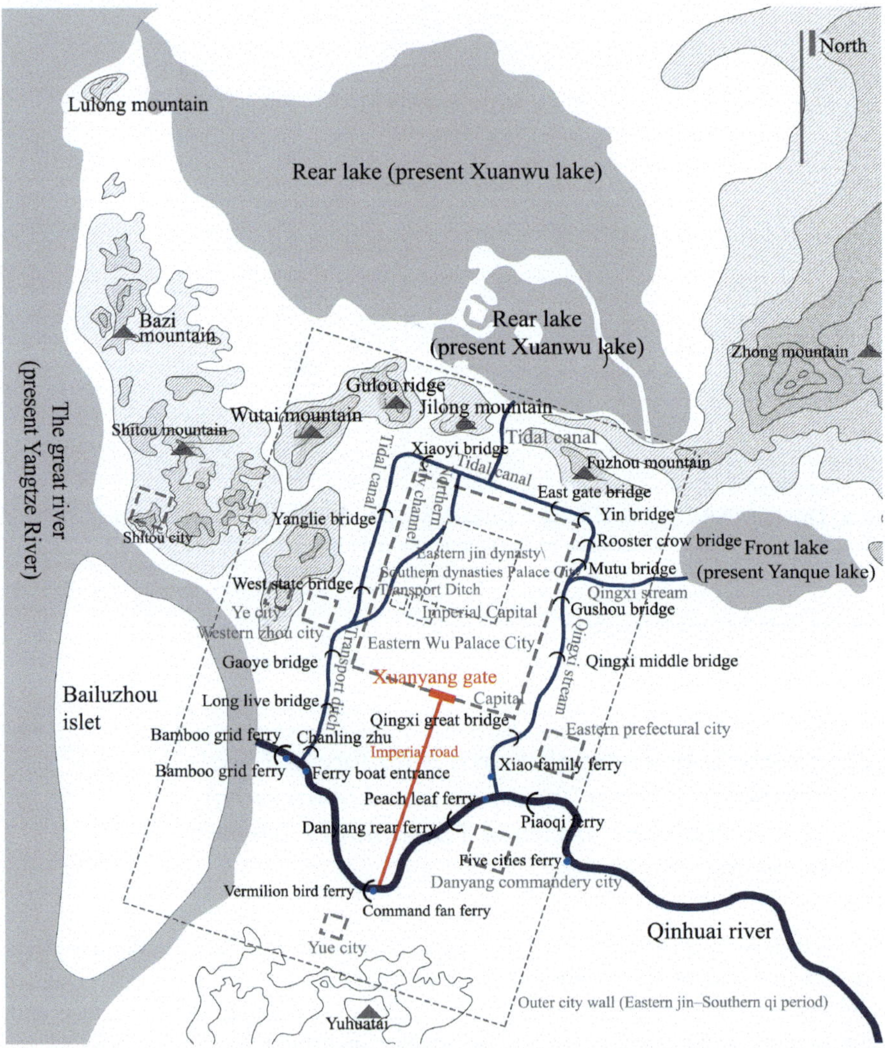

Fig. 6 Spatial Distribution Map of the Transportation Network of the Qinhuai River Water System. *Source* Redrawn by the author based on Notes 2, 34, 35, and 36

1. Ferries

The Qinhuai River served as a vital artery for the movement of people and goods from the Sanwu region into Nanjing City, with boats in constant passage and water–land exchanges in ceaseless activity. As spaces for the transition between river and road, ferries multiplied alongside the growth of commerce and the concentration of population (Table 1). Many ferries derived their names from places or events; romantic legends and the poetic verses of literati further enhanced their allure as sites of leisure and sightseeing. The Taoye Ferry was the most celebrated among them. In Wang Xianzhi's *Song of Taoye* from the Eastern Jin Dynasty, it is written:

> Taoye, O Taoye again!
> To cross the river, no oar you need.
> Cross the river and don't complain!
> I'll come to meet you, pay no heed [16].

According to tradition, "Taoye" was the beloved concubine of Wang Xianzhi, whom he often came to meet and escort at this very ferry along the Qinhuai. Thus the ferry was named in her honor.

Table 1 List of ferry crossings in the Qinhuai River water system

Ferry name	Location	Notes
Zhugedu Ferry	About two *li* southwest of the county seat, near today's Doumen Bridge	According to Annals of Jiankang, Wang Dun rebelled and crossed the river here from Zhugedu Ferry [4]
Taoyedu Ferry	West of Dongshuiguan Gate, at the confluence of Qingxi Stream and the Qinhuai River	Named after the story that Wang Xianzhi, a calligrapher of the Eastern Jin Dynasty, met and saw off his beloved concubine Taoye here [16]
Xiaojiadu (Yaodibu) Ferry	Near Dongshuiguan Gate	Named after the story that Wang Huizhi of the Eastern Jin moored his boat here and invited Huan Yi to play the flute [16]
Wucheng Ferry	East of Shangyuan County	–
Huishandu (Maogongdu) Ferry	Near Zhenhuai Bridge	When Chen Min occupied Jiankang, he led troops to the Great Avenue (Zhuque Avenue); Gu Rong waved his fan to signal, and Chen's army collapsed [17]
Duchaankou (Zhouzizhou)	Turn right east of the bridge, follow the canal south to the Huai River [18]	

Source Compiled by the author based on Notes 2, 22, and 35

2. Bridges

The Qinhuai River system contained numerous bridges, but due to the limitations of bridge-building technology and concerns over northern military threats, crossings of the Qinhuai and Qingxi Canal were primarily made in the form of floating bridges (fuhang), constructed by mooring boats together. In daily life, these vessels were linked to form a bridge for traffic; in times of war, the boats could be separated, creating a flowing barrier to repel enemies. The most famous of these were the Twenty-Four Floating Bridges, among which four are historically verifiable: Danyang, Zhuge, Zhuque, and Piaoqi [4]. From west to east, they were arranged as Zhuge Bridge, Zhuque Bridge, Danyang Bridge, and Piaoqi Bridge.

The Twenty-Four Floating Bridges not only served as crucial transportation nodes but also became celebrated venues for river viewing and leisure. They have been immortalized in poetry since antiquity. The *Jingding Jiankang Zhi*[7] notes:

> Du Mu[8] composed the verse:
>
> The green hills are vaguely seen and the stream is afar;
>
> In the south grass is not killed when autumn is o'er.
>
> The moon shines o'er the Twenty-four Bridges bright;
>
> Where is the jade-like belle teaching the flute at night? [4]

(It is worth noting that during the Six Dynasties, the administrative seat of Yangzhou was located in present-day Nanjing City; it was only in the Tang that it was relocated to today's Yangzhou. Hence, some argue that the "Twenty-Four Bridges" mentioned in the poem refers not to Yangzhou's later site but rather to the famous Twenty-Four Floating Bridges of Six Dynasties Nanjing City.) From these verses, one may discern not only the poetic beauty of the Qinhuai—its floating bridges, rippling waters, and luminous skies beneath stars and moon—but also the manner in which these bridges transcended their function as spaces of transportation, becoming vessels of lived experiences, joys, and sorrows.

3. Imperial Road (Yudao)

The imperial road, first constructed during the Eastern Wu, extended southward from the Xuanyang Gate of the imperial palace city to the Zhuque Gate, linking with the Zhuque Floating Bridge to form a vital transportation corridor connecting the northern palace city with the southern residential wards. This axis became one of the most prosperous spaces in the imperial capital. The imperial road was broad and level, flanked on both sides by imperial ditches and planted with locust and willow trees. From afar, one beheld rows of towering buildings aligned like fish scales; up close, continuous halls and pavilions rose in striking succession. The roadway bustled with imperial carriages and throngs of traffic, while on the Qinhuai River, light boats and pleasure barges accompanied by drums animated the entire scene.

Zuo Si of the Western Jin, in his *Rhapsody on the Capital of Wu*,[9] described:

Vermilion watchtowers stood in pairs; the Speedway was like a whetstone. It was planted with green locust trees and spanned by clear waterways. Their deep shade was massed and mighty; their pure flow, never-ending. For seven li, government offices lined the route, their pillars flanking the sunny road. Barracks stood close as the teeth of a comb; official bureaus were arrayed like a tapestry.

This portrayal illustrates the grandeur of the imperial road and the Zhuque Floating Bridge, where verdant waters, red towers, locusts, and willows framed a scene of bustling water–land transportation and thriving markets, capturing the flourishing urban imagery of the Six Dynasties capital.

(2) Environmental Resources and the Layout of Residential Wards (Lifang)

At the outset of the Eastern Wu capital, the city interior was filled with official offices, monasteries, and government precincts, leaving limited residential space, where only a few high-ranking officials and nobles resided. By contrast, the banks of the Qinhuai River, with their convenient access to grain transport, witnessed flourishing commerce, dense populations, and open, level terrain, thereby becoming the principal residential districts. As the Wu Capital Rhapsody (Wudu Fu) describes:

> Hengtang and Zhaxia,
> the dwellings there are lofty and splendid.
> Changgan stretches on and on,
> its soaring roofs interlocking.
> Those who live there are families of high rank and great prestige,
> the great and powerful,
> the brave and eminent:
> the descendants of the Yu and the Wei,
> the progeny of the Gu and the Lu [19].

The residents of Six Dynasties Nanjing City can broadly be divided into two categories: high officials and noble families, and commoners and merchants. The principal residential areas along the Qinhuai system were located in four main sites: on the southern bank near the river's confluence with the Yangtze at Changgan and Hengtang, on the northern bank along the imperial road, and on the banks of the Qingxi Canal. The distribution of these areas evolved in stages. During the Eastern Wu, commoners and merchants were mainly concentrated in Hengtang, while high officials and nobles resided primarily at Changgan; overall, however, the residential types were mixed, with no strict boundaries. After the Eastern Jin, with the turmoil of the "Five Barbarians" and the southward migration of aristocratic families, a large influx of northern gentry entered Nanjing City, necessitating the expansion of residential zones. Because the existing southern-bank districts were noisy, crowded, and inhabited by both officials and commoners, they were no longer ideal residences. Thus, the aristocratic elites shifted toward the northern bank of the Qinhuai and the banks of the Qingxi Canal, while the common populace remained concentrated on the southern bank (Fig. 7).

The transformation of residential distribution in the Six Dynasties was closely related to topography, landscape resources, and proximity to the imperial palace, rivers, and commercial districts.

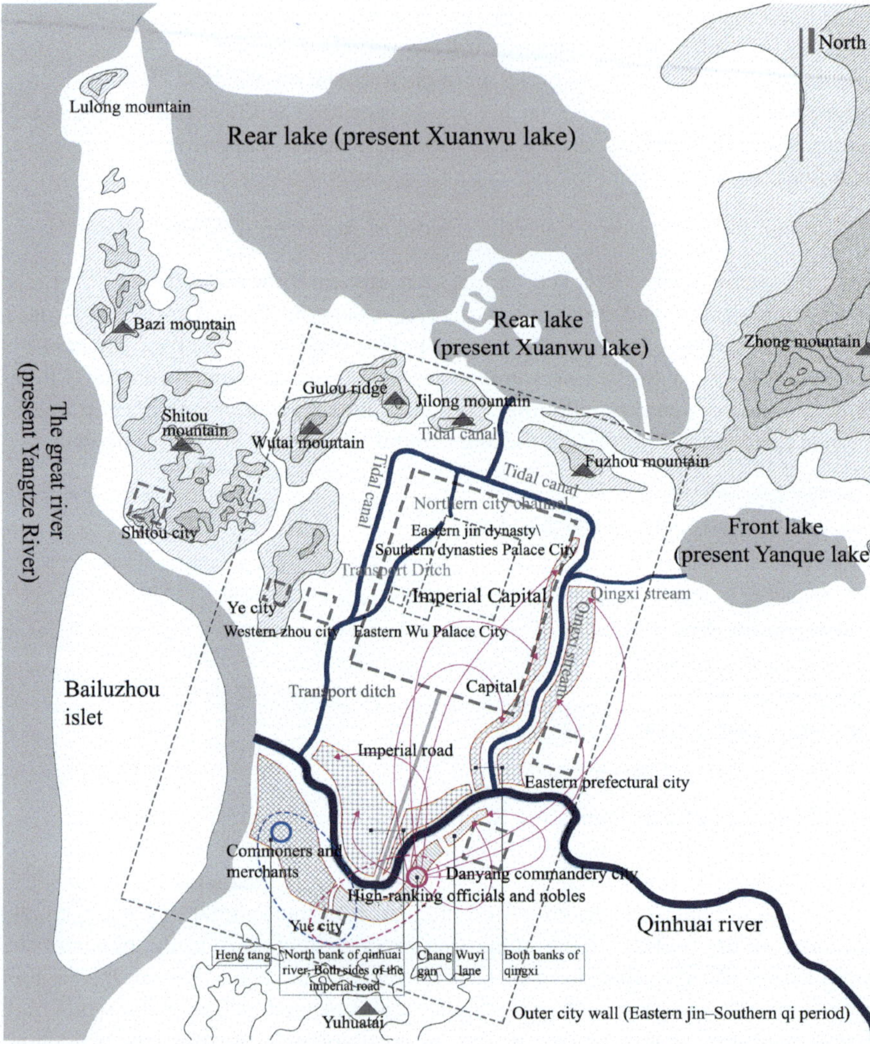

Fig. 7 Map of the distribution and transformation of residential spaces along the Qinhuai River Water System. *Source* Redrawn by the author based on Notes 2 and 34–36

(1) Hengtang: A long embankment built south of the Qinhuai during the reign of Sun Quan. Although the terrain was low-lying and the living environment poor, its position on the riverbank and near the Yangtze made it a hub of commerce and transport. Its residents were predominantly merchants, with ships moored side by side and the bustle of constant trade.

(2) Changgan: The site of the former Yue City, comprising the ridges between Great Changgan (Shizigang, the highest peak of Yuhuatai) and Lesser Changgan (Hualugang, the highest peak of Fengtai Mountain). Overlooking the Yangtze,

bordered to the north by the Qinhuai and to the south by Yuhuatai, its elevated terrain and scenic views, together with its relative closeness to the palace city, made it a preferred residential district for high officials and nobles. Notable residents included Zhang Zhao, Zhang Xiu, and the Lu brothers. The Wudu Fu describes this prosperity: "Wisdom succeeded to wisdom, and age to age; The hoofs of their horses were interlaced, the tracks of their vermilion wheels entwined."

(3) Wuyi Lane: One of the few aristocratic enclaves on Qinhuai's southern bank, which emerged after the Eastern Jin. According to Shan Qianzhi's Records of Danyang: "The origin of the 'Wuyi' (Black Garb) name comes from its status as the camp of the Black Garb Troops during the Wu Dynasty. In the early years of the Eastern Jin regime, it became the residence of the princes of the House of Langye [13]." Thus, aristocratic clans led by the Wang and Xie families, who had migrated south from Langya (modern Linyi, Shandong), settled here. Over time, more northern great families followed, and Wuyi Lane expanded into a major noble residential quarter second only to Changgan. Liu Yuxi later lamented in his poem: "By the Zhuque Bridge wild grasses sprawling,

At the Wuyi Lane the sun declining.
Swallows that skimmed by painted eaves in bygone days.
Are dipping now among the humble homes' doorways."

(4) Northern bank of the Qinhuai, along the Imperial Road: Owing to its proximity to the palace city and convenient access, many aristocratic mansions of prominent families were located here.

(5) Banks of the Qingxi Canal: Blessed with superior scenic resources such as Zhong Mountain and Yanque Lake, and being close to the palace, this area became a favored site for aristocratic estates and gardens. For example, the residence of Sima Daozi in the Eastern Jin's Dongfu City was described in the Jiankang Shilu as follows: "They constructed artificial hills and dug ponds, laid out rows of trees, bamboos, and other flora. The costs incurred reached an astronomical sum [9]."

In summary, the residential pattern of the Six Dynasties shifted from initial disorder to increasingly clear spatial differentiation: in the Eastern Wu, the southern bank of the Qinhuai was a mixed zone of officials and commoners; after the Eastern Jin, the southern bank became the primary quarter of commoners and merchants "markets and dwellings clustered along the southern road" [4] while the northern bank was dominated by the residences of high-ranking nobles—"on both sides of the imperial road, none but affluent households [20]." The banks of the Qingxi Canal: rich in scenic resources, were further developed into enclaves of aristocratic estates and gardens. Unlike the regimented ward system of northern capitals, however, the complex landscape environment of Nanjing City produced a more mixed and dispersed residential pattern, with looser boundaries and overlapping functions.

(1) Geographical Advantages and the Prosperity of Markets (Shisi)

"As the populace was already abundant, the common people were mostly engaged in trade, while the gentry relied on official stipends; the markets and stalls were arrayed no less than those of the Two Capitals [21]." During the Six Dynasties, Nanjing City emerged as the economic center of the southern regions, distinct from the earlier layout described in the *Kaogongji of the Zhouli*,[10] where the market was located behind the palace. Instead, in Nanjing City, most commercial spaces were concentrated along the southern banks of the Qinhuai River, for two principal reasons. First, the imperial palace city was encircled by mountains and lakes—bounded by Zhongshan, Fuzhou Mountain, Jilong Mountain, Lulong Mountain (today's Shizi Mountain), and Houhu Lake—leaving insufficient space for the development of markets. By contrast, the southern banks of the Qinhuai offered ample room for expansion. Second, the capital's commodities largely depended on supplies transported via Pogangdu Canal from the Sanwu region; through the canal transport system these goods circulated within the city. At the Qinhuai's confluence with the Yangtze, ships converged and people gathered, creating advantages of transport, location, and resources for commercial development.

Historically verifiable market types along the urban and suburban waterways included five categories (Table 2): major markets, minor markets, specialized markets, fodder markets, and garden markets. Major markets were large-scale, comprehensive centers of trade; minor markets were smaller, integrated trading sites; specialized markets dealt in specific commodities; fodder markets were suburban exchanges for fodder; and garden markets were set up within imperial or private gardens [22].

By the early Eastern Wu, major markets had already formed, with rows of shops lining both banks and merchant vessels ceaselessly traversing the waters. The Rhapsody on the Capital of Wu (Wudu Fu) vividly described the prosperity:

> They opened the market at dawn and all was received;
> It overflowed the crossroads and the hamlets.
> Goods of every kind were mixed in the same shop;
> The capital and the provinces were merged into one.
> Ladies and gentlemen lingered, gazing;
> Merchants and traders jostled side by side.
> In garments of fine and coarse linen,
> A bustling, teeming, converging throng.

As the economy developed, taxes levied on commercial activities in the major markets became a vital source of government revenue. In comparison, minor markets and specialized markets bore lighter fiscal burdens and thus proliferated along the Qinhuai's banks. As records state: "There were also minor markets, cattle and horse markets, grain markets, clam markets, and silk markets—ten in all—each arrayed

Table 2 List of markets along the Qinhuai River water system

Market name	Establishment period	Location	Type	Source
Jiankang Grand Market	State of Wu	In front of Jianchu Temple, south of the Qinhuai River	–	Taiping Yulan
Jiankang North Market	Reign of Emperor Jing of Wu, Yong'an period	North bank of the Qinhuai River	–	Taiping Yulan
Douchang Market (South Market, East Market)	Eastern Jin, Long'an period	Douchangli area, south bank of the Qinhuai River	–	–
Zhaojian Market	Eastern Jin	Area east of Qingxi Stream and north of the Qinhuai River, within the Dongfu City area	–	Jingding Jiankang Gazetteer
Zhuangyansi Small Markets (over ten)	Liu Song period	In front of Zhuangyan Temple, north of the Qinhuai River	Small markets	Book of Song
Cattle and Horse Market	Unknown	Along the Qinhuai River	–	Jingding Jiankang Gazetteer
Grain Market	Unknown	–	Specialized market	Jingding Jiankang Gazetteer
Clam Market	Unknown	–	Specialized market	Jingding Jiankang Gazetteer
Salt Market	Unknown	North of the Qinhuai River	Specialized market	Jingding Jiankang Gazetteer
Unknown	Eastern Jin period	Near Qingxi Stream, south of Xianggong Temple outside Qingming Gate	Fodder market	–
Residence of Sima Daozi	Eastern Jin period	Near Qingxi Stream	Garden market	Annals of Jiankang

Source Compiled by the author based on Notes 2, 22, and 35

along the Qinhuai with stalls where petty traders thrived [4]." In addition, commercial activity extended even into aristocratic estates and gardens.

Commercial markets in Six Dynasties Nanjing City were largely distributed along the Qinhuai River system, where markets, wards, and temples often intermingled. Examples include the major market in front of Jianchu Temple in Fotu Ward (Chusi

Ward), the Dou Market before Douchang Temple in Douchang Ward, and the fodder market before Xianggong Temple. This pattern stemmed from Nanjing's complex natural environment, which prevented standardized urban layouts. Moreover, the ward system (lifang) was underdeveloped, residential quarters were scattered, and—together with the flourishing of Buddhism and the proliferation of temples—commercial nodes naturally emerged in populous areas with available public space, such as temples and wards.

As one description records: "A modest cottage stood by the roadside, intermingled with a row of shops [23]." Another account depicts: "To the front, one gazed upon the bustling worldly dust enclosing the scene on all sides, observing the bustle and lull of the triple markets; To the rear, one glimpsed the ninefold tiers of purple pavilions, linked with the soaring pair of watchtowers [24]." (Fig. 8). These portrayals capture the lively interweaving of commerce, religion, and daily life along the Qinhuai, where markets formed integral nodes in the evolving urban landscape.

2.3 Constructing Temples Along the River and Establishing Gardens by Water: Economy and Culture as Catalysts for the Derivation of Six Dynasties Landscape

Following the southward migration of aristocratic families during the Eastern Jin Dynasty, Nanjing City emerged as the political, economic, and cultural center of the South. The abundant vegetation and diverse landscape environment of the Jiangnan region provided the natural conditions for garden construction. The economic growth of the capital laid a material foundation, while the concentration of literati and the flourishing of poetry, painting, and other arts created a fertile social atmosphere.

The gardens of Nanjing City during this period can be classified into three types: imperial gardens, private gardens, and temple gardens. Among them, the imperial and private gardens functioned as non-public landscape spaces, while temple gardens (primarily Buddhist temple gardens) served as public landscape spaces, constituting a distinctive cultural landscape of Nanjing City.

(1) Temple Gardens as a Distinctive Cultural Landscape

"The numerous temples of the Southern Dynasties, uncounted towers and terraces loom in the mist and rain." Buddhism was introduced to Nanjing City during the Eastern Han and rapidly flourished, reaching its zenith in the Six Dynasties. Magnificent temples and pavilions spread across the capital, making it the Buddhist center of the South. However, considering that Jinling was a metropolitan capital, the place upon which all paid allegiance depended, its temples and monasteries were like a forest, and its forums for doctrinal exposition were like a marketplace. From the five districts and six commanderies came a profusion of goods—fruits ripe and unripe, all mingled together."

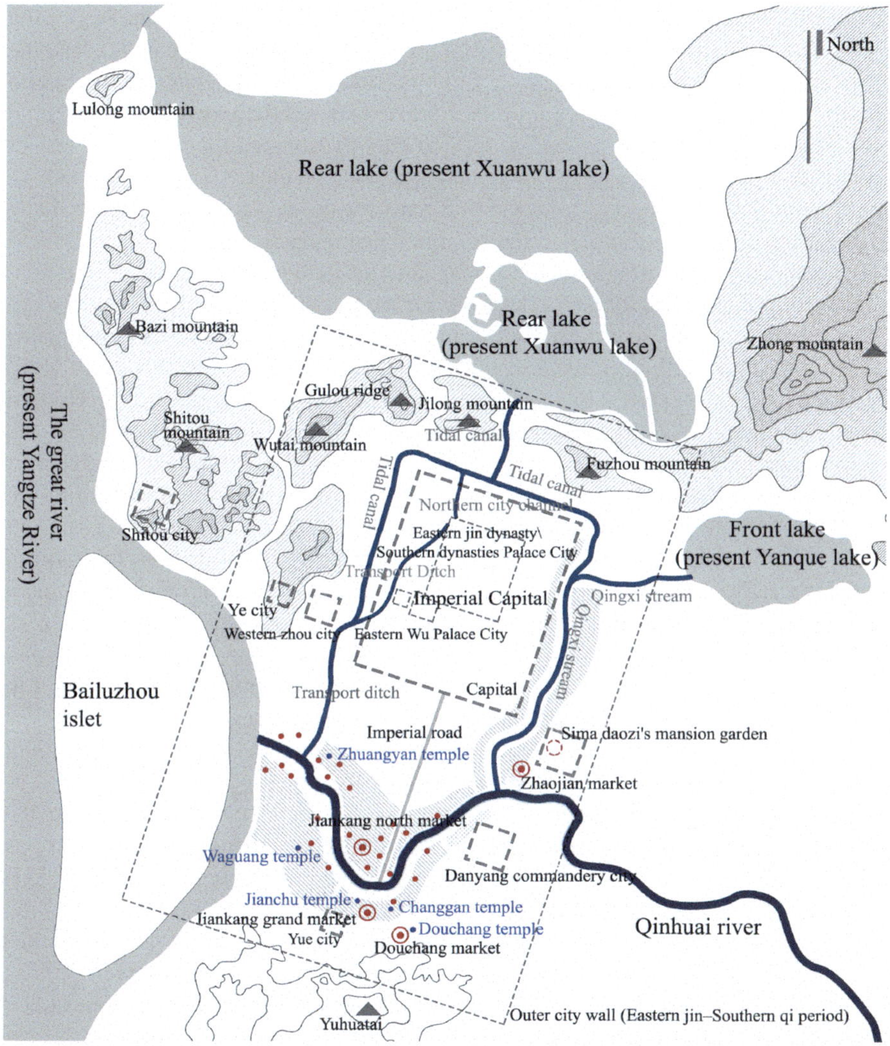

Fig. 8 Diagram of the mixed layout of markets, wards, and temples along the Qinhuai River water system. *Source* Redrawn by the author based on Notes 2 and 34–36

While the maxim "The world's famous mountains are largely occupied by monks" held true, these temples were not confined to hills in and around the city. They were also extensively distributed along the banks of the Qinhuai River system, producing an interwoven pattern of markets, wards, and temples. There were two main reasons for the location of Buddhist temples within residential and commercial quarters. First, it facilitated the propagation of Buddhism and accommodated the needs of urban residents. Officials or monks often established temples directly within wards (lifang). For example, Jianchu Temple—the "First Temple of Jiangdong"—was built in Xiaochanggan on the southern bank of the Qinhuai River, marking the beginning of temple construction in Nanjing City. Other examples include Changgan Temple in Changgan Ward, Douchang Temple in Douchang Ward, and Jianxing Temple in Jianxing Ward. These populous public spaces also fostered the growth of commercial activity. Second, high officials and nobles frequently donated their own estates for conversion into temples. Zhuangyan Temple, located "on the east bank of the Yundu, directly south of Zhuge Ferry [23]," was originally the residence of General Xie Shang in the Eastern Jin. Such temples often occupied prime residential areas.

Despite the mixture of markets, wards, and temples, the strong imperial and aristocratic patronage of Buddhism, coupled with the temples' symbolism of transcendence, meant that they were usually located in superior sites backed by mountains and facing water, with commanding scenic views. These temples also cultivated refined garden landscapes. For instance, Waguansi, constructed during the Eastern Jin on Hualugang of Fengtai Mountain on the southern Qinhuai, occupied a site overlooking the Yangtze. In the Liang Dynasty, the Waguan Pavilion was added: "A pagoda was built in the temple, soaring to a height of two hundred and forty *chi*, with the mountain itself serving as its foundation. At dawn, its shadow stretched across half the river. Ascending the pagoda and gazing out, one was met with a panoramic view of the river and the mountains." This made it not only a prominent public viewing platform in the suburban highlands but also a striking element of the city's skyline. Similarly, *Xianggong Temple Stele*, situated north of Zhongqiao Bridge on the Qingxi Canal, was formerly the residence of Emperor Ming of the Liu Song. The stele inscription of the Liang Emperor Jianwen reads: "Thus its eaves rise three zhang high, forming a hall dedicated to the worship of the gods. Surrounding it on all sides are interlinked pavilions, far from the dwellings of ordinary households. Grasses like those of the Snow Mountain of Forbearance and blossoms like the celestial Udumbara tree grace the grounds, their fragrance diffusing in all directions and their colors casting a wondrous radiance over the converging roads [26]." This testifies to the temple's splendor. By contrast, the Shanzhai Temple, built from the residence of Xie Ju in Wuyi Lane, exemplified the intimate beauty of natural springs and stones, so much so that it "seemed almost as if wrought by nature itself."

The flourishing Buddhist culture of the Six Dynasties thus gave rise to a multitude of temple landscapes. These not only provided the urban populace with spaces for religious devotion and leisure but also stimulated economic growth and advanced the development of public garden landscapes. In combining the vibrancy of urban markets and Buddhist ritual activity with the contemplative ideals of transcendence and reverence for nature, they formed the distinctive cultural landscape of densely clustered temples and communities of monks and nuns along the banks of the Qinhuai River system.

(2) Imperial and Private Gardens and Their Inclination Toward Natural Aesthetic Expression

In the Six Dynasties, beyond those gardens located within palace precincts, most imperial gardens and private gardens were distributed in areas of scenic beauty, particularly along the Qinhuai River and Qingxi Canal. Owing to abundant water resources, the mansions of nobles and private residences were typically furnished with ponds and lakes. As the saying went: "Of ten *mu* of land, nine were occupied by residences, and half of these by mountains and ponds [27]."

The imperial gardens of the Six Dynasties partly inherited the traditions of the Qin and Han, displaying magnificence and grandeur; at the same time, influenced by the culture of the literati (shiren), they also pursued natural charm, embodying an aesthetic taste that was fresh, refined, and elegant [28].

The Qingxi Canal, with its meandering "Nine Bends" and picturesque scenery, became a favored site of garden construction: "The foremost noble families of the capital were largely located to the left of the Qingxi Canal and north of the Chaogou Canal [9]." Among the imperial gardens established here was the Fanglin Garden, situated before Xianggong Temple and beside Zhongqiao Bridge. Converted from the former residence of Emperor Gao of the Southern Qi, it featured artificial hills, ponds, flowers, and trees. As recorded in the *Nan Shi*: "Furthermore, landscaping with lakes and hills was undertaken, planted with rare and exotic fruits and trees. The estate was adorned with the most exquisite carvings and decorations, achieving a splendor that rivaled the work of nature himself [23]."

On the southern banks of the Qinhuai River, imperial gardens were also located in areas of superior landscape setting. Examples include the Nanyuan Garden and the Jianxing Garden in Jianxing Ward. Nanyuan Garden, northeast of Waguan Temple, was first used as a private garden in the Southern Song and later became an imperial garden in the Southern Liang. The Northern Song poet Ma Yeting[11] reflected:

The Nanyuan Garden was a novel sight,
Surpassing other resorts east and west.
How many pavilions were left unvisited!
The labyrinthine paths led you far and nigh.
The breeze sweetened miles of lotus flowers fine;
The banks were shaded by thousands of willow lines.
I wonder who was so lucky as to live there, unaware.
That he'd mistaken for the Peach Blossom Stream divine.

The Jianxing Garden, meanwhile, was constructed by Emperor Wu of Liang for his cousin as a farewell. Ji Shaoyu, in his Excursion to Jianxing Garden, wrote:

This crimson hill embraces the imperial town;
This purple sea outvies the Forest Fine.
The silver towers hang a hundred feet high down;
The jade-like trees stand a thousand feet in line.
In the flowing water shimmered the reflections of carriages and canopies,
The wind wafts the sound of songs and piping clear.
We'd slow our steps to pick up a kingfisher's feather,
And look around for hairpin lost somewhere.
In setting sun the light veers in the hall;
Our square-canopied cabs shift their shadows on the ground.

At last we say our pleasure is not complete at all,
Nor do we care for gold to be found [29].

These accounts reveal the dual tendencies of Six Dynasties gardens: on the one hand, they inherited the opulence of earlier dynastic traditions; on the other, they expressed the refined literati pursuit of natural charm and poetic resonance, embodying the unique garden culture of Nanjing City.

The private gardens of the Six Dynasties can be divided by location into two types: urban private gardens and suburban estates. The urban private gardens were situated within the capital itself, characterized primarily by artificial landscapes, with grand and ornate architecture. By contrast, the suburban gardens were established on the city's outskirts, often incorporating farmland and residences, situated by mountains and rivers, and marked by a natural and rustic style [28].

Along the banks of the Qinhuai River, most private gardens belonged to the former type. For example, Ji Zhan's garden in Wuyi Lane was described as having "splendid halls and pavilions, with pools, bamboo groves, and woodlands sufficient for delight and appreciation [17]." Likewise, the garden residence of Sun Shuang on the eastern side of the Qingxi Canal was noted: "His personal dwelling verged on excess and extravagance, but its courtyards and chambers, with their artificial excavations, achieved the utmost charm of forests and springs [30]." Although limited in spatial extent, these urban private gardens, through the construction of artificial hills, ponds, tree groves, and flower plantings, succeeded in creating microcosms of mountain-and-water landscapes. They also served as venues for banqueting with guests, Buddhist discussions, and metaphysical debates, reflecting the open intellectual spirit of the Six Dynasties literati and their inclination to invest emotion in landscape.

Through the water-control projects of the Eastern Wu, the Qinhuai River was transformed from a primitive natural river into a water network permeating the entire capital. With the socio-economic development of the Six Dynasties, the Qinhuai River region evolved into a landscape complex, comprising functional spaces such as ferries, floating bridges, imperial roads, residential wards (lifang), and markets (shisi), alongside landscape spaces such as imperial gardens, private gardens, and Buddhist temple gardens. By day, the Qinhuai River bustled as depicted in the Rhapsody on the Capital of Wu: "Boats float on water, carts move on land; barges and carriages gather in multitude. The songs of oarsmen and the rumble of wheels persist from dawn till dusk." By night, it exuded the tranquility captured in *Twenty-Four Floating Bridges*[12]: "Ornamental pleasure boats, like bluebirds, cast their glow upon the nocturnal waves; under stars in multitude and clouds serene, the moonlight reigns abundant."

The functional spaces of the river system presented a mixed pattern of wards, markets, and temples. The Buddhist temple landscapes not only provided public spaces for communal activities but also introduced highland viewing platforms such as the Waguan Pavilion. Meanwhile, both imperial and private gardens embodied an artistic inclination toward "Rocks were gathered to channel water, trees were planted to create ravines. In a short time, it grew into a lush and intricate landscape, appearing as if created by nature itself [31]." (Fig. 9).

3 Riverside Landscapes of the Qinhuai River System in the Ming Dynasty

The Ming Dynasty marked the zenith of Nanjing's urban development. From 1368, when Emperor Zhu Yuanzhang established the capital there, the city became the foremost political, economic, and cultural center of China, its influence extending even overseas. At the same time, the landscape character of the Qinhuai River also reached its height. As vividly portrayed in The *Peach Blossom Fan*[13]:

> The pear blossoms look like snow, willows green mist make;
> Spring reigns along both sides of Qinhuai River.
> A row of bowers along the shore vies in splendor;
> The reflection of beauty vies with beauty in splendor.

3.1 Functional Installations—Unified Planning and Separated Residential Patterns

In order to rapidly strengthen national power and consolidate the new dynasty, the Hongwu Emperor Zhu Yuanzhang enacted a series of new policies. These included relocating populations, establishing prefectural schools, and constructing pavilions and halls to house official courtesans, thereby accommodating guests and envoys from within and beyond the realm. This ensemble of policies and construction activities was strongly reflected in the landscape character along the riverbanks.

(1) Transportation: Bridges Increasing Along with the Waters

In the Ming Dynasty, the city was interlaced with a dense water network, operating in a dual system of waterborne and land-based transport, which necessitated the construction of a great many bridges (Fig. 10). Some were revivals of ancient bridges from the Six Dynasties—for instance, the six bridges along the Yundu Canal were all renamed and preserved. The majority, however, were new constructions of the Ming period, concentrated especially along newly excavated waterways such as the Ming Imperial River and the Minor Canal. On the moats encircling the capital, bridges were commonly positioned near the city gates for the sake of convenient passage. This is evident from their names, such as Tongji Bridge, Zhengyang Bridge, Sanshan Bridge, and Shicheng Bridge, which directly indicate their relationship to11.3.1.2 Lifang System: The Eighteen Wards of the Southern City.

At the beginning of the Ming Dynasty, in order to replenish the population and stimulate economic growth, Emperor Zhu Yuanzhang relocated nearly one hundred

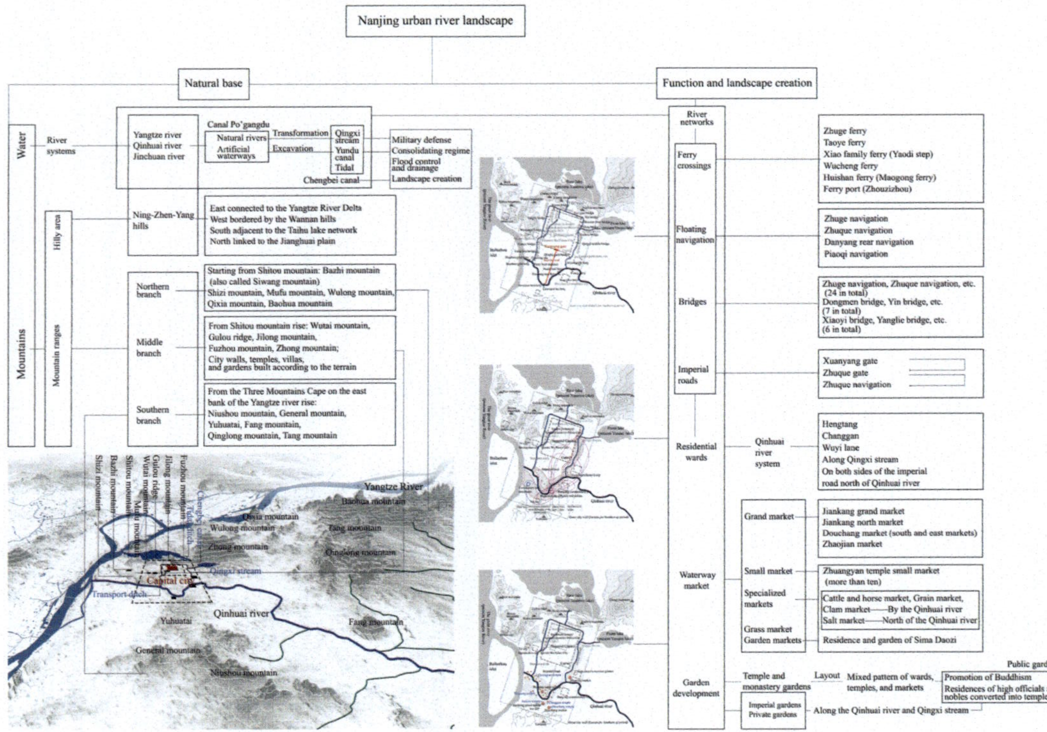

Fig. 9 Diagram of the relationship between the Qinhuai River water system and the development of riverine areas during the Six Dynasties

Fig. 10 Hypothetical distribution map of bridges along the Qinhuai River water system in the Ming Dynasty. *Source* Redrawn by the author based on Notes 37 and 38

thousand artisan households from across the empire to Nanjing City. These groups, representing diverse trades and crafts, were systematically organized and assigned to designated quarters along the southern Qinhuai River, thereby forming a clustered complex of workshops characterized by the front-shop and rear-residence pattern. This zone came to be known as the "Eighteen Wards of the Southern City" (Figs. 11 and 12).

As a consumer-oriented capital, Nanjing City had relatively few agricultural households but a substantial population of merchants and artisans. Consequently, the Eighteen Wards of the Southern City emerged as the principal residential districts of the capital. These wards were distributed along the Inner Qinhuai to the north and on both sides of the Imperial Thoroughfare, strategically chosen for their proximity to water resources and ease of transportation. The clustering of metalworking industries—such as silver- and iron-smithing workshops—further enhanced the effect of

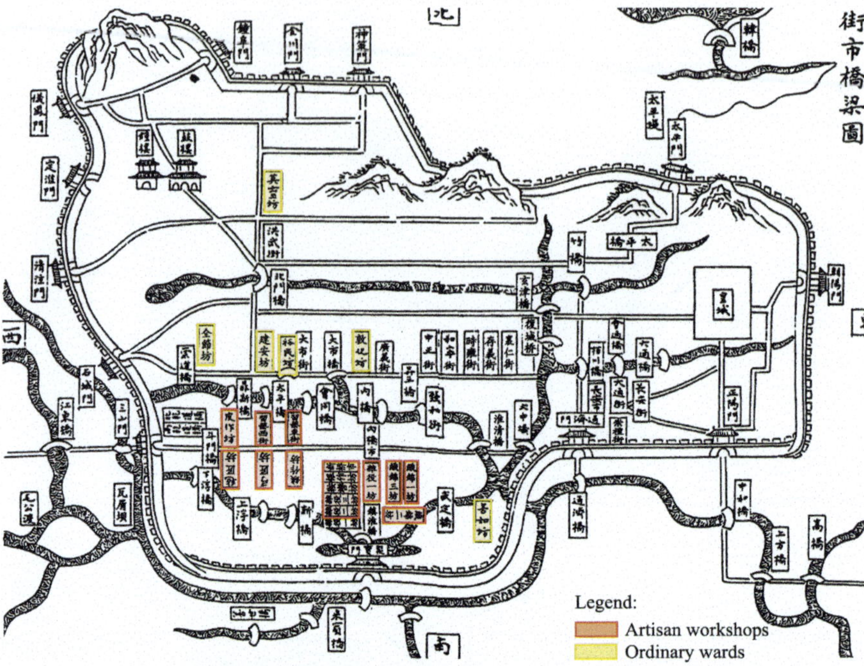

Fig. 11 Distribution map of wards along the Qinhuai River water system in the Ming Dynasty (1368–1644 CE). *Source* (Ming) Compiled by the Ministry of Rites, (Ming) Chen Yi, Hongwu Capital Map Records: Examination of Ancient and Modern Nanjing City—Street, Market, and Bridge Maps, Nanjing City Publishing House, 2017

industrial concentration, exemplifying the wisdom of segregated settlement by trade category in the practice of capital planning.

(2) The Ward System: The Eighteen Wards of the Southern City

At the beginning of the Ming Dynasty, in order to replenish the population and stimulate economic growth, Emperor Zhu Yuanzhang relocated nearly one hundred thousand artisan households from across the empire to Nanjing City. These groups, representing diverse trades and crafts, were systematically organized and assigned to designated quarters along the southern Qinhuai River, thereby forming a clustered complex of workshops characterized by the front-shop and rear-residence pattern. This zone came to be known as the "Eighteen Wards of the Southern City" (Figs. 11 and 12). As a consumer-oriented capital, Nanjing City had relatively few agricultural households but a substantial population of merchants and artisans. Consequently, the Eighteen Wards of the Southern City emerged as the principal residential districts of the capital. These wards were distributed along the Inner Qinhuai to the north and on both sides of the Imperial Thoroughfare, strategically chosen for their proximity to water resources and ease of transportation. The clustering of metalworking industries—such as silver- and iron-smithing workshops—further enhanced the effect of

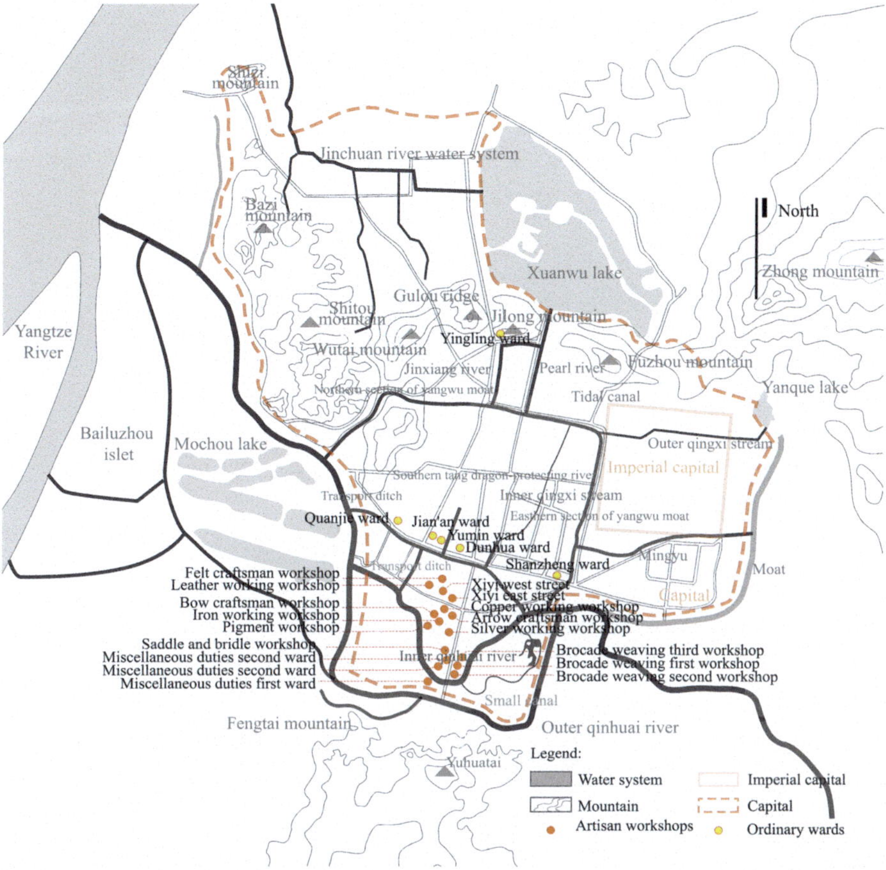

Fig. 12 Hypothetical distribution map of wards along the Qinhuai River water system in the Ming Dynasty. *Source* Redrawn by the author based on Notes 28, 39, and 40

industrial concentration, exemplifying the wisdom of segregated settlement by trade category in the practice of capital planning.

(3) Markets and Commercial Establishments: The Thirteen Markets of the Hongwu Era

During the Ming Dynasty, commodity exchange and commercial trade underwent further expansion, and the southern banks of the Qinhuai River emerged as the commercial hub of the capital, their prosperity exceeding even that of the Six Dynasties. Building upon the front-shop and rear-residence configuration of the Eighteen Wards of the Southern City, the government formally established the so-called "Thirteen Markets of the Hongwu Era," each specializing in distinct categories of goods (Fig. 13).

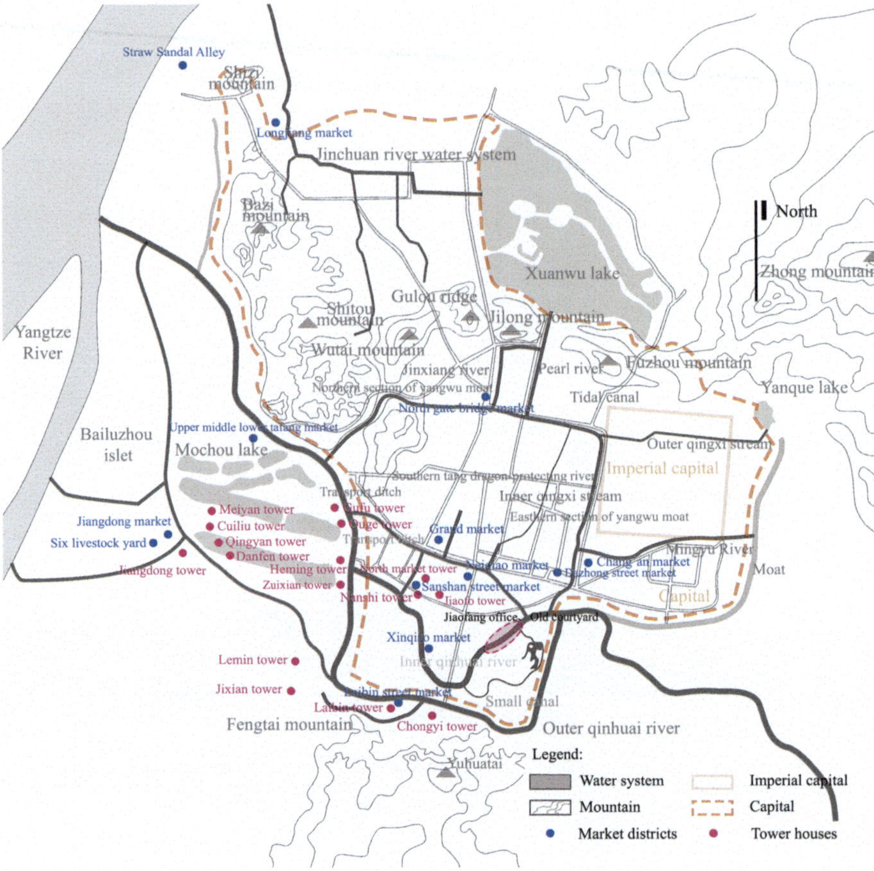

Fig. 13 Hypothetical distribution map of markets and pavilions along the Qinhuai River water system in the Ming Dynasty. *Source* Redrawn by the author based on Notes 28 and 39

Among these thirteen markets, eleven were situated directly along the waterways of the Qinhuai River system. Seven were located within the capital itself, distributed along the Inner Qinhuai, the Yundu Canal, the Qingxi Canal, and the Yangwu moat. Some of these, such as Xinqiao Market, Neiqiao Market, and Beimenqiao Market, were explicitly named after the nearby bridges. Six additional markets were positioned just outside the capital or at the periphery of the outer city wall, near city gates adjacent to river inlets that led into the Yangtze. Representative examples include Jiangdong Market outside Jiangdong Gate and the livestock markets at the city's edge.

This spatial pattern of distribution was closely tied to the consumer-oriented character of Nanjing City as a capital city. The daily necessities of its residents relied heavily on riverine transport, with supplies brought from surrounding regions via the Yangtze or the Qinhuai River waterways. As a result, numerous markets naturally developed both outside the city gates—serving as nodes where goods entered the capital—and within the city proper along the riverbanks.

(4) Entertainment: The Jiuyuan and the Sixteen Pavilions of Song and Dance.

To invigorate urban life, the Hongwu Emperor Zhu Yuanzhang, from the very founding of the Dynasty, placed great emphasis on urban entertainment. He established the Jiaofangsi, a governmental institution overseeing court music and official courtesans; constructed the Fuleyuan (later known as the Jiuyuan) as their residence; and founded the Sixteen Pavilions (Shiliu Lou) as venues for these courtesans to receive the many guests and envoys who frequently visited the capital (Figs. 13 and 14).

Jiaofangsi and Jiuyuan were located near Wuding Bridge in the eastern section of the Inner Qinhuai River. According to the *Baixia Suoyan*[14]: "In the early Ming Dynasty, the Court of Imperial Entertainment was established. Fuleyuan was first built at Qiandao Bridge and was later relocated to sites such as Wuding Bridge." Within the Sixteen Pavilions, three—namely the Nanshilou, Beishilou, and Jiaofolou—stood within the city, while the remaining thirteen were situated near the southwestern city gates.

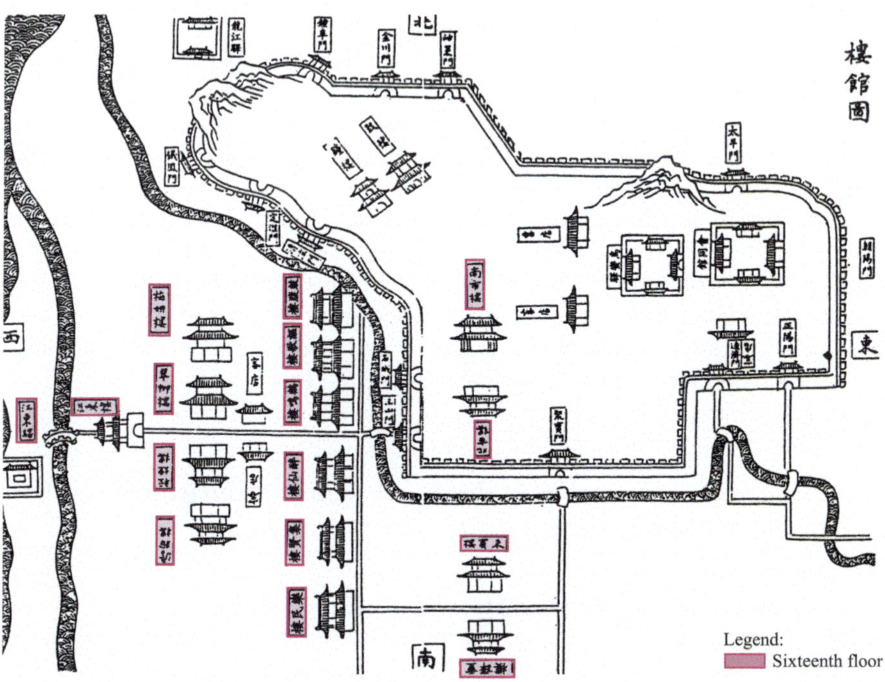

Fig. 14 Map of pavilions along the Qinhuai River water system in the Ming Dynasty. *Source* (Ming) Compiled by the Ministry of Rites, (Ming) Chen Yi, Hongwu Capital Map Records: Examination of Ancient and Modern Nanjing City—Map of Pavilions, Nanjing City Publishing House, 2017

This distribution followed practical and aesthetic considerations. First, foreign envoys and merchants customarily entered Nanjing City by ascending the Qinhuai River from the Yangtze, making the areas around Jiangdong Gate and Sanshan Gate inevitable points of passage. Second, the southwestern quarter of the city encompassed the scenic locales of Bailuzhou and Mochou Lake, where pavilions and halls set amidst natural beauty allowed the capital to display its cultural refinement and urban landscape character to visitors from afar.

(5) Education and Culture: The Confucius Temple Architectural Complex, a Gathering Place of Literati.

The Imperial Examination System[15] of the Ming Dynasty reached its peak, with the provincial, metropolitan, and palace examinations all concentrated in Nanjing City during the early Ming period. This greatly stimulated the development of the Confucius Temple educational district along the north bank of the Qinhuai River, stretching from Wende Bridge to Pingjiang Bridge [11].

The educational architectural complex comprised three main components, divided by Gongyuan West Street. To the west stood the Confucian Temple (Wenmiao) and the Imperial Academy (Xuegong), arranged in the traditional "Temple in front, Academy behind" pattern [13]. The Imperial Academy, also known as the Taixue, functioned as the state's national university for instruction and learning, while the Confucian Temple served as a ritual site where scholars venerated Confucius. The unique feature of the Nanjing City Confucian Temple lies in its Pan Pool, a semicircular ritual pond usually constructed before the main gate. Unlike standard forms, this Pan Pool was carved directly from the natural channel of the Qinhuai River, making it both the only such example integrated with a natural watercourse and the largest in scale.

To the east of Gongyuan West Street lay the Jiangnan Gongyuan, which served as the venue for the provincial and metropolitan examinations. Its densely arrayed examination cells (kaopeng) accommodated candidates during the preliminary tests. Each examination season, the area around the Gongyuan teemed with crowds of examinees, attendants, and merchants, creating a vibrant atmosphere that epitomized the cultural prominence of the Confucius Temple district (Fig. 15).

3.2 Gardens—Stagnation in the Early Ming, Flourishing in the Mid-to-Late Ming

During the first century of the Ming Dynasty, activities of garden construction were almost entirely suspended. Throughout the entire Ming period, no new imperial gardens were established, and although some temple gardens were constructed or renovated, their number and influence were far fewer than during the Six Dynasties. By contrast, in the mid-to-late Ming period, private gardens reemerged and experienced a rapid proliferation, marking a revival of garden culture along the Qinhuai River system (Fig. 16).

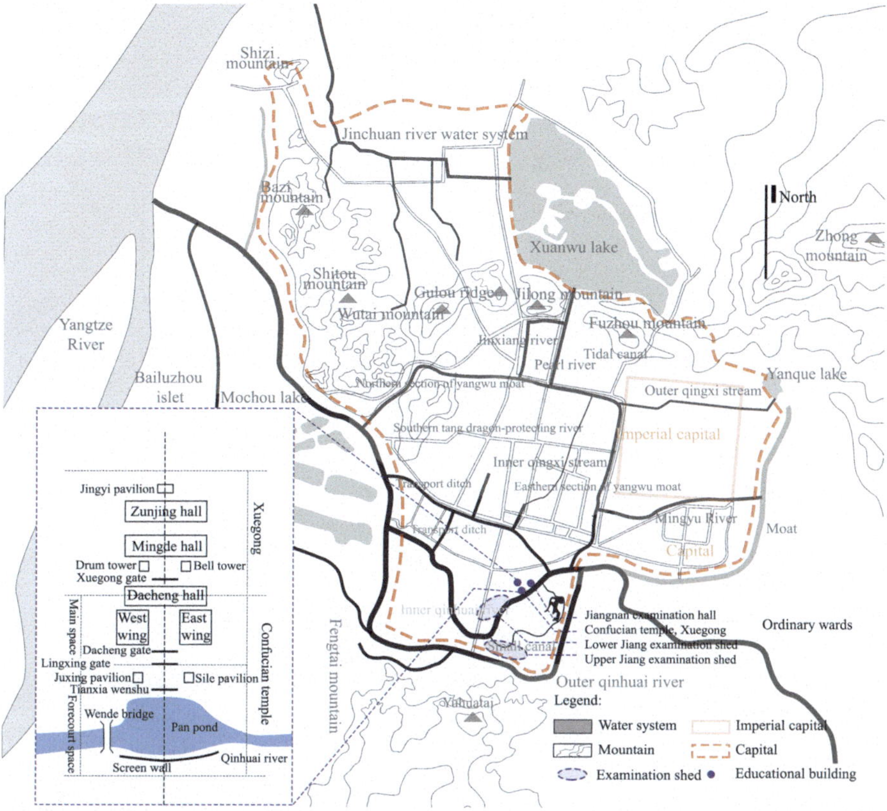

Fig. 15 Layout of educational and cultural buildings, including Confucian academies and temples of literature, along the Qinhuai River water system in the Ming Dynasty. *Source* Redrawn by the author based on Notes 28 and 41

(1) Imperial Gardens: Stagnation under the Frugality of the Founding Emperor

In the early Ming, Emperor Taizu (Zhu Yuanzhang) concentrated the state's principal financial and material resources on the construction of the capital city. Objectively, this left insufficient means to continue the building of imperial gardens. Subjectively, Zhu himself revered frugality, regarding the extravagance of garden-building as a fundamental cause of dynastic decline. As a result, during the first century following the founding of the Dynasty, activities of garden construction were virtually suspended. Although the emergence of capitalist tendencies in the mid-to-late Ming stimulated a gradual revival of garden-making, the political center had by then already shifted to Beijing. Consequently, garden construction in Nanjing City, as the secondary capital, was limited to private gardens of officials and aristocrats, with no new imperial gardens established.

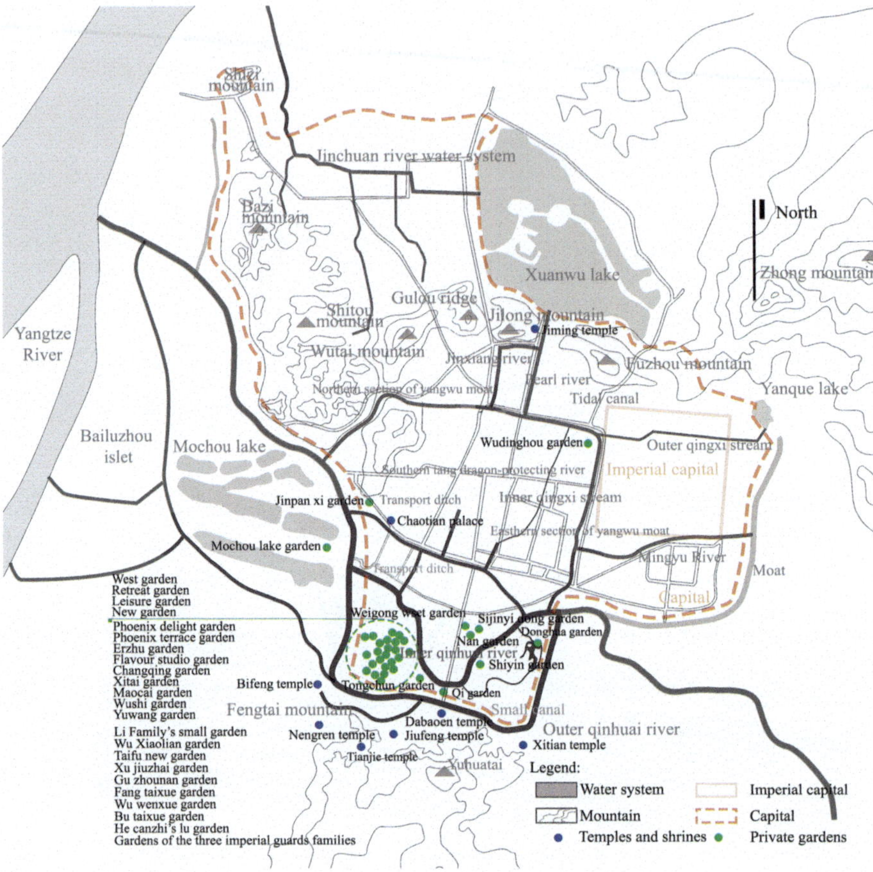

Fig. 16 Hypothetical distribution map of gardens along the Qinhuai River water system in the Ming Dynasty. *Source* Redrawn by the author based on Notes 16 and 38

(2) Private Gardens: The Flourishing After the Yongle Relocation

At the same time that Zhu Yuanzhang restricted the building of imperial gardens, he also curtailed the creation of private gardens belonging to high-ranking officials and aristocrats. In 1421 (the 19th year of the Yongle reign), Emperor Chengzu (Zhu Di) relocated the capital to Beijing, taking with him some 27,000 households of artisans. This caused large tracts of land along the Qinhuai River, particularly in the area of the "Southern Eighteen Wards," to fall vacant. With the economic revival of the secondary capital during the mid-to-late Ming, and the gradual relaxation of restrictions on garden-building, dozens of private gardens belonging to officials and aristocrats appeared along the southern Qinhuai River [13].

According to Wang Shizhen's *Record of Visiting the Various Gardens of Jinling*,[16] these gardens were densely concentrated in Xinghua Village, located along the south-western Qinhuai River. Two factors explain this clustering: first, the area's favorable setting, backed by hills and facing the river, offered ideal conditions for creating "urban mountains and forests"; second, many family estates were established there, such as Xu Tianci's Western Garden, whose division subsequently produced a number of subsidiary gardens.

In terms of style, compared with the refined elegance of Suzhou gardens and the mercantile character of Yangzhou gardens, the private gardens of Nanjing City embodied a dual quality of literati refinement and worldly sociability. Nanjing's political role endowed its gardens with an aristocratic character, while the city's commercial and courtesan culture lent them a pronounced entertainment function. Under the impact of the Ming's commercial economy, these gardens often functioned as commodities; their frequent change of ownership underscores their semi-industrial nature.

(3) Temple Gardens: Modest Growth Under the Supremacy of Confucian Orthodoxy

The Ming Dynasty's highly centralized political system enshrined Confucianism as the dominant ideology, while simultaneously fostering a tendency toward the syncretism of Confucianism, Buddhism, and Daoism. Because Zhu Yuanzhang had spent his youth as a Buddhist monk, he retained respect for Buddhism after ascending the throne, and thus the early Ming saw a number of projects to reconstruct or renovate Buddhist temples. However, during the reign of Emperor Jiajing (Zhu Houcong), with his promotion of Daoism and prohibition of Buddhism, overall development of both Buddhist and Daoist institutions remained limited. The temple gardens of this period were chiefly located along the Qinhuai River system, often situated at scenic foothills and elevated sites (Fig. 17).

3.3 *Scenic Culture—Excursions in Landscape, Appreciation of Famous Sights*

Although the pleasures of private gardens were confined largely to the upper strata of society, the natural beauty of Nanjing's mountains and waters was shared by all its inhabitants. In the early Ming, Emperor Taizu did not restrict the construction of public landscape. The city thus continued to possess numerous public scenic sites affiliated with or adjacent to the Qinhuai River system, where natural scenery predominated. Notable examples included Mochou Lake, Bailuzhou, Qingliang Mountain, and Yuhuatai.

The prosperity of the Ming economy and the flourishing of urban entertainment culture stimulated a vibrant culture of scenic appreciation. Its cultural culmination was the late-Ming compilation Forty Scenic Spots of Jinling Illustrated with Poems (*Jinling sishijing tuxiang shiyong*)[17] by Zhu Zhifan. Of these forty celebrated scenic spots, nineteen were situated along the Qinhuai River system, underscoring its paramount role in the urban landscape of Nanjing City.

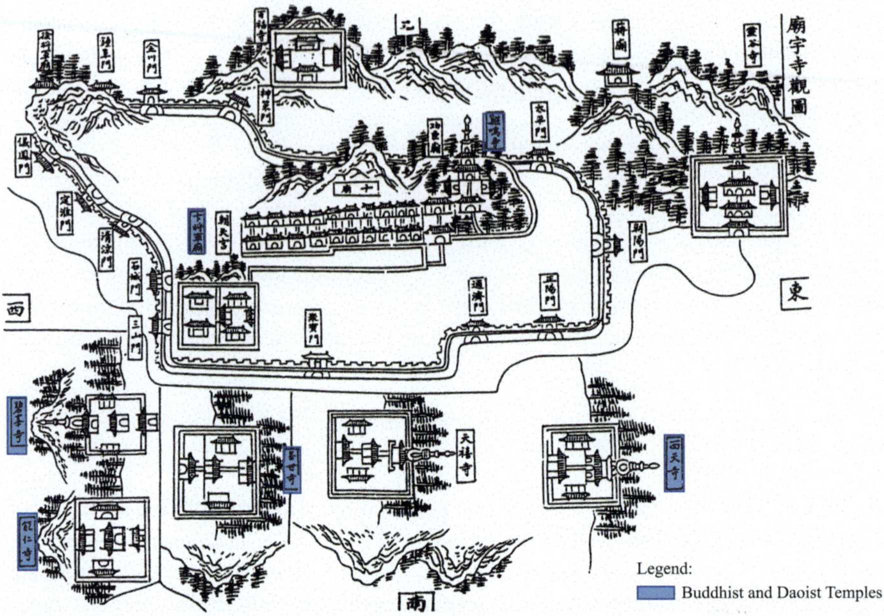

Fig. 17 Distribution map of temples and monasteries. *Source* (Ming) Compiled by the Ministry of Rites, (Ming) Chen Yi, Hongwu Capital Map Records: Examination of Ancient and Modern Nanjing City—Map of Temples and Monasteries, Nanjing City Publishing House, 2017

Based on thematic content and landscape type, these views may be broadly classified into four categories (Table 3).

Table 3 List of famous scenic spots along the Qinhuai River water system in the Ming Dynasty (1368–1644 CE)

Category	Scenic spot names
Lakes and Mountains, Water and Clouds Interconnected (2)	Mochou Panoramic View, Egret Spring Tide
Old Courtyards by the River, Oars and Lantern Shadows (4)	Changqiao Elegant View, Qinhuai Fishermen's Song, Taodu Riverside, Qingxi Boat Tour
Wards and Suburban Villages, Spring Outing Spots (3)	Wuyi Evening Glow, Changgan Spring Outing, Xingcun Wine Visit
Ascend Heights for Views, Visiting Temples and Listening to Sutras (10)	Bao'en Lighthouse, Tianjie Sutra Fish, Yuhua Leisure View, Fengtai Autumn Moon, Shicheng Snow Clearing, Qingliang Encircling Green, Yelu Secluded Pavilion, Xiedun Tranquil Enjoyment, Jilong Cloud Trees, Listening to Rain from the Heights

4 Comparative Analysis of Similarities and Differences

The Six Dynasties and the Ming Dynasty represent two pivotal periods in the rapid urban development of Nanjing City, and likewise two of the most illustrious stages in the evolution of riverside landscape character. The landscapes of these eras reveal both fundamental commonalities and distinctive features.

Overall, from the formative stage of the Six Dynasties to the mature stage of the Ming dynasty, the Qinhuai River system witnessed a sustained trajectory of riverside landscape development (Table 4).

Table 4 Comparative analysis between the six dynasties and the Ming dynasty

Aspect	Six dynasties	Ming dynasty	Cause	Landscape expression
Similarities				
Natural	Northern mountains widely distributed, Qinhuai River to the south	Same landscape pattern	Determines similar landscape structure	Landscapes clustered along both sides of the water system
Political	Successive changes of the legitimate Chinese dynasties	Unified Han Chinese empire	Similar political status	Six dynasties: Jinling ancient-themed poems Ming dynasty: Depictions of Qinhuai beauty in novels
Differences				
Political	Social turmoil, regime changes	Highly centralized	Political state affects layout planning	Six dynasties: Spontaneous growth Ming dynasty: Clearly planned
Economic	Commercial and consumer-oriented city	Highly prosperous commerce, early capitalism	Economic development enhances richness of landscape types	Ming dynasty: more diverse commercial forms, emergence of old courtyard restaurants and riverside houses
Social	Widespread suffering and unrest	Economic prosperity, luxury and extravagance	Social mood influences garden trends	Six dynasties: temples and monastic gardens flourished Ming dynasty: Private gardens thrived
Cultural	Hermit and recluse culture	Imperial capital culture	Cultural fashion stimulates public enjoyment	Six dynasties: Disheartened literati expressing emotion through wine Ming dynasty: Broad public participation, appreciation of scenic spots

4.1 Similarities in Landscape Structure and Influence

Although the Six Dynasties and the Ming Dynasty differed significantly in their political systems, cultural orientations, and social contexts, the capital city of Nanjing nevertheless exhibited analogous landscape structures and exerted comparably profound landscape influence.

(1) Shared Topographical Framework Determining Parallel Landscape Structures

Despite considerable changes in urban configuration between the Six Dynasties and the Ming Dynasty, including the eastward relocation of the imperial palace city and the expansion of the overall capital city—the principal elements of urban landscape such as residential quarters, commercial districts, transportation corridors, educational institutions, temples, and gardens remained densely concentrated along the banks of the Qinhuai River system. This spatial consistency stemmed from the constraining presence of mountain ranges in the north, which limited available urban space. The imperial palace city was constructed between mountains and rivers, leaving southward expansion toward the Qinhuai River as the primary option for urban growth. At the same time, the transportation advantages of the Qinhuai River accelerated the development of urban functions and the concentration of landscapes along its course.

(2) Comparable Political Status Yielding Parallel Landscape Influence

The countless poems commemorating Jinling that flourished after the Six Dynasties, together with the vivid depictions of the Qinhuai River scenery in Ming–Qing novels, attest that the landscapes along the Qinhuai River became deeply engraved in China's collective cultural memory, exerting enduring cultural influence. This phenomenon is inseparable from the city's significant political status.

The Ming Dynasty was a unified empire, whereas the Six Dynasties constituted a series of southern regimes sustaining themselves during times of fragmentation. Yet even amidst turmoil, Nanjing City remained an indispensable political stage. Both the Eastern Jin regime and the successive Southern Dynasties—Song, Qi, Liang, and Chen—were regarded by northern contemporaries as legitimate heirs of the Huaxia orthodoxy. They assumed a pivotal role in preserving cultural continuity and serving as the bridge between dynastic decline and renewal.

4.2 Differences in Landscape Distribution and Characteristics

In comparison with the Six Dynasties, the Ming Dynasty possessed a more stable political regime, a more developed economy, and a more flourishing culture. As a result, the overall landscape exhibited tendencies of growth and maturity. At the same time, prevailing social conditions led to divergent orientations in garden construction activities between the two periods, producing distinctive typological emphases (Fig. 18).

(1) Political Conditions and the Planned Nature of Spatial Layouts

The unification and stability of a political regime inevitably found expression in the planning of its capital city. Even though Sun Quan of Eastern Wu constructed palaces and fortified the city while also excavating canals to divert water, the residential and commercial functional spaces along the southern Qinhuai River banks lacked coherent planning, developing instead in a spontaneous and organic manner.

By contrast, in the Ming Dynasty, Zhu Yuanzhang, from the very founding of the dynasty, implemented a clear program of urban zoning and planning. At the metropolitan scale, the area south of the Qinhuai River was designated as a combined residential, artisanal, and commercial district. At a finer scale, both banks of the river were systematically organized into the "Southern Eighteen Wards" and the "Hongwu Thirteen Markets." In addition, along the western city gates, sixteen towers and pavilions were constructed on both sides of the water system; in the eastern stretch

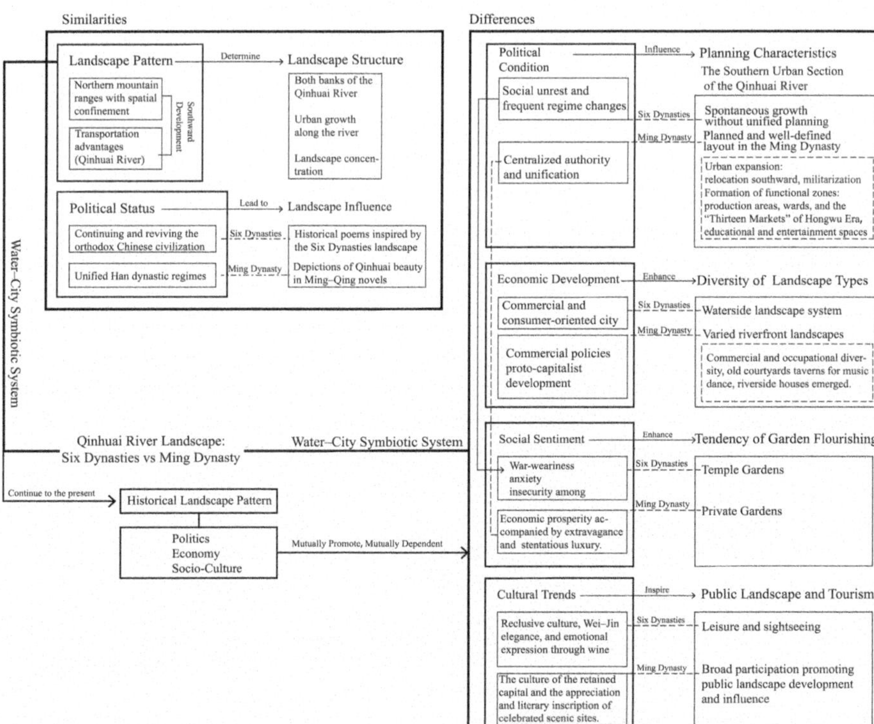

Fig. 18 Comparative analysis of landscape features along the Qinhuai River in Nanjing City during the six dynasties (220–589 CE) and the Ming dynasty (1368–1644 CE)

of the Inner Qinhuai River, the Jiangnan Gongyuan and the Jiuyuan entertainment quarter were established.

As recorded in the "Marketplace Volume" of *Kezuo zuiyu*,[18] "In the early founding of the state, streets and alleys were established, with artisans and merchants each assigned their own designated marketplaces," a vivid description of this prosperity organized within an orderly urban framework.

(2) Economic Development and the Enrichment of Landscape Typologies

Blessed with superior natural conditions and convenient water transport, Nanjing City enjoyed innate advantages for economic development. During the Six Dynasties, the city had already grown into an important commercial and consumption center. In the Ming Dynasty, however, a series of stimulus policies implemented by the rulers brought about an unprecedented flourishing of commerce, with the mid-to-late Ming even witnessing the embryonic stirrings of proto-capitalism. Compared with the Six Dynasties, Ming economic prosperity manifested along the Qinhuai River in the form of far more diverse landscape types.

First, the Ming era displayed a remarkable variety of industries and trades. This is vividly illustrated in the Prosperous Assemblage of the Southern Capital (*Nandu fanhui tu*,[19] Fig. 19), where a rough count identifies 109 distinct types of shop signs and banners—a level of specialization and differentiation not seen in the Six Dynasties' markets.

Second, the flourishing of entertainment quarters such as the Jiuyuan pleasure compound and the Sixteen Towers represented another striking feature. As products of Zhu Yuanzhang's official courtesan policy, these institutions were initially intended to project urban prosperity and secure the loyalty of subjects. Over time, however, Jiuyuan became a bustling commercial hub, where wealthy merchants congregated. Situated just across the river from the Jiangnan Gongyuan, it also drew the scions of aristocratic families and affluent clans, who gathered for feasts, judged courtesans, and indulged in leisure. The spectacle of lantern-lit boats and vibrant performances made the Qinhuai's pleasure quarters a unique riverside cultural landscape.

Finally, the Ming period saw the emergence of the distinctive riverside house typology. With wards densely arrayed along the banks of the Inner Qinhuai River and myriad trades prospering, nobles, officials, and wealthy merchants constructed

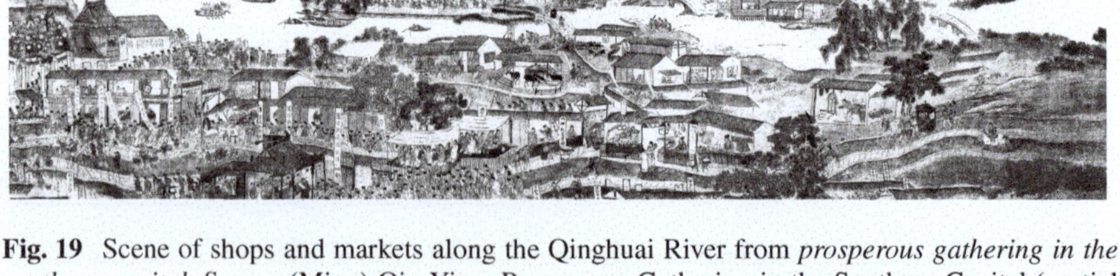

Fig. 19 Scene of shops and markets along the Qinghuai River from *prosperous gathering in the southern capital. Source* (Ming) Qiu Ying, Prosperous Gathering in the Southern Capital (partial view)

river halls and riverside residences. These structures extended a bay out over the water, with floorboards containing hinged panels that could be opened to exchange money and goods directly with passing merchant boats. Serving functions as varied as laundering, embarkation, shopping, cooling, and viewing the scenery, riverside houses lined the banks from the East Water Gate to the West Water Gate in continuous succession. Contemporary observers praised them with the verse: "Ten miles of embroidered spring breezes unfold, as a thousand doors and ten thousand households open upon the river."

(3) Social Sentiments and the Orientations of Garden Construction

The Six Dynasties and the Ming Dynasty unfolded under markedly different historical conditions, the former amid the chaotic fragmentation of competing powers, the latter during a highly centralized and prosperous empire These divergent contexts fermented distinct social sentiments, which in turn shaped different orientations and emphasis in garden development.

During the unsettled era of the Six Dynasties, the pervasive sense of insecurity, encapsulated in the phrase "everyone wearied of suffering, every household mindful of turmoil," fueled the flourishing of monastic gardens. The oft-quoted line, "four hundred eighty temples of the Southern Dynasties, how many towers amid the mist and rain," captures this unprecedented zenith of Buddhist temple landscapes.

By contrast, in the Ming Dynasty, it was economic prosperity coupled with a luxurious and ostentatious social ethos that catalyzed the efflorescence of private gardens. The contemporary expression, "Why bother seeking distant scenes while neglecting the nearby, when one can find serenity within the hustle and bustle? Seize the moment of leisure and embark on a journey, following the whims of your heart for an impromptu excursion.", vividly reflects the temperament of the age. It was this pursuit of refined yet worldly pleasures that made the private garden a favored form of landscape in the late Ming capital.

(4) Cultural Trends and the Stimulus of Public Landscape Aesthetics

In both the Six Dynasties and the Ming Dynasty, the striking mountain-and-river topography of Nanjing City consistently attracted the gaze and enjoyment of its residents. Yet the modes of urban sightseeing diverged across periods. In the Six Dynasties, landscape appreciation was often intertwined with the emotional release of disenfranchised literati, who expressed political disillusionment and personal melancholy through wine and poetry. By contrast, following the Yongle relocation of the capital, the political centrality of Nanjing City diminished, while its cultural characteristics became increasingly pronounced. In this context, the inscription, illustration, and cataloging of scenic sites—epitomized by works such as Illustrated Poems on the Forty Scenes of Jinling not only functioned as guides to local beauty but also as instruments of urban cultural promotion. This enhanced the construction, visibility, and impact of public landscapes, while simultaneously broadening the spectrum of participants in urban leisure.

From a structural perspective, rivers and cities constituted two interdependent life systems, together forming a "water–city symbiosis." Their relationship was inherently one of mutual reliance and reinforcement, articulated through the co-construction of a value chain integrating political stability, economic prosperity, social practice, cultural flourishing, and ecological resilience.

The Qinhuai River landscapes along its banks in both the Six Dynasties and the Ming Dynasty were shaped on the basis of context-sensitive design/planning, reflecting the interplay of political, economic, and socio-cultural factors. Thus, their similarities and differences can be traced to deeper causes rooted in the natural environment, politics, economy, and social culture. With the passage of time, some of the functions of the river may have gradually declined or disappeared; however, the overall historical landscape pattern of the region has persisted to the present, retaining significant cultural, landscape, and commercial potential. By comparing the similarities and differences between the political, economic, and cultural input conditions of these two major capital-building histories and the resultant transformations of the riverfront landscapes, one can better understand the traditional wisdom of urban construction based on waterscape systems and thereby grasp the enduring principles of water–city symbiosis [32].

A comparative reading of the inputs (political, economic, and cultural determinants) and outputs (landscape configurations and transformations) across the Six Dynasties and the Ming Dynasty illuminates the enduring logic of water-oriented urban planning. Ultimately, the Qinhuai River illustrates how a river, as the most crucial and continuously cultivated natural element of the city, could generate integrated functional and scenic complexes that fused natural resources with human construction. This process established the foundations for the celebrated Ming–Qing vision of "painted boats and music, unceasing day and night," while also offering insights into the sustainable development of river–city systems in historic capitals.

Notes

1. The "Six Dynasties" refers to the six successive regimes that established their capitals at Nanjing after the Jin court fled south following the collapse of the Western Jin: Eastern Wu (222–280 CE), Eastern Jin (317–420 CE), Liu Song (420–479 CE), Southern Qi (479–502 CE), Liang (502–557 CE), and Chen (557–589 CE).
2. "San Wu" denotes the Suzhou–Huzhou–Shaoxing region in the lower Yangtze, historically defined by shared water-based landscapes and integrated urban–rural systems rather than fixed administrative borders.
3. Hushu Culture (c. 3300–2200 BCE) refers to a Late Neolithic wetland-based culture in the lower Yangtze region, notable for rice farming, water-adapted settlements, and early landscape–hydrology integration.
4. *Spring and Autumn Annals of Emperor Xian*, a historical chronicle compiled during the late Han Dynasty (Emperor Xian's reign, 189–220 CE), traditionally attributed to court historians.
5. During the reign of King Wei of Chu (340–329 BCE), the ruler Xiong Shang established the settlement of Jinling (modern Nanjing). This early urban foundation shaped the regional landscape, strategic river crossings, and settlement patterns, providing a historical reference for the development of cities and urban waterways in the Lower Yangtze region.

6. *Six Dynasties Capital* documents the urban development, administrative organization, and landscape of Nanjing and other capitals during the Six Dynasties period (Eastern Wu 222–280 CE; Eastern Jin 317–420 CE; Liu Song 420–479 CE; Southern Qi 479–502 CE; Liang 502–557 CE; Chen 557–589 CE). The work provides historical records of city walls, streets, waterways, gardens, and market layouts, offering insight into early Chinese urban planning, riverine transport, and landscape use.
7. *Jingding Jiankang Zhi* (Gazetteer of Jiankang during the Jingding Era), compiled in the Southern Song Dynasty (Jingding era, 1260–1264 CE), is a local gazetteer describing the geography, urban layout, waterways, landmarks, temples, and administrative divisions of Jiankang (modern Nanjing).
8. Tang poet-official Du Mu (803–852 CE) captured urban layouts, waterways, and scenic landscapes in his writings.
9. Zuo Si (Western Jin Dynasty, 250–305 CE), a noted poet and scholar, authored *Rhapsody on the Capital of Wu*, a descriptive work celebrating the city of Wu (present-day Suzhou).
10. *Kaogongji Zhouli* (Rites of Zhou, Western–Eastern Zhou, c. 1046–256 BCE) describing craftsmen's organization, workshop management, and technical production.
11. Ma Yeting of the Northern Song Dynasty (960–1127 CE) was a scholar-official whose writings describe local cities, waterways, gardens, and scenic landscapes, offering insights into Song-era urban and landscape planning.
12. *Twenty-Four Floating Bridges* (Ming Dynasty, 1368–1644 CE) describes the design and use of 24 floating bridges in Jiangnan, revealing historical engineering, water management, and urban-landscape integration in canal-based cities.
13. *Peach Blossom Fan* (Qing Dynasty, 1699 CE; set in late Ming 1627–1644 CE) is a historical drama by Kong Shangren portraying the fall of the Ming Dynasty, vividly describing Jiangnan gardens, urban spaces, and elite residential landscapes.
14. *Baixia Suoyan* (Qing Dynasty, eighteenth century CE) is a local gazetteer from Nanjing documenting streets, gardens, temples, markets, and waterways, offering detailed views of urban planning, social life, and scenic landscapes.
15. The Imperial Examination System in the Ming Dynasty (1368–1644 CE) was a state-run meritocratic system for selecting officials through competitive exams, primarily testing knowledge of Confucian classics, literary composition, and administrative skills.
16. Written by Ming Dynasty (1368–1644 CE) literatus Wang Shizhens (1526–1590 CE), *Record of Visiting the Various Gardens of Jinling* describes his visits to multiple private and imperial gardens in Jinling (modern Nanjing). The text records garden layouts, water features, plantings, pavilions, and scenic views, reflecting Ming-era landscape aesthetics, urban leisure spaces, and the integration of natural and constructed elements within the city.
17. *Jinling sishijing tuxiang shiyong*, Wang Shizhen's Ming-era (1526–1590 CE) record of Jinling gardens details garden layouts, water features, and pavilions, revealing Ming landscape aesthetics and urban leisure planning.
18. Written by Ming Dynasty (1368–1644 CE) literatus Wang Shizhen (1526–1590 CE), *Kezuo Zuiyu* is a collection of essays and commentaries on literature, art, and cultural practices.
19. Created in the mid-Ming Dynasty (1368–1644 CE), Nandu Fanhui Tu is an illustrated scroll depicting the bustling urban life, markets, and festive gatherings of Nanjing (then known as Nandu).

References

1. Editorial Committee of Jiangsu Provincial Gazetteers. (2001). *Jiangsu Provincial Gazetteer: Hydraulic Works Gazetteer*. Jiangsu Ancient Books Publishing House.
2. Lu, H. (2002). *The six dynasties capital city*. Nanjing Publishing House.
3. Quan, W. (2007). *A study on the Interaction between the topography of mountains and rivers and urban construction in early Ming Nanjing*. Shaanxi Normal University.

4. Zhou, Y. (Song). (2012). *Gazetteer of Jiankang, Jingding Era*. Nanjing Publishing House.
5. Guo, L. (1984). The status and role of the Qinhuai River in Nanjing's History. *Journal of Nanjing Normal University (Social Sciences Edition), 4*, 80–85.
6. Chen, Z, Chen, Y. (Late Qing–Early Republic). (2008). *Nine treatises on the miscellaneous records of Jinling: Local products and customs of Jinling*. Nanjing Publishing House.
7. Zheng, E., She, L., & Zhang, Y. (2016). The historical changes of the Qinhuai River. *Jiangsu Water Conservancy, 5*, 60–62, 72t.
8. Han, P., & Han, W. (2004). *A historical account of the Qinhuai River*. Nanjing Publishing House.
9. Xu, S. (Tang). (1987). *Veritable records of Jiankang*. Shanghai Ancient Books Publishing House.
10. Yang, G., & Wang, Z. (2008). *Gazetteer of the Nanjing City Wall*. Phoenix Publishing House.
11. Wu, T. (2011). *Planning of Jiankang in the six dynasties*. Tsinghua University Press.
12. Guan, Z. (Spring and Autumn). (2016). *Guanzi. Annotated by Li Shan*. Zhonghua Book Company.
13. Li, F. (Song). (1960). *Taiping imperial encyclopedia*. Zhonghua Book Company.
14. Chen, S. (Jin), Pei, S. (Song, annotated). (2011). *Records of the Three Kingdoms*. Zhonghua Book Company.
15. Kong, S. (Qing). (1959). *The Peach Blossom Fan*. People's Literature Publishing House.
16. Office of Local Gazetteers of Qinhuai District, Nanjing. (1996). *Ten miles of Qinhuai Gazetteer*. Local Chronicles Publishing House.
17. Fang, X. (Tang). (1974). *Book of Jin*. Zhonghua Book Company.
18. Chen, Z., & Chen, Y. (Late Qing–Early Republic). (2008). *Nine treatises on the miscellaneous records of Jinling: A brief record of Yundu bridges and roads*. Nanjing Publishing House.
19. Xiao, T. (Liang), & Li, S. (Tang, annotated). (1977). *Selections of refined literature*. Zhonghua Book Company.
20. Yao, S. (Tang). (1973). *Book of Liang*. Zhonghua Book Company.
21. Wei, Z. (Tang). (1997). *Book of Sui: Treatise on geography*. Zhonghua Book Company.
22. Lin, G. (2004). *A preliminary study on the urban form of Jiankang in the six dynasties*. Shaanxi Normal University.
23. Li, Y. (Tang). (1975). *History of the Southern dynasties*. Zhonghua Book Company.
24. Yan, K. (Qing). (1999). *Collected Writings of the Sui Dynasty*. The Commercial Press.
25. Zhang, X. (Yuan). (1991). *New Gazetteer of Jinling in the Zhizheng Era*. Nanjing Publishing House.
26. Ge, Y. (Ming). (2001). *Gazetteer of Buddhist Monasteries in Jinling*. Nanjing Publishing House.
27. Yifeng, Y. (2010). Exploring the landscape architecture of Nanjing during the six dynasties. *Chinese Landscape Architecture, 26*(07), 57–61.
28. Shengbo, G. (2009). *A study of the gardens of the six dynasties capital*. Nanjing Agricultural University.
29. Guo, M. (Song). (2016). *Anthology of Yuefu Poetry*. Shanghai Ancient Books Publishing House.
30. Yao, S. (Tang). (1972). *Book of Chen*. Zhonghua Book Company.
31. Shen, Y. (Southern Dynasties). (1974). *Book of song*. Zhonghua Book Company.
32. Wu, J. (Qing). (2002). *The Scholars*. People's Literature Publishing House.
33. Yao, Y. (2008). The geographic pattern of "crouching tiger and coiled dragon" and the urban landscape of Nanjing. In *Proceedings of the Academic Exchange Conference of the Landscape and Environmental Planning Committee of the Urban Planning Society of China* (2008)
34. Yifeng, Y. (2009). The evolution of Nanjing's urban water system and research on its modern landscape. *Urban Planning, 33*(11), 39–43.
35. Tingting, Z. (2015). *A study on the spatial layout of Jiankang in the six dynasties*. Shaanxi Normal University.
36. Weiwei, D. (2012). *A preliminary study on the evolution of the urban landscape along the inner Qinhuai River since the Ming Dynasty*. Southeast University.
37. Xiangmei, C. (2016). *A study on bridges in Ming Dynasty Nanjing*. Shaanxi Normal University.
38. Zhu, X. (2015). *Illustrated investigation of ancient sites in Jinling*. Zhonghua Book Company.

39. Ministry of Rites (Ming, compiled), & Chen, Y. (Ming, authored). (2017). *Illustrated Gazetteer of the Capital in the Hongwu Era: A comprehensive record of Jinling, Ancient and Modern.* Nanjing Publishing House.
40. Xinfu, Y. (2003). *Ten records of Jinling.* Wu Guxuan Publishing House.
41. Rong, P. (2008). *A preliminary study of Confucian temples in China.* Beijing Forestry University.

Chapter 12
Urban Subarea—Landscape Evolution of the Urban Inner-Port Wharves Under the Influence of the Grand Canal

Abstract This Chapter investigates the landscape evolution of urban inner-port wharf subareas shaped by the Grand Canal, situating these districts within the broader framework of canal-based landscape systems and urban development. As key spatial nodes linking the main canal with inland urban fabrics, inner-port wharves (An inner-port wharves is a cargo and passenger docking area embedded within a city's internal river or canal system, functioning as a node of transport, commerce, and urban life.) historically functioned as interfaces for transportation, commerce, and hydraulic regulation, giving rise to distinctive waterscape-oriented urban landscapes. From the perspective of landscape architecture and urban history, this chapter examines how canal infrastructure, water networks, and surrounding landforms jointly structured the spatial morphology and functional organization of inner-port districts. Through comparative case studies—including Shichahai in Beijing, Changmen in Suzhou City, and the eastern canal area of Liaocheng City—the chapter traces the long-term transformation of these subareas from transport-oriented port landscapes to multifunctional urban spaces. The analysis demonstrates that although the logistical role of the Grand Canal diminished over time, the landscape structures established by canal-based water systems persisted and continued to influence urban form. Through adaptive reuse, these inner-port landscapes were gradually integrated into contemporary urban systems as cultural landscapes, public waterfronts, and carriers of historical memory. By focusing on the interaction between hydraulic infrastructure, landscape form, and urban subareas, this chapter reveals how canal-driven landscape systems shaped urban growth at the micro scale, complementing basin-scale regional development discussed in subsequent chapters.

The Grand Canal's grain transport system[1] connected inner-port wharfs with cities along the route, fostering the commercial prosperity of the urban areas where these wharfs were located and shaping urban areas characterized by a landscape pattern centered on water. Although these areas, once flourishing under the impetus of hydraulic infrastructure, have lost their waterborne transport function, their landscape pattern endures, and, following functional transformation, they continue to serve as vital conduits for the prosperity of the urban area.

X. Zheng, *Terrapoetic Urbanism*, https://doi.org/10.1007/978-981-95-8676-9_12

The relative spatial relationship between the canal and the city determined the classification of inner-port wharfs into three major types: those located within the city, at the urban edge, and beyond the city. This study selects three representative sectors—the Shichahai subarea in Beijing, the Changmen subarea in Suzhou City, and the eastern subarea of the ancient city of Liaocheng along the Grand Canal—to examine the formation and transformation of landscapes in inner-port wharf sectors shaped by the ancient Grand Canal, and to reveal how a historic landscape structure centered on water has evolved over time and continues to hold significance in regional development.

"By the tavern pavilion they hail a fisher's boat; ten-thousand-acre glass and ten-thousand-acre sky. About to cross the creek and ferry eastward, the sound of flutes and songs reaches all the way to the Drum Tower."

The poem "*Shuiguan Bamboo Branch Song* (*shuiguan zhuzhi ci*)[2]" by the Ming poet Gao Heng vividly portrays the enchanting scenery and thriving prosperity of the Shichahai subarea in his time. Even today, visitors to Shichahai can still experience the vibrant waterscape depicted in his verses. The enduring prosperity of the Shichahai subarea fundamentally stems from its waterborne environment, and the most significant driving force behind its development lies in its historical role during the Yuan Dynasty (1271–1368) as an inland port wharf of the Beijing-Hangzhou Grand Canal.

The Grand Canal, as the lifeline of grain tribute transport, spanned the long feudal history of China, facilitating trade and cultural exchange between the northern and southern regions. It profoundly influenced the development of cities along its course in terms of urban pattern, urban scale, culture, and landscape. In particular, the inland port wharf subareas that connected the main channel of the Grand Canal with the cities played a pivotal role; their formation and evolution in aspects such as commercial trade, landscape framework, and historical connotation were inextricably linked to the grain tribute transport of the Grand Canal.

1 Transshipment Hub—Inner-Port Wharves Landscapes and Their Impetus to Surrounding Areas

The Beijing-Hangzhou Grand Canal, extending over 1700 km as an artificial waterway, traces its origins to the Han Gou Canal[3] of the Spring and Autumn and Warring States Periods (770–221 BCE). It was during the Sui Dynasty (581–618 CE) that full connectivity was first achieved, and by the Yuan Dynasty, its configuration had essentially taken shape. Stretching from north to south, the canal linked more than twenty cities and regions across the six provinces of Beijing, Tianjin, Hebei, Shandong, Jiangsu, and Zhejiang (Fig. 1).

An inner-port wharf refers to a nodal point along waterborne transport routes adjacent to the urban fabric, serving as a hub for vessel access and berthing. Cities along the grain tribute transport routes utilized inner-port wharf areas for cargo loading

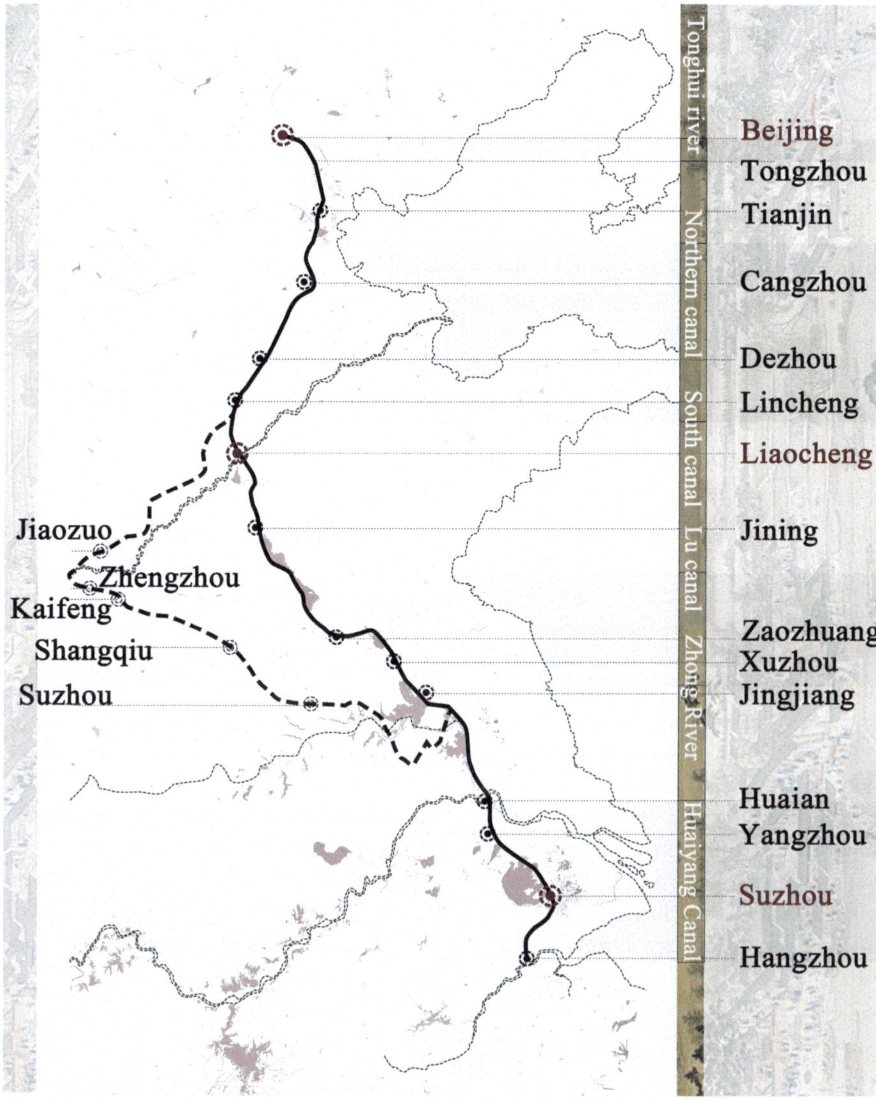

Fig. 1 Distribution of cities along the grand canal (Beijing–Hangzhou)

and unloading, passenger embarkation and disembarkation, commercial trade, logistical provisioning, and engaged in landscape construction along the riverbanks. Such districts also evolved into prominent landmarks that attracted visitors and gatherings, while being equipped with facilities for recreational activities, commerce, services, and warehousing, thereby constituting a historically water-centric and functionally integrated city hub.

The Grand Canal main navigation channel traversed urban territories without directly interfacing with the city itself, instead linking to it through inner-port wharves. These inner-port wharves simultaneously connected to the main navigation channel by water and interfaced with the urban domain, thereby facilitating two-way integration between the canal and the city and functioning as pivotal nodes for water-borne transport and logistics. Goods moving in both directions—via the Grand Canal and the city—were loaded, unloaded, and engaged in commercial exchange within these inner-port wharf precincts. Under the impetus of grain tribute transport, the areas surrounding the inner-port wharves gradually urbanized, emerging as urban prosperity districts with the inner-port wharf as their landmark.

1.1 Inner-Port Wharves and Their Waterscape Transformation

Since the excavation and navigation of the Grand Canal, the establishment of inner-port wharves, and the linkage between cities and the main channel of the Grand Canal through Hydraulic Infrastructure, Waterborne Transport and the resulting Waterscape System have endowed the Canal City with its distinctive Waterscape Structure.

During the period of flourishing Waterborne Trade, Inner-Port Wharves and their associated Hydraulic Infrastructure primarily served transport functions, with the surrounding urban area thriving as a hub for logistics transshipment. However, as the Grand Canal experienced cycles of prosperity and decline, and as alternative modes of transport emerged, the transport role of these Inner-Port Wharves gradually diminished. Over time and through environmental transformations, the Inner-Port Wharves and their adjacent areas acquired profound cultural connotations and significant historical value. Evolving from a traditional commercial zone centered on the Waterscape, they have transformed into a participatory Historic Cultural Landscape, a vibrant Urban Commercial District organized around the Waterscape, and, as a Landmark Landscape, they remain an integral component of the Waterscape System of the Canal City.

1.2 Hydraulic Landscape Infrastructure and Its Promotion of Regional Development

In canal cities, inner-port wharves served not only as hubs for trade and exchange but also as scenic waterside areas where various aspects of production and daily life depended on their presence. Consequently, the emergence of inner-port wharves significantly accelerated population growth in the surrounding districts, stimulated diverse commercial activities, and facilitated the rapid development and prosperity of local commerce, thereby advancing the urbanization process of adjacent areas.

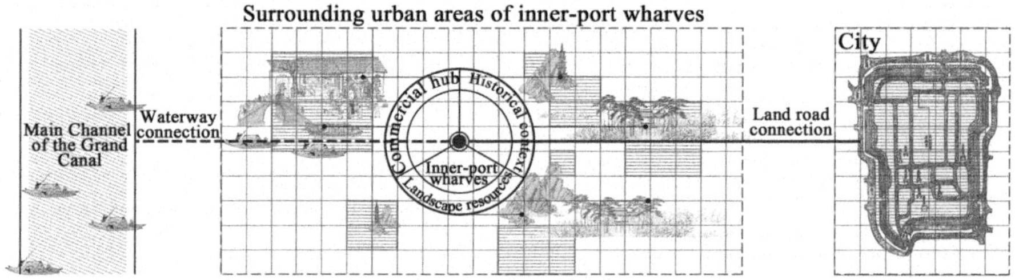

Fig. 2 Schematic diagram of inner harbor wharves connecting the grand canal and the city

As the inner-port wharf districts, once thriving under the impetus of waterborne trade, underwent functional transformation to become vibrant areas combining historic character and waterscape characteristics, the landscape structures shaped by the intervention of hydraulic infrastructure persisted as vital drivers of regional development, enabling these areas to sustain prosperity in a new adaptive form.

In summary, the landscape foundation inherent in the hydraulic infrastructure of the inner-port wharves, the historical context acquired through their long historical evolution, and their role as the core of transportation and trade have jointly driven the formation, development and transformation of the urban area surrounding the inner-port wharves, sustaining its vitality to the present day (Fig. 2).

2 Differentiated by Geographical Position—Typological Classification of Canal Inner-Port Wharves

The inner-port wharves situated along the Grand Canal have played a consistently significant role in driving the development of their surrounding urban districts, yet this influence exhibits both universality and context-specific variability. Among the critical factors shaping this dynamic are the manner in which the Grand Canal traverses a city and the relative spatial relationship between the inner-port wharves and the urban fabric itself.

When the Grand Canal traverses a city, the influence of the water-transport wharves radiates outward from the urban core, exerting a transformative impact on intra-urban development, as seen in cities such as Beijing, Wuxi, and Hangzhou. Conversely, when the Grand Canal merely skirts the urban periphery, the inner-port wharves located at the city's edge may exert limited influence on the central urban fabric, instead serving as pivotal drivers of peripheral growth, exemplified by Suzhou City and Yangzhou. In instances where the Grand Canal bypasses the city entirely, water-transport wharves situated beyond the city limits wield even less direct influence over its interior; yet, they often catalyze the emergence of new commercial port towns, as evidenced in Liaocheng, Jining, and Linqing.

Accordingly, based on the locational relationship between inner-port wharves and the city, three representative typologies can be identified: inner-port wharves situated within the urban core, at the urban periphery, and beyond the city proper. The developmental trajectories of the Shichahai canal subarea in Beijing, the Changmen canal subarea in Suzhou City, and the eastern canal subarea of the ancient city of Liaocheng serve as exemplary cases through which these respective forms are analyzed and interpreted.

3 Prospering by Water: Typical Cases of Urban District Landscape Development Under the Influence of Inner-Port Wharves

3.1 Inner-Port Wharves Located Within the Urban Core: The Case of the Shichahai Subarea in Beijing

Shichahai, comprising the Front Lake (Qianhai), Back Lake (Houhai), and West Lake (Xihai), is situated to the northwest of the Central Axis of Beijing's urban core and represents the largest water surface within the city's inner precincts. The Shichahai subarea has undergone a historical evolution process characterized by three distinct phases.

(1) Exploration of Landscape Resources and Establishment of Hydraulic Infrastructure

The Shichahai area, comprising a chain of lakes left in the wake of the rerouting of the Yongding River, underwent a profound transformation with the establishment of the Daning Palace during the Jin Dynasty (1115–1234). This intervention elevated the locale into a refined scenic landscape park, thereby laying a superior landscape resource foundation for the capital site selection and the configuration of the subsequent landscape pattern of the imperial city.

During the Yuan Dynasty, Shichahai transitioned from a closed enclave to an open urban space. Under the direction of Guo Shoujing,[4] an extensive thirty-year program of hydraulic engineering was undertaken, linking the Grand Canal and establishing the foundational structure of Beijing's fluvial network and its emergent waterfront landscape. This transformation rendered Jishuitan an inland port of the Grand Canal within Dadu City. Concurrently, in the metropolitan planning of Yuan Dadu, the area was designated as a commercial center, in accordance with the principle of "court in front, market behind" as stipulated in *the Kao Gong Ji of the Zhou Li.*[5] Thus, the Shichahai subarea emerged as a dual nexus of grain tribute transport and commerce [1].

During this period, fluvial and maritime transport operated in tandem, with Jishuitan serving as a principal harbor of Dadu City. Grain barges and merchant vessels from across the empire converged at the Shichahai grain tribute transport wharf, facilitating large-scale distribution and exchange. On the northeastern bank, the Xiejie Market, together with the bustling precinct before the Bell and Drum Towers, constituted the most vibrant commercial district of the capital, where merchants gathered in great numbers. The Shichahai area thus experienced an unprecedented apex of prosperity (Fig. 3).

(2) Development of Public Spaces and Formation of the Landscape Pattern

By the Ming Dynasty (1368–1644), with the transformation of the city's water network, Shichahai relinquished its role as a grain transport wharf, evolving instead into the sole public open water area accessible to Beijing's residents. The scenic interplay of lake and hills was integrated into the core of the city's spatial organization, giving rise to an urban waterscape system centered on Shichahai. Although the dense flotillas that once obscured waters receded during the Ming and Qing Periods (1644–1912), the regional prosperity catalyzed by waterborne transport endured, manifesting in new forms within the urban fabric.

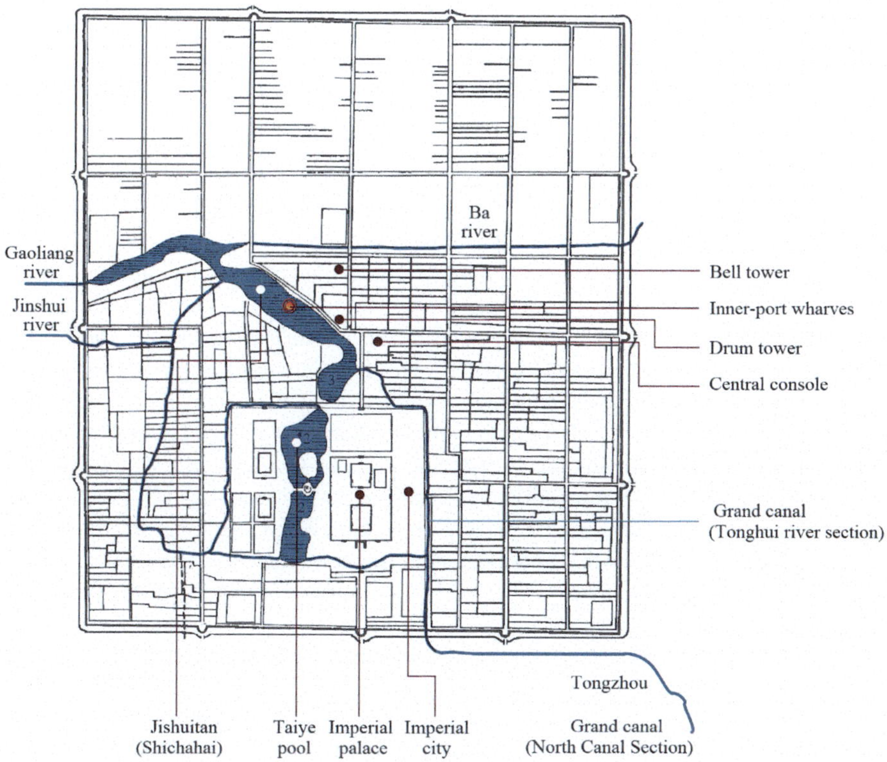

Fig. 3 The Grand Canal and Beijing City in the Yuan Dynasty (1271–1368). *Source* Redrawn by the author based on Note 1

1. Natural Landscape

The description of Shichahai in *the Record of Scenic Spots of the Imperial Capital* evokes the distinctive natural environment of the area: "A single watercourse enters the gate; its breadth extends three or four li. Where the water is deep, fish inhabit it; where it is shallow, there are lotuses, water caltrops, and gorgons. Even if there were neither lotus nor water caltrop, the waters would still be clothed with reeds and cattails—such are the inherent endowments of these waters [2]." This vivid depiction conjures a tableau of aquatic vegetation intermingling in harmonious profusion, with waterfowl gliding and swimming fish darting freely amidst the lush waterscape.

Concurrently, the authorities undertook extensive trees around Shichahai and Jishuitan, planting groves of trees and cultivating expansive paddy fields, thereby crafting an evocative watertown landscape reminiscent of the Jiangnan region. A contemporary account records: "East of Desheng Gate lie several hundred mu of paddy fields; irrigation ditches and channels course through them, with rows of willows planted along the embankments. Upon the ridged plots, the dew on the seedling rice mingles and drifts like a light mist [2]." This deliberate transformation imbued Shichahai with a sensibility evocative of the Jiangnan region, rendering it a place where émigrés from the south could indulge in nostalgic recollection of their homeland.

2. Monastic Complexes and Private Residences

Beyond the evolution of its natural landscape, the Shichahai subarea during this period witnessed the construction of a remarkable array of temples and monasteries as well as private residential gardens. As recorded in the Record of the *Scenic Sites of the Imperial Capital*: "From the Buddhist monasteries drift the sounds of bells and stone chimes; from the lakeside pavilions and villas come songs and melodies. The true charm lies in sights seen and sounds heard from afar—appreciated across the water. Standing at the gate of Jingye Monastery, one's gaze takes in the waters to the south; seated at the Grand Preceptor's Park, the Horse-drying Ground, Mirror Garden, Lotus Nunnery, or Liu Maocai's Garden, one's view extends to the waters to the north. Looking eastward, there is Fang Garden, best admired at dusk; looking westward, there are Man Garden, Shi Garden, Yang Garden, and Wang Garden, and beyond them the Western Hills, best viewed in the morning [2]." Following the dynasty's collapse, the Qing Dynasty saw imperial prince's gardens supplant these religious and residential complexes. Among them, the enduring Prince Gong's Mansion and Prince Chun's Mansion stand as prominent remnants of this historical transformation.

These temples and monasteries together with private residences, which initially flourished owing to the presence of water, in turn elevated the urban area of Shichahai by attracting high-ranking officials and imperial aristocracy. This reciprocal relationship fostered a refined synergy between the natural environment and the cultural context, achieving a harmonious equilibrium that epitomized the aesthetic ideals of the period.

(3) Reconstruction of Landscape Character and the Integration of Classical and Modern Elements

Entering the twenty-first century, densely clustered bars emerged along the shores of Front Lake (Qianhai) and Back Lake (Houhai), gradually giving rise to a bustling bar street that attracts a steady stream of patrons. Here, the natural scenery, historical sites, and folk culture collectively forge a profound historical and cultural ambience, which stands in striking contrast to yet harmoniously coexists with the vibrancy of modern urban life. This juxtaposition and integration epitomize the dynamic interplay between Beijing's identity as both a historic city of enduring legacy and a thriving international metropolis.

Fueled by waterborne transportation, Shichahai has, over centuries of rise and decline, evolved into a historic cultural landscape area that intricately integrates urban fabric, cultural customs, architecture, natural scenery, and intangible cultural heritage. Situated at the heart of Beijing, it endures as a waterway axis that preserves and narrates the historic character of the old city.

3.2 Inner-Port Wharves Situated at the Urban Edge: The Case of the Changmen Subarea in Suzhou City

Changmen, the western gate of the ancient city of Suzhou City, flourished as one of the most prosperous commercial districts in the nation, owing to its advantageous proximity to the Grand Canal and its convenient transportation links—earning it the epithet "Changmen in its Golden Age." The evolution of the Changmen subarea unfolded across three distinct stages.

(1) Interlaced Water Systems and Connected Water Transport Routes

Since antiquity, Suzhou City has been a quintessential watertown, where waterborne transport constituted an indispensable mode of communication. Following the opening of the Grand Canal, its waters were introduced into the city from the vicinity of Fengqiao Town on the western outskirts (Fig. 4), entering through the Changmen Gate in the west and the Qimen Gate in the north, and exiting via the eastern Loumen Gate and southern Panmen Gate, where they once again merged with the main channel of the Grand Canal. These city gates were constructed as dual-purpose portals for both land and water traffic. The diverted canal waters coursed along the outer city moat and inner city moat flanking the city wall, penetrating the urban fabric and aligning broadly with the terrestrial road network to form a rectilinear matrix of three longitudinal and three transverse waterways. From these principal channels proliferated countless subsidiary water lanes, intricately interlacing the entire city [3].

The establishment of interconnected waterborne transport corridors fundamentally transformed the spatial form of ancient Suzhou City, engendering an extensive

Fig. 4 Connection between the Grand Canal and Tang River in Suzhou City—Fengqiao Bridge

and fluid water network of intersecting channels. Functioning as a pivotal transshipment hub for land–water transport, Suzhou City emerged as a principal point of origin for goods and materials shipped northward via the Grand Canal (Fig. 5).

(2) Commercial Prosperity and the Formation of an Axis-Core Spatial Configuration in the Urban Sector

Since the establishment of Helü Great City, Changmen Gate has held preeminence among Suzhou's city gates, serving as a potent political symbol. During the Yuan Dynasty, despite the presence of canal grain transport system, maritime shipping remained dominant; consequently, the regional landscape of the Changmen Gate area

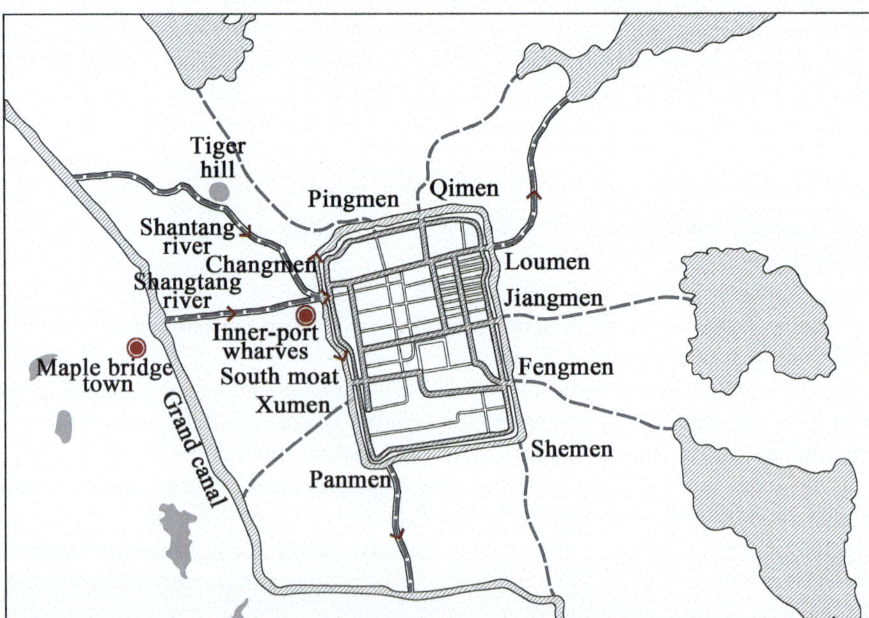

Fig. 5 The Grand Canal and Ancient Suzhou City in the Qing Dynasty (1644–1912). *Source* Redrawn by the author based on Note [4]

was scarcely developed, retaining the character of a suburban area defined largely by natural scenery.

By the Ming and Qing Dynasties, the re-dredged and fully navigable Grand Canal had become the principal artery of waterborne transportation. Flowing past the western perimeter of Suzhou's ancient city, the canal intersected at the strategically significant Changmen Gate, where five waterways converged, affording exceptional water-land connectivity. Goods entering and leaving Suzhou City were largely consolidated at the Nanhao Wharf beyond Changmen Gate, where fleets of merchants moored in profusion. This commercial vitality catalyzed the urban transformation of the Changmen precinct from a natural periphery into a fully integrated urban district.

Radiating from Changmen Gate, five principal waterways extend respectively toward Xumen Gate, Qimen Gate, Fengqiao Bridge, Tiger Hill, and the inner city, articulating a distinctive axial-core spatial pattern within the precinct. Among them, the Shangtang River and Shantang River served as the primary conduits, channeling trade from Changmen Gate through the Grand Canal to destinations across north and south China. This connectivity engendered not only the flourishing of the Changmen area but also the prosperity of Suzhou City in its entirety [5].

The Shangtang River, extending from Changmen Gate to Fengqiao Bridge, constituted the principal waterway linking the city with the main navigation channel of the Grand Canal. Running parallel to Shangtang Street, it formed the region's most vital transportation artery. Along its banks, guild halls clustered densely, accommodating the congregation of myriad merchants, while the passage of a thousand sails animated the waterborne trade. This bustling activity shaped a distinctive riverine landscape. As recorded in *the Gazetteer of Hanshan Temple (Qing Dynasty)*[6]: "From Changmen Gate to Fengqiao Bridge spans ten *li*; the skyline was crowded with fleets of masts, voices of boatmen chanting at a distance, uninterrupted from dusk till dawn. Boats were coupled bow to stern in procession, extending all the way to Xushu Villa."

Another critical conduit linking Changmen Gate to Tiger Hill was the Shantang River, running parallel to Shantang Street, a renowned commercial district that gracefully extended for seven li. While the Shantang River functioned primarily as a transportation corridor integrating the city with the main navigation channel of the Grand Canal, it simultaneously operated as a transitional axis from urban landscape to natural landscape. The vicinity of Duseng Bridge and Shantang Bridge near Changmen Gate was lined with densely packed shops, forming a vibrant commercial front. Progressing westward toward Bantang Bridge, the scenery gradually transformed into picturesque natural vistas, culminating in a suburban recreational area at the city's edge.

The development of waterborne transport catalyzed the transformation of the Changmen Gate precinct from an area of scenic natural landscapes into a vibrant commercial center. Over time, Changmen Gate evolved from a mere political symbol to the urban core of Suzhou City, establishing a distinctive radial urban waterscape pattern (Figs. 6, and 7).

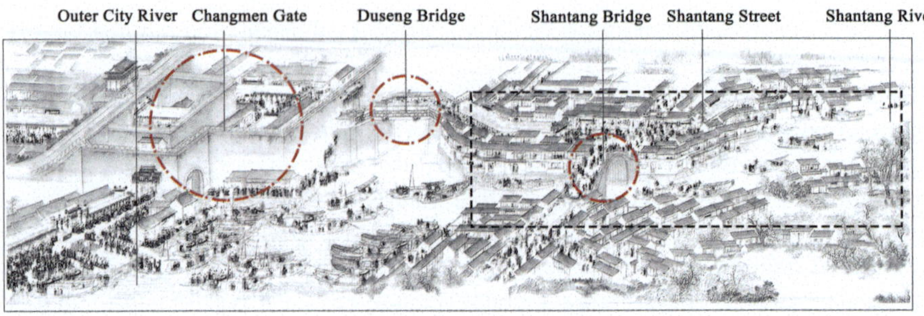

Fig. 6 Overview of Changmen and Shantang Street in Suzhou City. *Source* (Qing) Wang Hui, Yang Jin, Kangxi Southern Inspection Tour, Volume 7

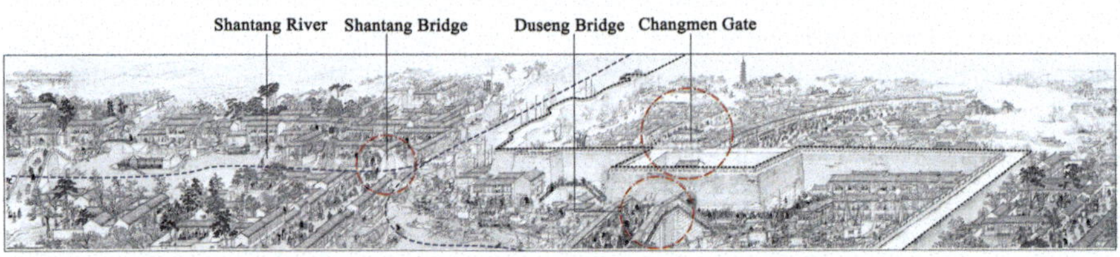

Fig. 7 Area outside Changmen. *Source* (Qing) Xu Yang, Gusu Prosperity Map

(3) Contemporary Changmen Gate: The City Landscape of a Historic City

The Changmen Gate area retains its distinctive confluence of five waterways and the illustrious legacy of Prosperous Changmen, rendering it today a unique spatial locus for uncovering historical and cultural traditions and showcasing the vibrant city life of Suzhou City. These enduring qualities have shaped a city landscape imbued with the unmistakable character of Suzhou's historic city.

In recent years, the preservation and planning initiatives in the Changmen Gate area have focused on the reconstruction of several core landscape nodes, including the city gate itself and the drawbridge. The district has been re-envisioned as a commercial district, integrating the Shantang Street scenic tourism belt and featuring Suzhou's traditional Jiangnan brands, folk handicrafts, and distinctive local delicacies. Furthermore, a comprehensive tourist transportation system has been established, incorporating boat terminals, parking facilities, and public transport stops, thereby enhancing connectivity between waterborne and land-based routes and fostering the development of historic city loop tourism [6].

Over the course of time, the Changmen Gate area has evolved from a once-flourishing commercial center into a significant historic urban landscape of the ancient city. Although the tides of history have shifted, the confluence of five waterways remains one of the most dynamic and vibrant districts within Suzhou city.

3.3 Inner-Port Wharves Located Outside the City: The East Canal Subarea of the Ancient City of Liaocheng

The area along the canal to the east of the ancient city of Liaocheng, driven by grain transport system, developed a thriving commercial trade and evolved into a new bustling commercial district in the urban periphery. The transformation of this area unfolded through the following three stages.

(1) Encircling the City with a Lake and Bypassing It with a Canal: A Unique Urban Water System Pattern

Located in the western plain region of Shandong, Liaocheng's ancient city, also known as Dongchang Ancient City, developed a distinctive water-city model. Historically, frequent flood disasters caused by breaches of the Yellow River compelled ancient inhabitants to excavate artificial waterways, resulting in the formation of Dongchang Lake surrounding the square-shaped ancient city. This created a unique urban form—a city within water, encircled by a lake.

The earliest intersection between the Grand Canal and the Ancient City of Dongchang dates back to the Sui Dynasty (581–618 CE), when the Yongji Canal traversed the northwestern part of the ancient city. Following the extensive re-excavation and dredging of the Grand Canal during the Yuan Dynasty, the Huitong River section flowed north–south through the eastern side of Dongchang Ancient City. With the opening of the canal, it became connected to Dongchang Lake, forming an integrated water system around the ancient city (Figs. 8, and 9).

(2) The Prosperity of Grain Transport System and Its Impetus to the Development of the Canal Sector

The impact of the canal on Liaocheng became particularly pronounced during the Ming and Qing Dynasties, when the Grand Canal functioned as the primary waterborne transportation artery linking north and south. During this period, Dongchang Prefectural City emerged as one of the nine major commercial ports along the ancient canal. Beyond the city walls, canal wharves of varying scales extended in a continuous belt, serving as the principal hubs for cargo distribution. The scene was one of unceasing activity: merchant vessels converged in great numbers, their masts forming a dense forest along the riverbank, while goods-laden boats queued for unloading, creating a vibrant and bustling spectacle stretching for several *li*.

The city district surrounding the wharves to the east of the ancient city, functioning as a commercial port district of its time, exhibited a scene of remarkable prosperity—"Boats crisscrossed like shuttles, people surged like tides, shops stood shoulder-to-shoulder, workshops were arranged like stars across the sky." This flourishing of commerce gave rise to a newly vibrant city center beyond the city walls, fundamentally dependent on the Grand Canal. The network of market streets in this area was laid out along both banks of the canal, adapting to the natural topography and following the river's course. All primary and secondary market streets were

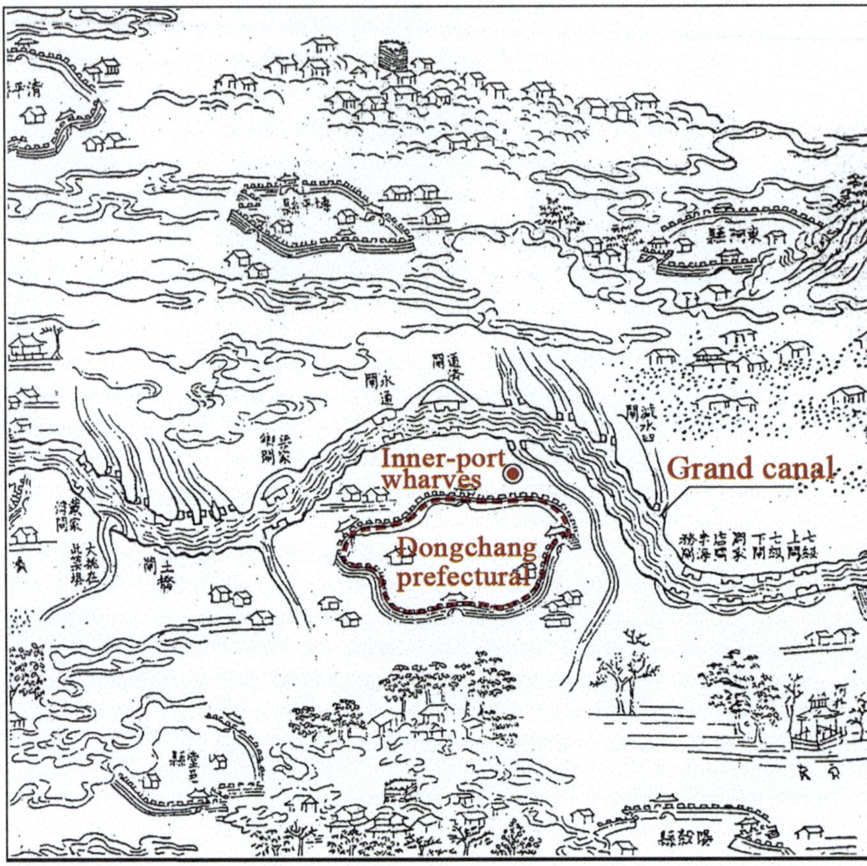

Fig. 8 Canal surrounding the Ancient City of Dongchang. *Source* (Qing) Fu Zehong, Zheng Yuanqing, Xingshui Jinjian—Canal Map of Dongchang Prefecture

connected to the canal, forming a distinctive radial framework, and the names of many of these streets remain in use to this day [7].

This enduring prosperity was sustained by the flourishing of grain tribute transport via the Grand Canal for over four centuries, earning Liaocheng the epithet: "A vital nexus of canal grain transport system, the elbow at the capital's flank, and the foremost metropolis north of the Yangtze River."

(3) The Decline of Water Transport and the Emergence of a Historic and Cultural Landscape Area

During the Xianfeng Period (1851–1861),[7] a breach of the Yellow River caused siltation of the Grand Canal in the Liaocheng section, gradually depriving it of its transport advantages. Although the bustling scenes of boats and ships have long faded, the canal-side structures—such as sluice gates, inner-port wharves, bridges, streets,

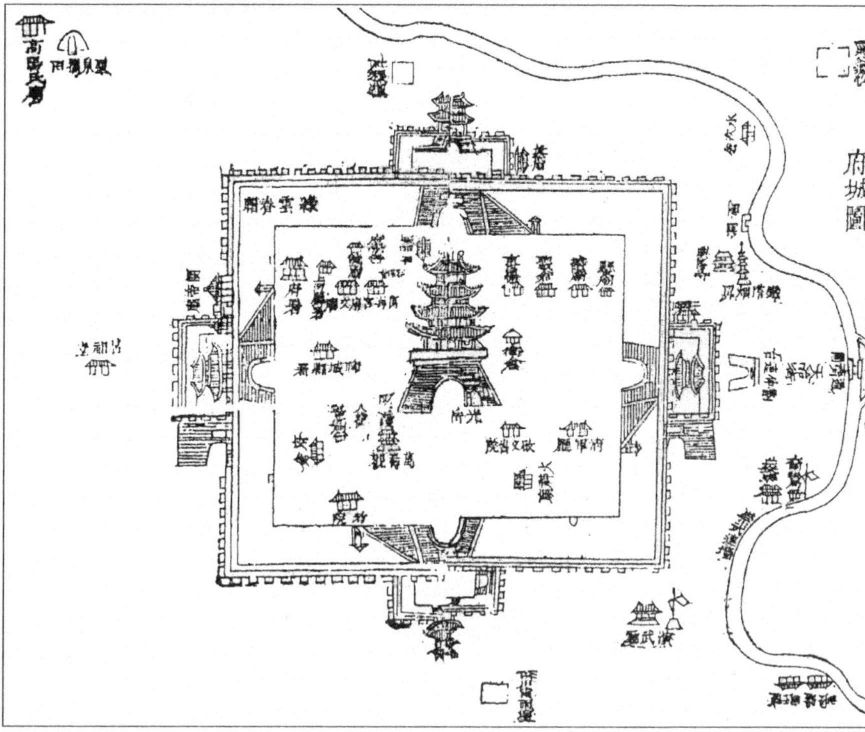

Fig. 9 Relationship between the Grand Canal and Ancient Dongchang City in the Qing Dynasty. *Source* (Qing) Dongchang Fuzhi, Volume 1, Illustrations of the Prefecture City, 13th Year of Jiaqing

and lanes—were not abandoned when the canal lost its function as a route for grain transport system. Instead, the historic urban fabric along the Grand Canal, including ancient towns, old neighborhoods, historic alleys, and traditional architecture, has retained its original layout, evolving into a valuable historical and cultural landscape. Furthermore, to this day, the surrounding Dongchang Lake and the canal-linked water system of Liaocheng have been preserved, continuing to shape the city's waterborne culture and aquatic scenic area (Fig. 10).

4 Tracing the Origins—The Influence of Ancient Canal Inner-Port Wharves on the Urban Subarea Landscape Development

The Grand Canal's completion and the establishment of hydraulic infrastructure at the inner-port wharves catalyzed the development of both the inner-port area and the city in multiple dimensions. While the impacts exhibited considerable similarities

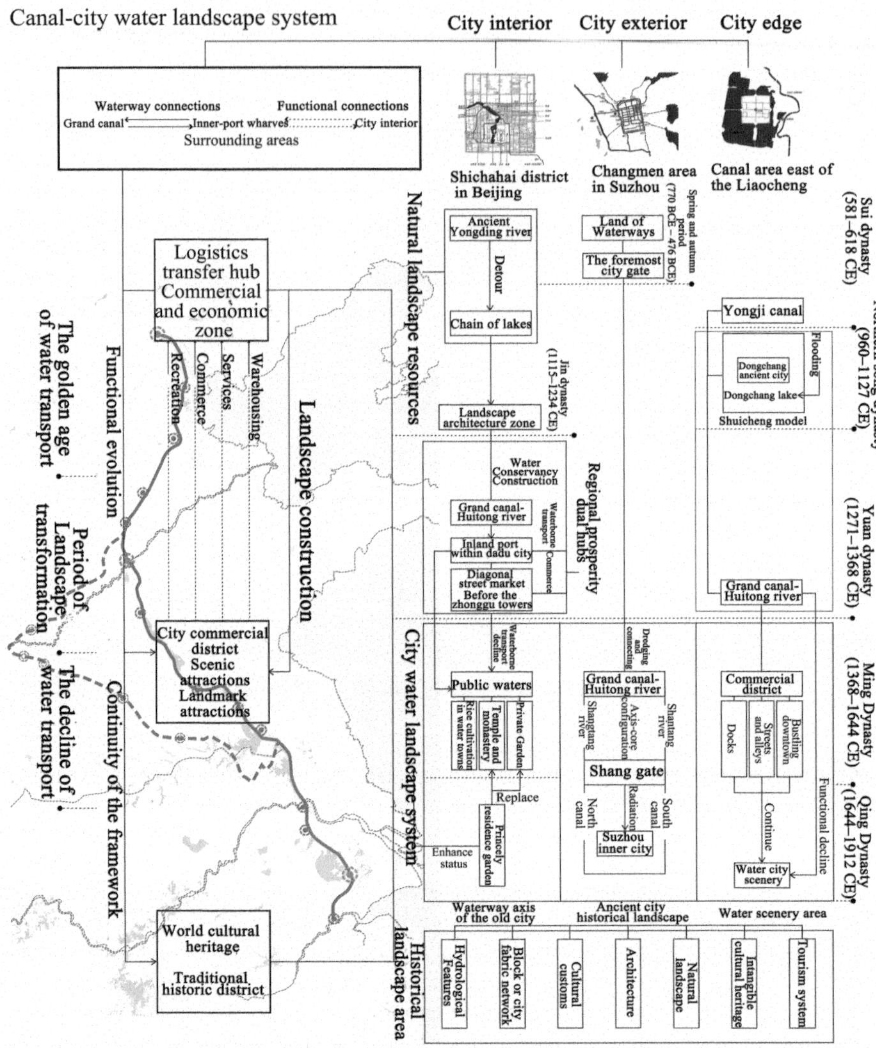

Fig. 10 Schematic diagram of the waterscape system in canal cities

across regions, they also manifested notable variations contingent upon the specific spatial location and typology of the wharves in relation to the city (Fig. 11).

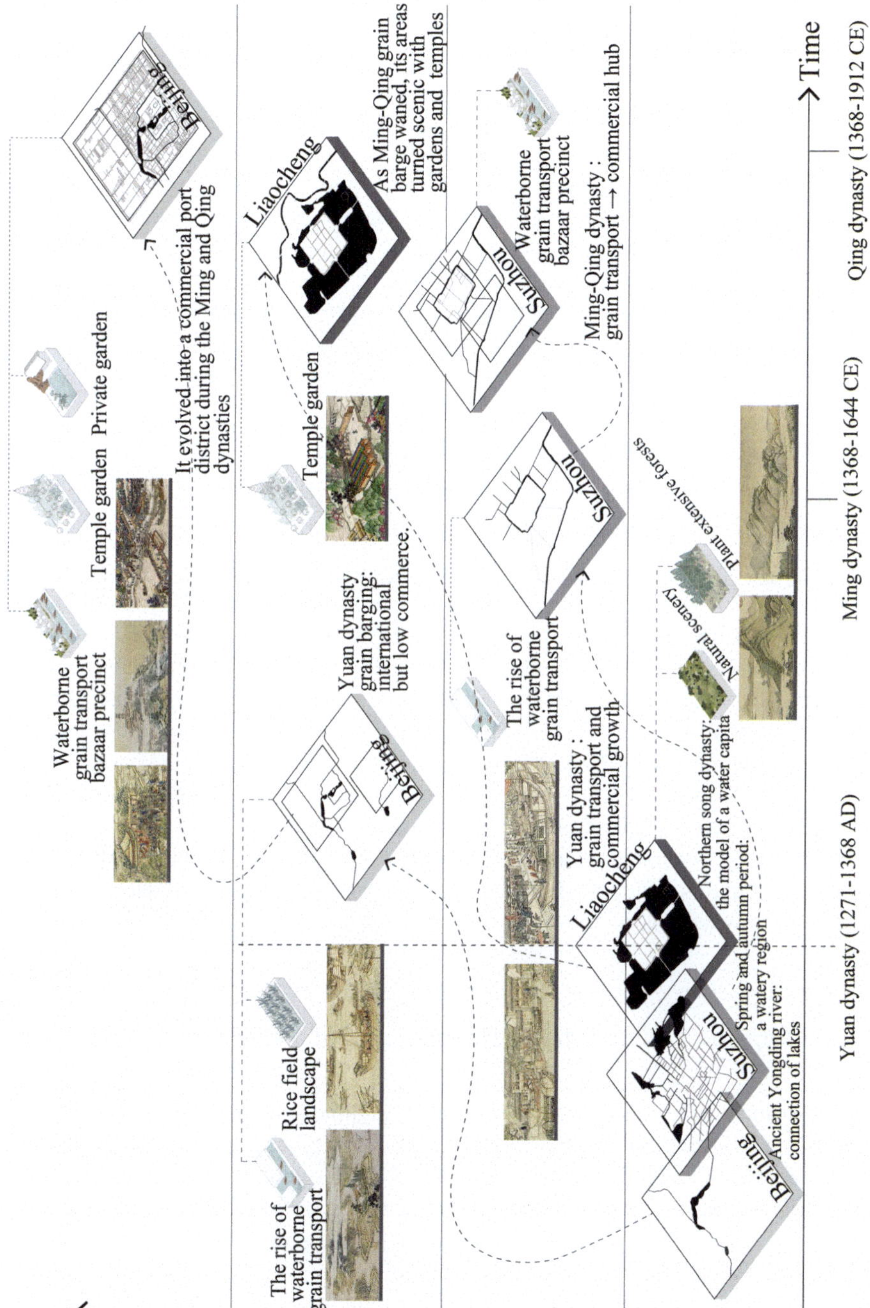

Fig. 11 Comparative diagram of the relationship between city and canal in Beijing, Suzhou City, and Liaocheng based on time and landscape type

4.1 *The Strategic Location of Water Transport Hubs as Catalysts for the Emergence of Prosperous City Commercial Districts*

The circulation of cargo via water transport facilitated the trade exchange between the city and the outside world, profoundly stimulating the economic development of the entire city. Particularly, port areas often emerged as strategic transport hubs, attracting merchants from across regions. The influence of these commercial centers on the wider city varied according to the spatial location of the wharves relative to the city. When the inner-port was located within the city center, it significantly propelled urban commerce; conversely, when the wharves were situated on the city's periphery, their impact waned, with only adjacent urban areas benefiting from their presence.

4.2 *Shaping a Distinctive Regional Waterscape Pattern and Its Transformation for Continuity into the Present*

The axis-core configuration of the Changmen Gate area in Suzhou City, defined by the confluence of five waterways, remains evident to this day. Numerous celebrated sites, such as the famed Tiger Hill and the Seven-Li Shantang Street, owe their renown to the canal. This area has successfully transformed from a commercial center into a historic landscape district, contributing to Suzhou City's status as a prominent tourist destination in Jiangnan. Similarly, Shichahai has evolved into the largest public open water area within Beijing's city, where a landscape pattern interweaving history and modernity captivates visitors from across the nation. Meanwhile, in the east canal area of Liaocheng Ancient City, the original network of ancient streets and hydraulic infrastructure remains intact, having been repurposed into a cultural landscape of the ancient canal, now celebrated as a significant scenic area.

4.3 *Driving the Urbanization Process of the Subarea*

The establishment of hydraulic infrastructure at the inner-port wharves transformed the area into a commercial district of remarkable city prosperity, drawing population inflows that, in turn, further stimulated local development. This mutually reinforcing dynamic significantly accelerated the urbanization of the district. Moreover, the trajectory of urbanization was shaped by the locational context of the inner-port wharves. When the wharves were situated away from the city center, they catalyzed urbanization in the urban periphery, driving the outward expansion of the city. Conversely, when positioned outside the city itself, the wharves—leveraging their connection to the canal—could even give rise to emergent bustling central districts beyond the traditional city core.

The opening of the Grand Canal and the construction of hydraulic infrastructure engendered distinctive spatial forms and developmental dynamics in the cities along the canal and their inner-port wharves. Over time, as the waterborne transport function of the Grand Canal diminished, the commercial functions of the inner-port districts associated with grain tribute transport gradually waned. Nevertheless, the prosperity once brought by the waterborne network and the historic landscape pattern it fostered have endured. Through adaptive functional transformation, these areas continue to sustain and invigorate the urban vitality of the city to this day.

Notes

1. Grain Transport System (Caoyun) was the state-organized system of moving tax grain along rivers and canals—especially the Grand Canal—to supply capital cities and armies. It shaped waterway infrastructure, port facilities, and settlement patterns across imperial China.
2. Gao Heng's Shuiguan Zhuzhi Ci (sixteenth century, Ming dynasty) uses the folk-style Zhuzhi form to depict everyday life around urban water gates—multifunctional infrastructures combining hydraulic control, transport, and city management in canal cities.
3. The Han Canal (Hanguo) is an early artificial waterway constructed in the Spring and Autumn period to link the Yangtze River with the Huai River system. It functioned as a precursor segment of the Grand Canal, enabling military movement, grain transport, and the formation of canal-based urban landscapes.
4. Guo Shoujing (1231–1316) was a Yuan-dynasty state scientist whose work combined astronomy and large-scale water management, including urban water-supply and river regulation for the Yuan capital (Dadu, present-day Beijing).
5. The Kaogong Ji (late Warring States period), incorporated into the Rites of Zhou, is an early Chinese technical text that sets out normative principles for city layout, infrastructure, and state-controlled spatial order.
6. The Hanshan Temple Gazetteer (Qing dynasty) is a monastic gazetteer documenting the temple's history, spatial layout, gardens, and surrounding waterways in Suzhou. It provides insight into the integration of Buddhist sites with canal landscapes and urban infrastructure in late-imperial Jiangnan.
7. The Xianfeng reign period (1851–1861) refers to a decade in the late Qing dynasty marked by political instability, warfare, and fiscal strain. These conditions profoundly affected urban management, water-control systems, and landscape maintenance, often leading to the decline or adaptive reuse of traditional infrastructures.

References

1. Hou, R. (1997). A preliminary discussion on the planning and design of the Yuan Capital (Dadu). *City Planning, 3*, 10–13.
2. Liu, T. (Ming Dynasty), & Yu, Y. (Ming Dynasty). (1963). *Record of scenic sites of the imperial capital*. Beijing Publishing House.
3. Yin, L., & Wen, Y. (2011). *The wise enjoy water (XVI)—Notes on cultural relic investigations by the Grand Canal Historical Investigation Group* [2015-02-01]. http://www.sdmuseum.com/show.aspx?id=2844&cid=83

4. Li, Y. (2007). *A study on the landscape culture of the ancient canal in Suzhou*. Suzhou University.
5. He, F. (2010). The formation and development of the urban landscape in the Western Changmen area of Suzhou. *Journal of Chinese Historical Geography, 1*, 71–79.
6. Fu, X., & Xie, A. (2009). Landscape planning of the green space on the Western section of Suzhou's Ancient City Ring. *Anhui Agricultural Science, 10*, 4757–4759.
7. Wang, H. (2007). *A study on the canal economy of Dongchang during the Ming and Qing Dynasties*. Nanchang University.

Chapter 13
Basin: The Regional Landscape and Development of the Fuchun River Basin

Abstract This Chapter focuses on the Fuchun River Basin and examines the formation of its regional landscape and its decisive influence on regional urban and settlement development from a basin-scale perspective. Through an analysis of topography, hydrological systems, and ecological foundations, the chapter elucidates the coupled landscape structure formed by mountains, plains, and dense water networks. On this basis, case studies of historical cities such as Suzhou City, Shaoxing City, Hangzhou City, Nanjing City, and Yangzhou City are employed to illustrate the adaptive strategies through which ancient Chinese cities developed in response to basin-scale natural conditions. The chapter emphasizes the role of natural landscapes as integral components of regional defense systems, transportation networks, and ecological frameworks, while also examining the regulatory function of human interventions, such as city wall construction, hydraulic engineering, and water network management in the evolution of the basin landscape. By synthesizing the spatial relationships between mountains, water, and cities, it argues that urban form is not an isolated product, but rather the outcome of long-term interactions within a broader regional landscape system. The study demonstrates that the regional landscape of the Fuchun River Basin not only provided the essential natural foundation for urban emergence and prosperity, but also profoundly shaped the spatial configuration, boundaries, and landscape structures of ancient Chinese cities, offering important insights into basin-scale regional development and urban formation.

Dwelling in the Fuchun Mountains[1] is one of the Ten Greatest Ancient Chinese Paintings. The scenery it depicts is situated within the Fuyang area of Hangzhou City, in the Fuchun River Basin (Fig. 1). Within this basin, the dynamic interplay of mountain–water–objects–people manifests as "each in its proper place, vivid and measured," thereby sustaining a favorable human settlement environment that has enabled continuous habitation across generations. As a precious cultural legacy of the agrarian civilization era, *Dwelling in the Fuchun Mountains* offers more than a visual feast; the painting, as an imprint of the mind, embodies profound ecological insights. The wisdom of harmonizing human activity with nature resonates with the pursuit of an ecological human settlement in the process of terrapoetic urbanism. The beauty

X. Zheng, *Terrapoetic Urbanism*, https://doi.org/10.1007/978-981-95-8676-9_13

of the Fuchun River's banks derives from the synergy between its unique natural landscape and carefully moderated human interventions. Together, they shaped the regional landscape system along the river, fostered the development of local settlements, and gave rise to the pictorial and cultural representation that is *Dwelling in the Fuchun Mountains.*

"In the early autumn, when the air and mist are utterly clear, the sky and mountains share the same hue. On both sides of the river, hills rise and fall in a winding path that follows the turning peaks. The river nourishes fertile soil, with sandy shallows and level fields. Villages and cottages are half-hidden amidst the clouds and haze, while fishing boats appear and disappear on the shimmering waves. Nearby, the trees are lush, deep green, spaced in harmonious density. In the distance, streams and mountains recede into the depths, with cascading waterfalls hanging like inverted ribbons. Pavilions and small bridges each find their perfect place; figures and birds are rendered with lively appropriateness." This sentence provides a vivid description of the Fuchun River Basin landscape as depicted in *Dwelling in the Fuchun Mountains.*

The Fuchun River is the name given to the section of the Qiantang River that extends from Tonglu in Zhejiang Province to Wenyan in Xiaoshan. The middle and lower reaches of the Fuchun River Basin lie within the Fuyang district of Hangzhou City, Zhejiang Province. The region is distinguished by its picturesque landscape and abundant natural resources, and has long enjoyed the reputation of being a "land of fish and rice." Scattered among the riverine landscape are renowned towns and ancient villages with distinctive Jiangnan regional characteristics. Throughout history, the area has been economically prosperous and culturally flourishing. It has given rise to eminent figures such as Sun Quan[2] of Eastern Wu during the Three Kingdoms Period (220–280 CE), the poet Luo Yin[3] in late Tang Dynasty (c. 830 s–907 CE), the Yuan Dynasty (1271–1368) painter Huang Gongwang,[4] the Qing Dynasty (1644–1912) father-and-son prime ministers Dong Banda and Dong Gao,[5] as well as the modern literary master Yu Dafu.[6] At the same time, the region attracted the visits of distinguished literati, including Li Bai,[7] Bai Juyi,[8] and Ji Xiaolan,[9] who left behind numerous poems, essays, and calligraphic works extolling the scenic beauty of the Fuchun River's banks (Fig. 2).

According to the research of Professor Wang Bomin of the China Academy of Art and Mr. Jiang Jinle, a scholar who focuses on the works of Huang Gongwang, the inspiration for Huang Gongwang's creation of *Dwelling in the Fuchun Mountains* originated from the regional landscape system along both banks of the Fuchun River within the Fuyang region [1]. After experiencing the verdant mountains, lucid waters, secluded settings, and profound historical depth of this region, Huang Gongwang depicted its spiritually endowed natural landscape, local settlements, and the simple, tranquil rhythms of agrarian life. The work reflects the wisdom of harmonious coexistence between humans and nature, encapsulated in the principle of unity of heaven and humanity.

The scenery depicted in *Dwelling in the Fuchun Mountains* constitutes a valuable resource for investigating the human settlement environment of the Fuchun River Basin during the agrarian civilization period, and may be interpreted as a form of documentary material of its time. Employing the method of pre-iconographic

Fig. 1 County map of the Fuyang area during the Daoguang period (1821–1850) of the Qing Dynasty (1644–1912)

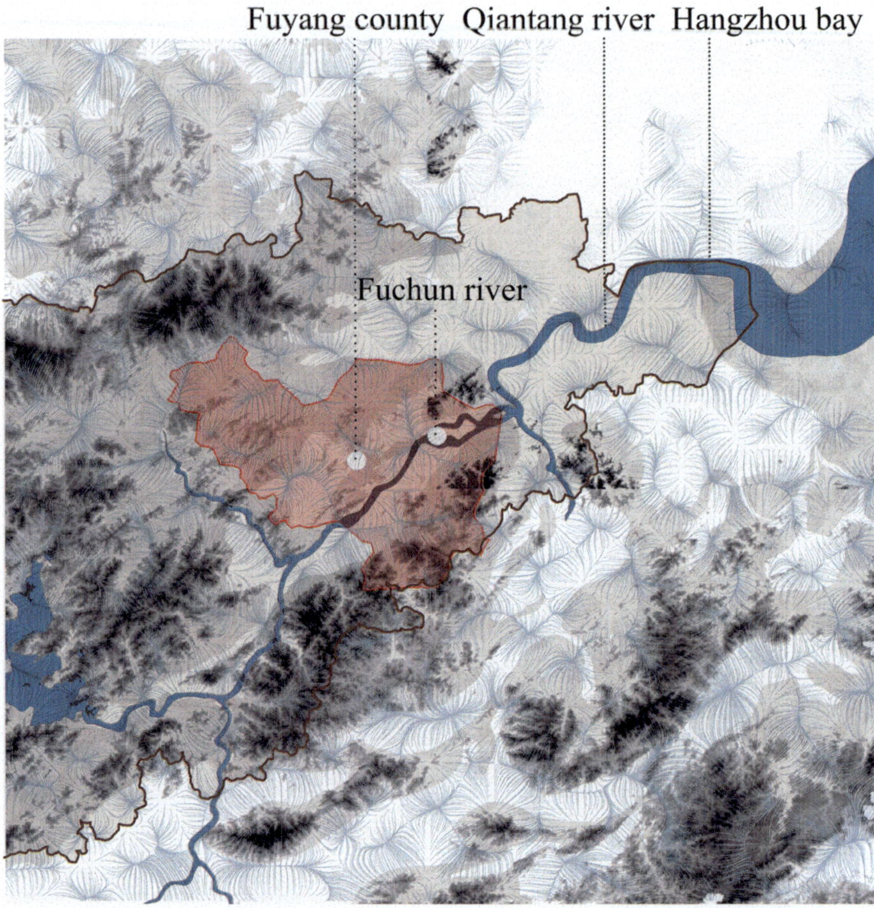

Fig. 2 Location of Fuyang County

description within the framework of iconology [2], this study seeks to explain the physical and visual facts embedded in the artistic formal elements of the literati landscape painting genre. In doing so, it aims to explore the landscape morphology and evolutionary process of the ecological human settlement environment in the Fuchun River Basin during the agrarian civilization period [3].

The scope of this study is primarily focused on the core area of the Fuchun River Basin—namely, the Fuyang district of Hangzhou City. *Dwelling in the Fuchun Mountains* was completed during the Yuan Dynasty. In light of the scarcity of systematic documentary sources from the Yuan period, and given that pre-industrial society was characterized by relatively slow processes of environmental formation and transformation, this research draws comprehensively upon documentary records from the Yuan, Ming (1368–1644), and Qing dynasties.

The regional environment of the Fuchun River Basin is structured upon a grand framework of natural landscape. Through the intervention of hydraulic facilities, a human-modified water landscape was created, which in turn facilitated the development of farmland landscapes and settlement landscapes. Within this process of formation, an agrarian cultural landscape characterized by "fishing, woodcutting, farming, and reading" emerged. Accordingly, this study examines the formation and transformation of the human settlement environment in the basin through the interconnected dimensions of natural environment, artificial water landscapes, farmland landscapes, settlements, and agrarian cultural landscapes, while also analyzing their correspondences with *Dwelling in the Fuchun Mountains*.

1 Painting as the Imprint of the Mind—Re-envisioning the Artistic Conception of Unity of Heaven and Humanity

Huang Gongwang lived during the late Yuan and early Ming Dynasty (c. 1350 s–early 1400 s). Having become disillusioned with his official career, he withdrew into the landscape for seclusion. In his seventies, he resided beneath Shaoji Spring in Miaoshanwu, on the banks of the Fuchun River. Following Daoist principles and likening himself to an immortal, he named his dwelling Xiao Dongtian ("Small Grotto-Heaven"), alluding to the Daoist ideal of a blessed abode of the immortals.

Along both banks of the Fuchun River, the landscape alternates between towering mountains enclosing dangerous shoals and rapids, and stretches of low hills, distant peaks, and wide, gently flowing waters. The local populace lived in harmony with this environment, settling amidst mountains and waters. Fertile plains and farmlands extended across the riverbanks, interlaced with paths and irrigation channels; villages dotted the foothills and shores, where human activity intermingled with the sounds of chickens and dogs. Thus, natural environment, farmland landscapes, and settlements coalesced seamlessly into an integrated whole (Fig. 3).

Huang Gongwang, long nurtured by the landscape of this region and subtly influenced by a rural life of farming labor, absorbed the spirit of "learning from nature externally and drawing from the heart internally." Through a process of refinement, he distilled the authentic landscape features and artistic conception of the Fuchun River Basin into a heightened representation. Responding to phenomena through form and emphasizing ink over color, he ultimately devoted approximately seven years to completing the monumental handscroll *Dwelling in the Fuchun Mountains* (Fig. 4).

From the perspective of pictorial composition, the scroll begins and concludes with light-ink renderings of distant mountains and islets, extending the spatial field toward both ends. In the foreground and middle ground, the mountain ridges are vertically prolonged, thereby enhancing spatial depth. Strategic blank spaces are left throughout the painting, while detailed brushwork is concentrated on rocky outcrops at mountain summits, slopes, and foothills, as well as on settlements located at

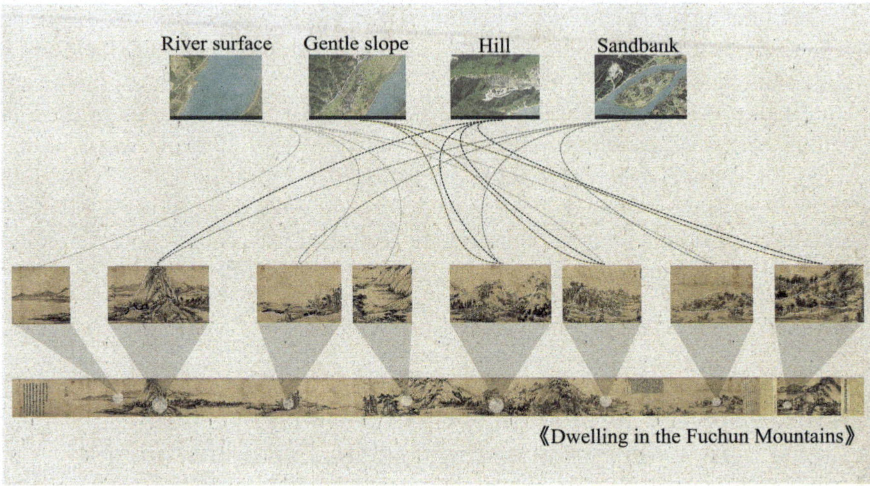

Fig. 3 Landscape types of the Fuchun River Region. *Source* Yuan Dynasty (1271–1368) Huang Gongwang, *Dwelling in the Fuchun Mountains*

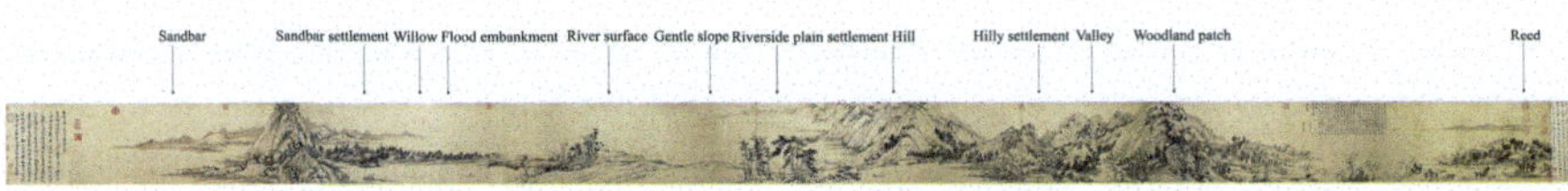

Fig. 4 *Dwelling in the Fuchun Mountains*, Huang Gongwang

flat plains or gentle slopes, mountain rocks and vegetation, riverine pavilions and stone bridges, and fishermen in boats. The scenery gradually fades from near to far, accentuating contrasts between foreground and background, and imbuing the composition with a heightened sense of spatial layering.

The pictorial scenes appear at times majestic and vast, as if within arm's reach, and at other times simple, ethereal, and expansively profound. By employing techniques of visual advancement and recession, Huang Gongwang revealed the spatial configuration of encircling mountains and embracing waters within the Fuchun River Basin, while also depicting the expansive river surface with its boundless mists and waves. At the same time, with low village dwellings half-hidden amid forested hills, delicate pavilions poised along the riverbank, and rustic villagers portrayed in a state of tranquil composure, he sketched an approachable regional landscape that conveys the gentle local character of Jiangnan's rural atmosphere. It was precisely this plain yet genuine, naturally adorned beauty of Fuchun, long nurtured in Huang Gongwang's mind, that ultimately gave rise to the creation of this timeless masterpiece [4].

2 Harmonizing Humans and Nature—The Constituent Elements of the Landscape Within the Ecological Human Settlement

In *Guanzi · Wuxing*,[10] it is stated: "Only when human beings are in harmony with Heaven, can the beauty of the world emerge." The scenic beauty of the Fuchun River Basin is inseparable from its natural landforms, yet it is equally indebted to the wisdom of the ancients, who respected nature and lived in symbiosis with it. Through the gradual evolution of agrarian civilization over millennia, this landscape was progressively formed.

2.1 Extraordinary Natural Configurations Shaping a Unique Landscape

> The stream is bluish green,
> And so clear that you can see the bottom at a thousand feet,
> Where swimming fish and tiny stones are visible to your eyes.
> The swift current is like an arrow;
> The raging billows seem like galloping horses.
> The steep banks flanking the stream are covered with trees that thrive in coldness.
> They seem to strive upwards for greatness,
> Stretching skyward,
> And forming thousands of peaks.
> The spring water splashes on the stones with a gurgling sound;
> The fine birds sing to each other with a warbling voice.
> The cicadas' chirping is unending,
> And the monkeys' cry is ceaseless.

This is how the Southern Liang Dynasty (502–557) writer Wu Jun, in his *Letter to Zhu Yuansi*,[11] described the magnificent landscape of the Fuchun River (Figs. 4 and 5).

(1) Hill Landscapes

Shaped by the incision of the Fuchun River, the terrain along its banks divides into two principal blocks: low mountains to the southeast and hills to the northwest. The remaining ridges of the Tianmu Mountains stretch across the northwest, while the residual ranges of the Xianxia Ridge extend into the southeast. Mountains and hills together account for 78.61% of the total municipal area. The undulating hills along both riverbanks possess a pleasing scale and gentle topography, forming tranquil and secluded valleys.

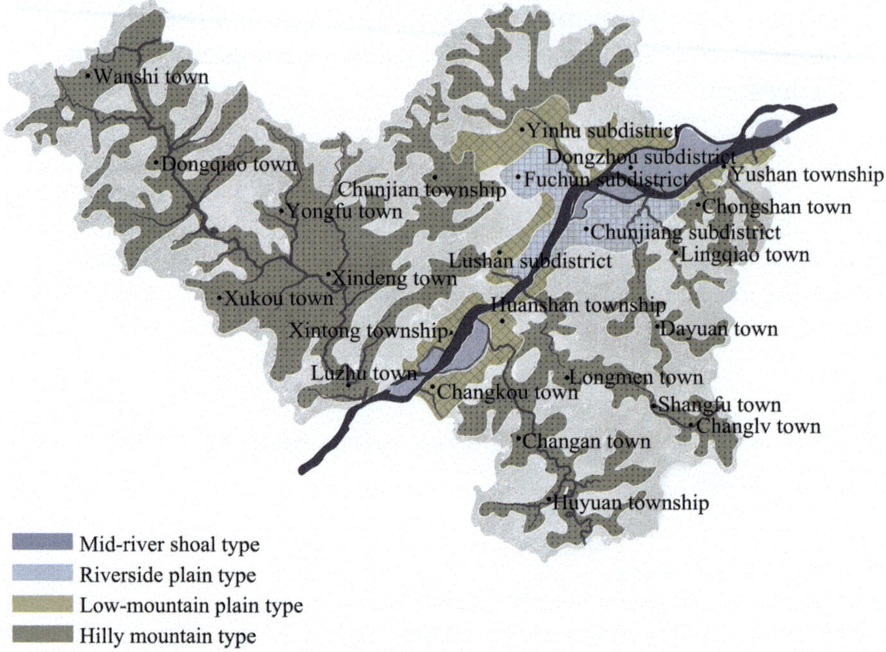

Fig. 5 Map of landscape types in the Fuyang Region

In the northwest, the remnants of the Tianmu Mountains descend step by step southeastward into the basin, presenting wave-like undulations that gradually extend to the riverbanks, with hilltops reaching elevations of 400–500 m. Numerous ridges lie between the hills, with deep and fertile soil layers.

In the southeast, the ridges of the Xianxia Mountains extend from the southeast toward the northwest, pressing close to the riverbank and submerging beneath the riverbed. Here the terrain is lofty, with hilltops exceeding 400 m in elevation, and the mountain masses form continuous layers of overlapping ridges. One branch enters the region from Xiaoshan, extending northwestward to form Yushan Mountain and Lishan Mountain; another branch enters from Zhuji, stretching northwestward to join Yushan Mountain and Lishan Mountain, together constituting the entire southeastern mountainous area.

As a recluse, Huang Gongwang also projected his emotions onto the hilly landscapes along the banks of the Fuchun River. Based on field investigations and scholarly research on Huang Gongwang, it has been inferred that the mountain forms at the beginning of *Dwelling in the Fuchun Mountains* reference the scenery around Guanshan Mountain within the basin. The dynamic clusters of hills, ravines, dense forests, and sand islets depicted in the central portion may reflect landscapes near Zhongbu, Tangjiabu, and various sandbars within the basin. The solitary peak at the end of the scroll corresponds to "Zhuren Peak" behind Miaoshanwu—where Huang Gongwang resided in seclusion. By adopting the sobriquet Yifeng Daoren ("Daoist of the Lone Peak"), Huang metaphorically embedded his own persona within the solitary peak at the end of the scroll [1].

(2) Streams and River Islets

The Fuchun River Basin is endowed with abundant water resources, characterized by the system of "one river and ten streams" [5] (Fig. 6). The "one river" refers to the Fuchun River, while the "ten streams" are its primary or significant tributaries, all of which converge into the Fuchun River within the basin (Fig. 7).

The *Geography Chapter of the Fuyang County Gazetteer*[12] *(Qing Dynasty)* records: "According to the locals, the tidal bore surges upstream, reaching as far as Fuyang before receding, and some say it even goes all the way to Tonglu. However, based on my observations, Fuyang and Tonglu do not actually experience the tidal bore. Why is this? The upper reaches of the river (around Tonglu) are fed by three tributaries. The terrain is elevated, and the river's flow is powerful and swift. Having charged over a hundred *li* from the narrow estuary between two mountains to the boundaries of Fuyang, the sea tide, like an arrow at the end of its flight, has exhausted its momentum and can no longer contend with the river current. Thus, it has no choice but to retreat." This indicates that when the tidal bore of the Qiantang River advances upstream, it encounters the downstream flow of the Fuchun River, resulting in a confrontation of forces. At such moments, the originally placid river surface turns turbulent.

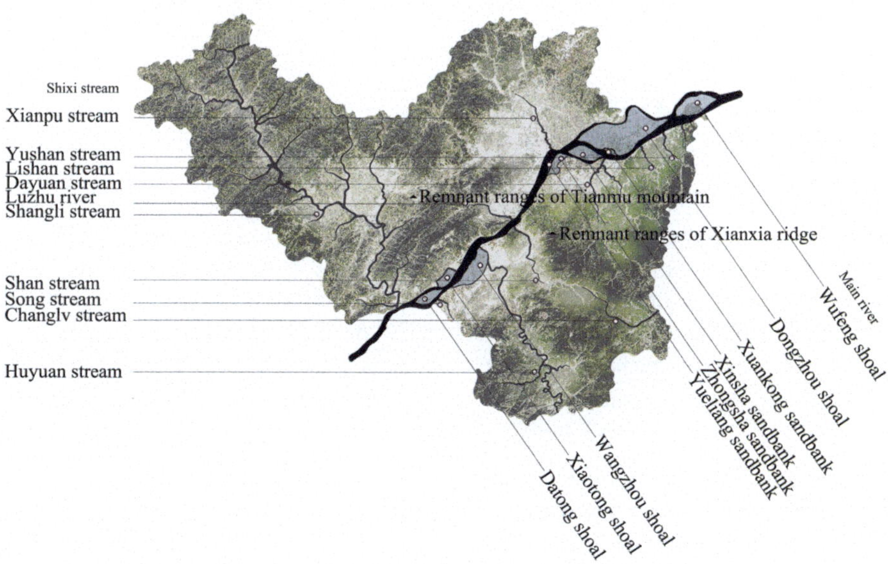

Fig. 6 Map of mountains and water systems in the Fuyang Area. *Source* Redrawn by the author based on the 2007–2020 Fuyang Area Master Plan's elevation factor analysis and water system planning

Fig. 7 Diagram of the landscape configuration of the Fuchun River

Meanwhile, under the action of river flow, silt carried by the waters gradually deposits in the river under the combined influence of tidal surges and salinity, eventually forming multiple sand islets. These constitute a unique fluvial spectacle within the basin: from Tonglu in the upper reaches to Xiaoshan in the lower reaches, more than ten large and small sand islets are distributed. With favorable ecological conditions, these islets gradually gave rise to the celebrated landscapes of "Egrets on the Islets of the Fuchun River" and "Wild Geese Descending on Zhongsha."

(3) Riparian Plains

Along the banks of the Fuchun River, extensive alluvial plains have developed, primarily consisting of fluvial floodplains formed by alluvial deposits. These include the plains on both riverbanks as well as the lacustrine depression called "Huangtianfan." In their original state, these areas had an average elevation of 7 m, accounting for 19.46% of the total territory. The plains are characterized by deep, fertile soils. However, during periods of heavy rainfall, the rising waters of the Fuchun River frequently backflow into the plains, resulting in flooding disasters. Despite these risks, the broad, flat terrain with fertile, deep soil provides abundant arable land [4].

The gently undulating hills along the riverbanks possess a pleasing scale, with soft mountain forms and secluded, tranquil valleys. *The Letter to Zhu Yuansi* describes this vividly: "Those who seek high positions

Their hearts will cease to strive for fame;

Those who are engaged in state affairs.

Will give up their ambition and remain." This passage illustrates how viewing the graceful peaks and serene valleys of the Fuchun River landscape allows visitors to feel relaxed and enchanted, lingering in harmony with the refined elegance of its mountain forms. This natural aesthetic resonates with the literati and recluses' pursuit of inner peace and contentment.

The riparian plains extend from the southwest toward the north, with varying widths on either side, eventually reaching the outer terraces of the low hills and shallow mountains. These areas later became the principal farmlands and grain-producing zones of the region.

The Huangtianfan Plain is encircled by mountains on its east, west, and north sides, while its southern edge borders the Fuchun River. With a gradient of only 1°–2°, it was originally formed as a lacustrine depression through long-term incision and erosion by the river. Later, as the Biezimen channel in Haining connected with Hangzhou City Bay, the water level of the Fuchun River dropped. Marine silt deposits accumulated to significant depth, and with the river's course changes, the lagoon gradually evolved into a riparian plain. Within the Fan, rivers meander through low-lying terrain, giving the area a shape reminiscent of an iron cauldron.

(4) Hills and Valleys

In the painting, the Fuchun River traverses between the hills on both banks, with the river surface appearing at times broad and profound, at times narrow and elongated. Judging from the rippling patterns of the river's surface along the mountain foothills depicted in the painting, one can infer that the waters alternated between tranquil expanses and swift currents eroding the banks. Between the mountains, coves and inlets were formed, with various large and small sand islets sketched within the river, where waterfowl are shown resting and playing near the shores.

Through field investigations, landscapes corresponding to the scenes depicted in the painting have been identified. Based on visual analysis, coordinate positions of the photographic viewpoints were annotated, the distances between the shooting points and the actual landscape sites were measured, and a viewshed diagram was produced (Fig. 8).

2.2 *Hydraulic Interventions Reshaping the Water Landscape*

Water is the foundation of agriculture, yet the natural water environment does not fully meet the requirements of agricultural production. Thus, it became necessary to construct hydraulic works in order to regulate and redistribute surface and groundwater, thereby both mitigating water-related hazards and enabling the productive use of water resources. Each region possesses its own distinctive history of water conservancy development. Through the management and utilization of water by means of hydraulic systems, marshes, depressions, woodlands, and sloping lands within the natural landscape were gradually transformed into fertile fields interlaced with irrigation networks, and settlements accordingly took shape in situ [6].

While the Fuchun River nourished the populations along its banks, it also frequently produced water-related calamities. During floods, "houses were submerged, and people nearly became fish"; during droughts, "grains and trees withered, and the people suffered from hunger." [5] Consequently, the forebears relied

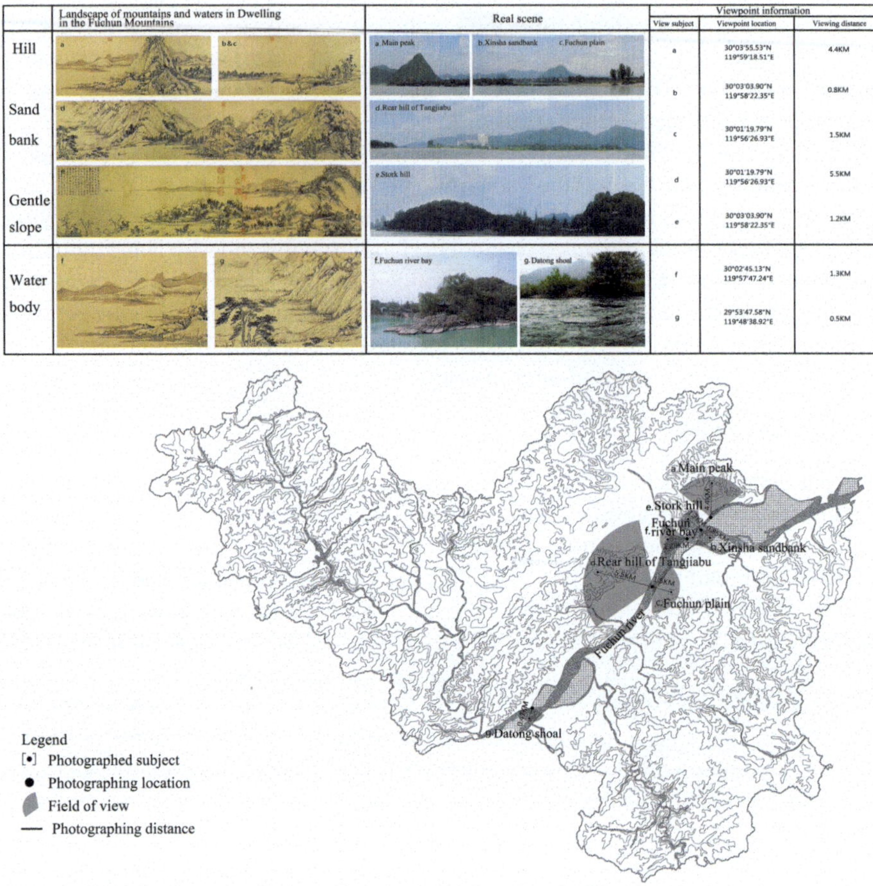

Fig. 8 Top: distances from the real scene to the corresponding photograph locations marked on the map; Bottom: analysis diagram comparing the natural landscape in the painting with the actual landscape. *Source* (Yuan Dynasty, 1271–1368 CE) Huang Gongwang, Dwelling in the Fuchun Mountains

on long-term modes of human intervention in nature to reshape the environment, creating conditions suitable for both production and habitation. This integration of artificial and natural forms of landscape modification influenced not only the siting of rural settlements but also their internal spatial patterns. In turn, it laid the foundation for a harmonious human settlement environment while simultaneously enriching the regional landscape system.

(1) Artificial Water Landscapes Formed for the Purpose of Flood Control

Because the Fuchun River historically maintained high water levels, seasonal rises frequently caused backflow into farmlands and settlements along its banks. To mitigate this threat, flood-control levees were constructed along riversides and at the periphery of settlements. In addition, agricultural levees were built to protect farmland, which in turn promoted the development of polder fields. Flood-control engineering in the region began as early as the Tang Dynasty (618–907), with the construction of the protective urban levee known as the "Fuchun Dike." This structure not only safeguarded the settlement but also laid a stable foundation for the permanence of the urban base.

To reinforce the levees and prevent soil erosion, groves of trees were commonly planted along the riverbanks, willows were placed at the water's edge, and extensive reed beds were cultivated at the land–water interface. Similarly, in the Taihu Basin, it was a widespread practice to plant trees on earthen embankments of rivers and canals, while reeds were cultivated in the water. The root systems of these plants strengthened the embankments and stabilized the soil. This illustrates a typical soil-conservation method of the agrarian civilization in the Jiangsu–Zhejiang region. *Dwelling in the Fuchun Mountains* also captures such interventions, depicting tree groves and willow plantings along the banks, thereby presenting a riverside landscape shaped by human mediation of nature.

With the growing need for safety, additional flood-control levees were constructed along the banks of the Fuchun River, incorporating sluice gates to regulate the inflow and outflow of water. Flood-control engineering in the region began in the Tang Dynasty, when the earliest protective urban levee, known as the "Fuchun Dike," was built. Extending approximately 1000 m from Guanshan Mountain in the east to Xianpu River in the west, it not only safeguarded the settlements but also laid a stable foundation for the permanence of the city base. In the same year, the Xianpu Dike was constructed on the left bank of Xianpu River, stretching about 500 m from Hailu Lake in the east to Xianpu River in the west, and connecting with the Chunjiang Dike to protect the northern bank's settlements and farmland. As settlements expanded along the river, this levee later evolved into the substructure of a major internal transportation road within the settlement. In addition, agricultural levees specifically designed to protect farmland were first established in the Qing Dynasty. With the increase in available labor, large-scale construction continued until the 1950s, facilitating the development of polders and providing crucial safeguards for the further growth of riverside settlements based on agrarian livelihoods (Fig. 9).

(2) Artificial Water Landscapes Formed for Irrigation and Diversion

The region is mountainous, with dense mountain streams characterized by short courses and rapid flows, which in the rainy season easily result in flash floods. Residents of the hilly areas often constructed levees, built dams, and excavated canals to divert and channel water, thereby dredging river courses. The *Fuyang County Gazetteer*[13] *(Qing Dynasty)* records that between the Tianshun era (1457–1464) of Ming Dynasty and the first year of the Tongzhi reign (1862),[14] the Gexi stream was successively dredged and diverted eight times, directing its waters into the surrounding settlements' moats and irrigation channels. In addition, ancient practices frequently involved stream-channel modification projects, whereby certain obstructive meanders hindering flood discharge were transformed into straight channels. This not only shortened the course length but also broadened the floodway, thereby

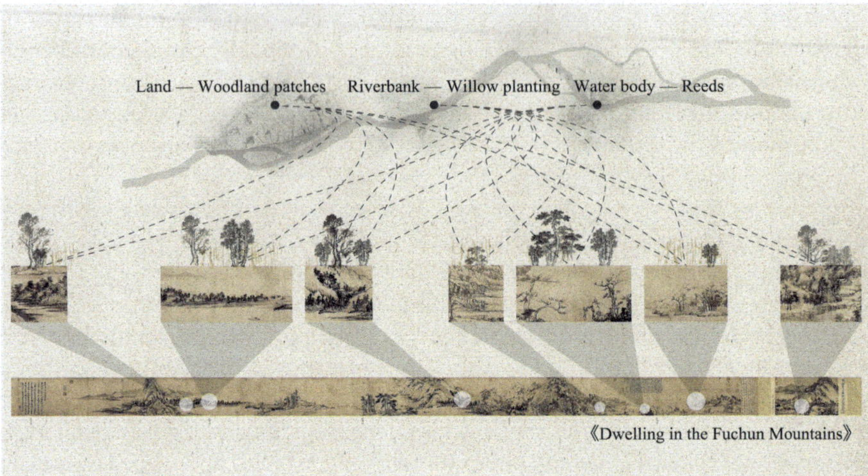

Fig. 9 Riverside interventions. *Source* (Yuan Dynasty) Huang Gongwang, Dwelling in the Fuchun Mountains

regulating the downstream velocity of mountain flood peaks during the rainy season, while simultaneously improving irrigation conditions.

During the dry season, when the water level of the Fuchun River dropped, natural irrigation from rivers alone could not meet the needs of farmland production. At the same time, fields located in mountain depressions often faced irrigation difficulties due to their distance from water sources. To address this, the ancients constructed ponds and embankment dams near farmlands to store water for drought prevention and irrigation, while also providing water for daily domestic use. In addition, they built complementary hydraulic facilities, such as canals to divert water, sluice gates on dams to regulate water levels, and weirs to channel river water into irrigation ditches. These weirs were commonly built with stone, while floating weirs (made of grass bundles)[15] were sometimes constructed across small streams. From the Yongchun era (682–683) of Tang Dynasty to the eleventh year of the Chenghua reign (1475) of Ming Dynasty, a total of seventy-three weirs and dams were recorded within the county. Waterwheels, often made of bamboo or wood, were also widely used, relying on human power to lift water from lakes, bays, ponds, springs, inlets, and wells within rural settlements for both agricultural production and daily life [5]. Together, the lakes, bays, ponds, and associated diversion facilities became some of the most approachable and life-sustaining elements within the regional landscape system.

(3) Complementary Facilities

1. Water-Lifting Devices Historically

Historically, the principal water-lifting tools in the region were waterwheels, which relied on human and animal power to draw water from rivers, bays, ponds, and springs. In hilly and semi-mountainous areas where water sources were scarce, wells were excavated, and devices such as ju dou (wooden dippers) and ju gao (counterpoised levers) were employed to draw water from them.

The ju gao, also known as the "well sweep," was an agricultural implement used for irrigation. It was widely employed in areas such as Shangguan, Longmen, Huanshan, and Xindeng, where it played an important role in irrigation during dry seasons.

The ju dou was a traditional irrigation tool shaped like a dustpan, used to lift water for fields. However, because of its high labor intensity and low efficiency, it was not widely adopted in the region, except in the hilly areas of Longmen and Huanshan where it saw more frequent use.

The waterwheel, also referred to as the "dragon-bone wheel," was used both for lifting and draining water. Initially, hand-operated waterwheels were common, but due to their low efficiency, they were later replaced with treadle waterwheels. The treadle waterwheel harnessed the inertia of rollers and bearings to raise water. When higher lifts were needed, several treadle wheels could be connected in succession, thereby reducing manual labor. Subsequently, animal power was introduced to replace human labor, which further increased agricultural efficiency.

2. Ditches

Canals and ditches were generally used in conjunction with the aforementioned hydraulic facilities, linking various water-storage installations into an integrated water environment regulation system. They often served the dual functions of diverting water for irrigation and draining excess water. In the hilly areas, ditches were frequently excavated along the edges of mountain slopes; while connecting to the broader water system, they also played a role in preventing mountain flood disasters.

Within the hydraulic history of the Fuyang area, several canals are particularly noteworthy, including the Huangtianfan South Canal, the Dongzhou Irrigation Canal, the Hong Canal, and the Dafeifan Canal.

2.3 Context-Sensitive Design/Planning in Land Reclamation Led to the Formation of Agricultural Landscapes

Agriculture is closely bound to specific geographical settings: it must not only adapt to the productive capacity of the land but also respond to the challenges of local ecosystems. In order to secure sources of food, people, through long-term accumulation of experience, transformed their environments and adopted farming methods attuned to natural conditions within different geographic contexts. In doing so, they shaped farmland landscapes, which occupy the largest share of the earth's surface and most directly reflect the interactions between humans and land.

Day after day, tilling and raising livestock upon the land constituted the core of production for villagers in agrarian society, embodying the "everydayness" and "vernacularity" of the regional landscape system. The farmland landscape thus emerged as a central component in the constitution of the regional landscape.

Agriculture in the Fuyang region has a long and continuous history. Archaeological discoveries indicate that as early as 5000 years ago, the area was already inhabited and engaged in primitive farming and pastoral activities. The construction of hydraulic facilities provided essential support for the development of settlements and farmland cultivation. With subsequent population growth and the increase of labor resources, people gradually shifted from small-scale farming or fishing activities to more systematic agricultural and fishery production adapted to the local geographical conditions. By the Tang Dynasty, agriculture had become highly developed. This advancement fostered both the improvement of hydraulic infrastructure and the diversification of farming practices, thereby giving rise to distinct types of farmland landscapes such as polder fields, fishponds, level fields, and terraced fields.

(1) Plain Farmland Landscapes

In the broad plains, large expanses of level fields were systematically arranged. Depending on the shapes of different cultivation units, the level fields were divided into strip fields and square fields. The ridges that separated these units functioned, during the period of the small-scale peasant economy,[16] both as boundaries demarcating individual households' plots and different crop varieties, and as narrow paths facilitating field access and agricultural operations. Complementary irrigation ditches were excavated alongside the field ridges, channeling water from nearby sources to irrigate the farmland.

(2) Waterfront Polder Farmland Landscapes

In riparian zones prone to flooding and on sand islets within the river, people sought to protect arable land by constructing embankments or ring dikes from earth and stone around the fields. Sluice gates were installed along these embankments to regulate the inflow and outflow of water, thereby forming polder fields.

(3) Hilly Terraced Farmland Landscapes

In the hilly areas, level terraces and sloped terraces were constructed along the mountain contours. Water stored in reservoirs at the mountain summits was channeled into ditches excavated along the slopes, or mountain streams were utilized through their natural gradients to irrigate the terraces successively, step by step.

2.4 Settlements Oriented by Mountains and Waters: The Formation of Vernacular Settlement Landscapes

(1) Settlement Layout

Settlements are the places where humans gather and live; they are products entirely shaped by human adaptation to, and utilization of, nature, and they most clearly manifest the cultural dimension of the regional landscape system. The forms and organizational patterns of settlements are closely tied to the natural environment, the regulation of the water environment through hydraulic facilities, and the practices of farmland cultivation and production. When the research scale is expanded from the individual settlement to the larger grouping of settlement clusters, and considered as an integrated whole, studying their forms and organizational patterns within the broader environment yields insights of greater practical significance [6].

Traditional settlements in the region were predominantly Han Chinese lineage villages where farming and learning were transmitted across generations. Informed by the traditional ecological worldview of unity of heaven and humanity, the siting, layout, orientation, and vertical scale of these settlements were carefully adapted to their environmental contexts. Consequently, different types of settlements emerged in response to differing geographical conditions.

Rural settlements in river-basin regions were constrained by topography; their formation was closely related to the orientation of natural rivers, valley lines, and ridgelines, while also maximizing the use of terrain in consideration of agricultural production. The depictions of village dwellings in the painting fully reflect how the forebears-built homesteads are in harmony with the surrounding environment.

Within the Fuchun River Basin, settlements can be classified according to their geomorphological contexts into riparian plain settlements, sandbar settlements, and hilly mountain settlements (Fig. 10).

1. Riparian Plain Settlements

Riparian plain settlements were primarily located near rivers with abundant water sources or distributed along foothill plains, where large, contiguous tracts of farmland could be preserved. Because plain settlements were less constrained by spatial limitations, they tended to be larger in scale. The settlements were relatively dispersed and most often exhibited irregular forms.

2. Sandbar Settlements

Sandbar settlements were typically sited on elevated terrain, where thicker and more stable soils reduced the risks of subsidence and dampness, thereby favoring construction. Given the region's frequent rainfall, higher ground also prevented water accumulation; during periods of rising tides, elevated siting reduced the likelihood of flooding. These settlements generally extended along the longitudinal axis of the sandbar, often taking on elongated forms. By adopting this spatial layout, aligned with both the landform and the water system, the settlements not only adapted to the

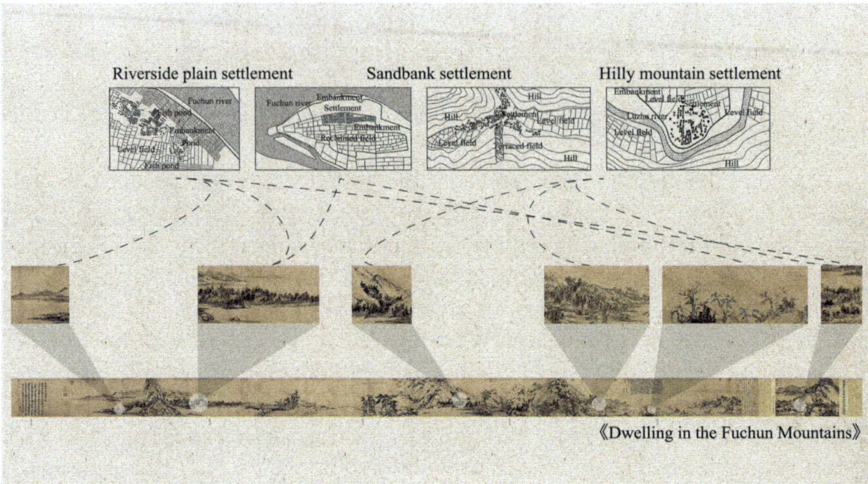

Fig. 10 Correspondence between settlement layout and real-world sites. *Source* (Yuan Dynasty) Huang Gongwang, *Dwelling in the Fuchun Mountains*

local geomorphology but also maximized their orientation toward the river surface, thereby ensuring open and favorable visual corridors.

3. Hilly and Mountain Settlements

Hilly and mountain settlements were largely distributed along the main tributaries of the Fuchun River, with their layouts concentrated near riverbanks. In some cases, settlements extended along mountain foothills or were situated upon low hills. Because the surrounding mountains constrained available space, these settlements were generally linear in form and smaller in scale. Only in broader valley areas did larger, more block-like settlements emerge.

(1) Architecture of the Settlements

1. Residential Dwellings

Residential dwellings in the scroll painting, the residential architecture of the settlements mostly appears as single-household, single-building units. Each building was composed of three main parts: the platform base, the main structure, and the roof.

The base was constructed of brick and stone, raised above ground level to support the entire house. The main structure rested on this base, formed with a post-and-beam timber frame system and typically rising one or two stories high. Doors and lattice windows were installed between the columns, with decorative mullions most often rendered in the form of wicker patterns.

Above, the roof was created with a wooden structural frame, covered with pitched gable roofs that were commonly used in vernacular residences. This roof form was particularly effective for drainage during the rainy season [7].

2. Recreational Architecture

Recreational architecture primarily took the form of pavilions, terraces, and towers, most often situated along the riverbanks or on mountain slopes to facilitate the viewing of river scenery, while also providing scholars with favorable environments for reading and contemplation.

The Bànjiāng Pavilion, built at the river's edge, featured a wooden platform extending over the water, exemplifying strong waterside accessibility and intimacy. In contrast, the Bànshān Pavilion (Half-Mountain Pavilion), located midway up a slope, offered superior views. On elevated slopes with expansive river vistas—such as Guanshan Mountain—multi-storied towers (two to three stories in height) were often constructed, with railings installed on the uppermost level to allow outdoor viewing of the river scenery or observation of potential flood hazards.

The four pavilions and one hillside tower depicted in the scroll painting faithfully reflect the actual landscape of recreational architecture within the region.

2.5 *The Material Environment and the Formation of Agrarian Cultural Landscapes of Fishing, Woodcutting, Farming, and Learning*

(1) Fishing: Rational Utilization of River Resources

The Fuchun River, situated near the East China Sea, possesses clear waters and abundant fishery resources. It is one of the principal distribution regions for freshwater fish in China and is also a well-known spawning ground for the migratory shad, which made the local fishery particularly prosperous. In addition to direct fishing in the river, aquaculture was another important production method. According to the *Record of Escaping the Heat* (Bishulu Hua, Song Dynasty),[17] by the late Song Dynasty (960–1279) aquaculture in ponds had begun in Zhejiang: fry purchased from markets were transported back in wooden buckets and released into ponds for cultivation. By the Yuan Dynasty, households located near water sources were commonly engaged in pond fish farming. Some practices combined agriculture, forestry, and fishery, such as cultivating clams in paddy fields or developing orchards integrated with fishponds. These practices produced alternating layouts of ponds and paddy fields, or ringed the edges of ponds with fruit trees, forming distinctive agrarian cultural landscapes.

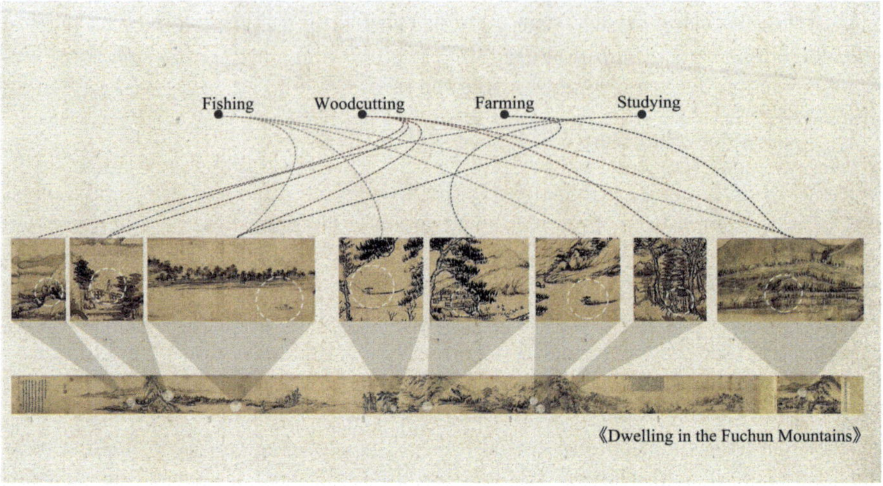

Fig. 11 Cultural landscape of agriculture and fishery. *Source* (Yuan Dynasty) Huang Gongwang, *Dwelling in the Fuchun Mountains*

Dwelling in the Fuchun Mountains depicts the autumn season, just after villagers had completed the harvest of their rice paddies. At this time, fish populations in the river, having grown and reproduced through the spring and summer, were at their most abundant and flavorful, marking the peak fishing season. Rural labor was thus redirected to fishing, providing households with an additional source of income. The scroll painting vividly portrays this scene: two fishermen are clearly shown, while another pair of villagers seems to be engaged in fishing together. Wearing bamboo hats and rowing small boats across the river, they embody the picturesque imagery of "fishing boats at sunset" characteristic of the Jiangnan waterside landscape (Fig. 11).

(2) Woodcutting: Moderate Extraction of Forest Resources

During the agrarian civilization period, everyday needs for food, housing, and transportation all demanded large amounts of firewood and timber, making people's daily lives highly dependent on the surrounding forests. The Fuchun River Basin was richly forested, with abundant species, including many fast-growing varieties such as Chinese fir and pine. The ancients opened mountain paths to facilitate the gathering of timber and medicinal herbs, giving rise to traditional occupations that relied on selling firewood or herbal medicine.

Typically, villagers would first cut the branches of shrubs or trees, and only secondarily harvest fast-growing species. Moreover, felling was carried out in a dispersed and non-concentrated manner, ensuring that forests remained lush and regenerative over time.

In the scroll, mountain paths are depicted as if they encircle the mountain's base in winding curves. With meticulous ingenuity, Huang Gongwang painted a woodcutter carrying a heavy load of firewood along one such path, while on a nearby bridge two other figures, holding shoulder poles, appear to be setting out to cut wood.

(3) Farming: Cultivation of Crops According to Local Conditions

In the basin, extensive level fields were devoted to the concentrated cultivation of high-yield and high-demand crops such as rice, rapeseed, and maize. In the hilly areas, terraced fields were often planted with tea, forming tea gardens of considerable scale and creating regional specialties. On the hillsides, large peach orchards were cultivated for fruit production. The fertile and loose soils of the river sandbars were particularly suitable for mulberry and Chinese tallow tree cultivation. According to the *Fuyang County Gazetteer (Qing Dynasty)*, the islands primarily produced mulberry for sericulture and Chinese tallow for fuel and oil pressing. The widespread planting of mulberry trees stimulated sericulture techniques and thereby promoted the development of handicrafts within the basin.

In Fuyang, mulberry trees grew tall with large leaves and were locally referred to as "lotus-leaf mulberry." The Chinese tallow trees bore broad crowns and blazed red with foliage in autumn. Tea trees remained evergreen throughout the year, while peach and plum trees blossomed in the spring with graceful beauty, scattering petals in profusion. The county magistrate Niu Huan[18] of the Kangxi reign period (1662–1722) composed a poem:

The river winds its mist-veiled course past sandy shores;
The riverside is dotted with winsome cottages.
With hoes, we move green bamboos after snow outdoors;
With baskets, we pick green tea leaves before spring showers.
Where is the afloat brewed in spring day?
There's an oriole warbling for miles of flowers.
The land is rich in peace of a prosperous day,
We've planted mulberry and hemp as in the Land of Peach Blossom Spring [8].

Huang Gongwang selected the polder-field landscapes of the sandbars, most characteristic of the basin, as inspiration for his art. With expressive brushwork, he depicted fertile island farmlands where trees with ornamental qualities and broad, shading crowns were planted along the ridges. These trees not only prevented soil erosion but also provided cool resting places for farmers during fieldwork. In this way, the scroll evokes an idyllic pastoral life, akin to a worldly Peach Blossom Spring.

(4) Learning: Emotional Attachment to Landscape and Spiritual Pursuits

In the Yuan Dynasty, Han Chinese scholars often faced limited opportunities for official appointment, leaving their career prospects bleak. As a result, many redirected their aspirations toward Daoist spirituality, seeking seclusion in the simplicity and authenticity of the natural landscape. Huang Gongwang was no exception. The early

Qing Dynasty painter and calligrapher Wu Li[19] once observed of him: "Through painting he escaped from fame, living freely and at ease, growing old amidst the forests and streams."

The scroll likewise depicts the presence of the recluse: a figure sits contentedly within a riverside thatched pavilion, leaning against the railing and gazing at waterfowl freely resting upon the river. Immersed in the scenery, he seems to forget both self and world, embodying the unity of subject and object. It is quite possible that Huang Gongwang was here portraying an image of himself.

3 Preservation and Inheritance—Constructing the "New Dwelling in the Fuchun Mountains" on the Foundation of Mountains and Waters

Praised by later generations as the "Lanting Pavilion within a painting,"[20] *Dwelling in the Fuchun Mountains* presents the beauty of the ecological human settlement environment within the Fuchun River Basin. This beauty was nurtured by the natural landscape environment and, beginning in the Qin Dynasty, took form through the continuous interaction and reciprocal adaptation between humans and nature. By exercising appropriate interventions in the natural environment, the ancients created habitable and productive settings within the landscape, thereby enabling settlements to take root across the basin. In this way, the lives and livelihoods of the people were sustained over generations, forming a harmonious regional landscape system where humans and nature were in concord, and where beautiful ecological habitation flourished.

This unique beauty, nurtured over thousands of years, remains a valuable landscape resource of the Fuyang region. Urbanization in Fuyang should therefore preserve the landscape framework of the basin, draw upon the accumulated wisdom of traditional building practices, apply gradual and context-sensitive interventions into nature, and make rational use of both natural and cultural resources (Fig. 12).

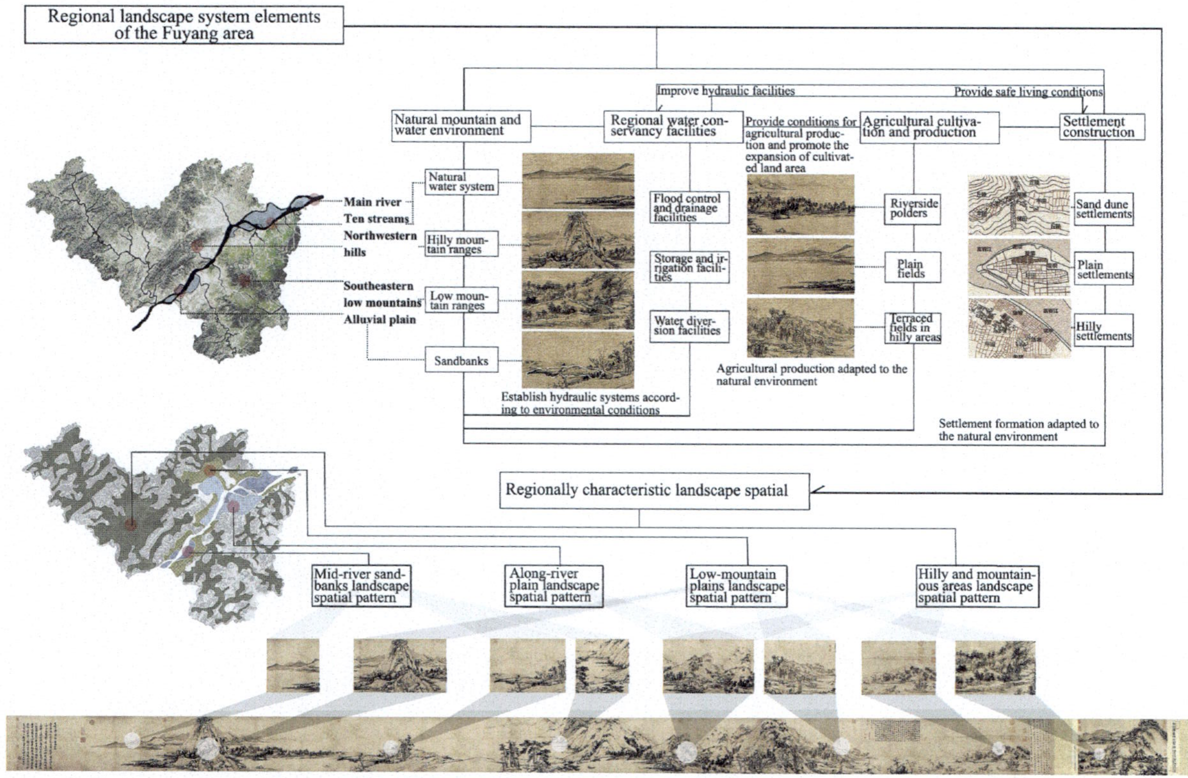

Fig. 12 Components of the regional landscape system in the Fuchun River Basin and their relationship with settlement development. *Source* (Yuan Dynasty) Huang Gongwang, *Dwelling in the Fuchun Mountains*

Notes

1. Dwelling in the Fuchun Mountains (c. 1347–1350), painted by Huang Gongwang in the late Yuan dynasty, is a handscroll landscape depicting the Fuchun River basin.
2. Sun Quan (182–252), founder of the Eastern Wu state during the Three Kingdoms period, ruled the lower Yangtze region from Jiankang (modern Nanjing), where river systems and terrain played a central role in urban defense and regional governance.
3. Luo Yin (833–909) was a late Tang poet whose works often depict everyday urban life and regional landscapes in Jiangnan, offering insights into lived environments beyond court culture.
4. Huang Gongwang (1269–1354), a leading Yuan-dynasty literati painter, articulated a model of landscape integrating mountains, rivers, and human dwelling, most famously in Dwelling in the Fuchun Mountains.
5. Dong Bangda (1699–1769) and his son Dong Gao (1740–1818) were Qing dynasty grand secretaries and court painters, closely involved in the visual and administrative construction of imperial landscapes and garden aesthetics.
6. Yu Dafu (1896–1945) was a modern Chinese writer whose prose foregrounds personal experience and emotional responses to landscapes, cities, and place.
7. Li Bai (701–762), a Tang-dynasty poet, is renowned for landscape poetry that emphasizes rivers, mountains, and travel, shaping an early poetic imagination of natural scenery as expansive and mobile space.
8. Bai Juyi (772–846), a Tang poet and local official, wrote extensively on urban life, waterways, and gardens, linking poetic expression with governance and landscape management.
9. Ji Xiaolan (1724–1805) was a Qing scholar and editor whose writings reflect literati perceptions of cities, landscapes, and cultural memory in late imperial China. Ji Xiaolan (1724–1805) was a Qing scholar and editor whose writings reflect literati perceptions of cities, landscapes, and cultural memory in late imperial China.
10. The "Five Phases" (Wuxing) chapter of the Guanzi (Warring States period) articulates a theory linking natural cycles with state governance, emphasizing seasonal timing, environmental regulation, and the management of land, water, and resources. It reflects an early systemic view of human–environment relations in political planning.
11. "Letter to Zhu Yuansi," written by Wu Jun in the early sixth century (Southern Liang dynasty), is a prose letter describing the Fuchun River landscape. It presents an experiential view of rivers, mountains, and movement through space, emphasizing perception and spatial sequence rather than formal planning.
12. The Geography Section of the Fuyang County Gazetteer (Qing dynasty) documents the region's topography, river systems, settlements, and transportation routes. As an official local gazetteer, it provides structured information on the spatial organization and environmental setting of the Fuchun River area.
13. The Geography Section of the Fuyang County Gazetteer (Qing dynasty) documents the region's topography, river systems, settlements, and transportation routes. As an official local gazetteer, it provides structured information on the spatial organization and environmental setting of the Fuchun River area.
14. The Tianshun era of the Ming dynasty (1457–1464) represents a mid-Ming period of political consolidation and routine urban and water management. In contrast, the first year of the Tongzhi reign (1862) marks an early late-Qing recovery phase following widespread warfare, when damaged cities, waterways, and landscapes entered a period of reconstruction and adaptive repair.
15. A floating weir (or vegetated weir) refers to a low, temporary water-control structure made of earth, grass, brushwood, or bundled vegetation. Widely used in traditional agrarian societies from early imperial China onward, it diverted water for irrigation while allowing seasonal adjustment and flood passage, forming a flexible interface between rivers and farmland.
16. The smallholder agrarian economy refers to the dominant rural economic system in imperial China (roughly from the Qin–Han period to the late Qing), characterized by family-based

farming on small plots, intensive land use, irrigation dependence, and close integration of settlement, fields, and waterways. This system shaped dispersed villages, fine-grained agricultural landscapes, and everyday water–land management.

17. Bishu Lu Hua (early twelfth century), written by Ye Mengde, is a Song-dynasty miscellany that records observations on landscapes, climate, urban life, and travel. It offers firsthand insight into how natural settings and built environments were experienced and discussed in late Northern Song China.
18. Niu Huan, a county magistrate during the Kangxi reign (late seventeenth–early eighteenth century), was a local Qing official responsible for water control, infrastructure maintenance, and land administration. Such magistrates played a key role in shaping everyday landscapes through practical governance rather than formal design.
19. Wu Li (1632–1718) was an early Qing dynasty painter and poet associated with Jiangnan literati culture. His landscape paintings emphasize travel, real topography, and river–mountain settings, reflecting an experiential understanding of regional landscapes rather than purely symbolic composition.
20. "Lanting Pavilion within a painting" refers to a recurring literati painting motif, developed mainly from the Yuan period onward, that visualizes the site and atmosphere of the Lanting Gathering associated with Wang Xizhi (353 CE). Rather than a literal depiction, it represents an idealized landscape combining watercourses, pavilions, vegetation, and scholarly gathering, functioning as a cultural model of literati landscape space.

References

1. Jiang, J. (2011). *Huang Gongwang and Eight Essays on Dwelling in the Fuchun Mountains.* Cultural Relics Press.
2. Iconological research is divided into three dimensions: the pre-iconographical description, which focuses on artistic formal elements and explains the physical and visual facts of primary or natural subjects; the iconographical analysis, which provides knowledge-based interpretation of the identities of figures, narrative content, and historical context; and the iconological interpretation, which elucidates the cultural, political, economic, religious, and social background of image-making and reveals the specific worldview reflected in the artwork.
3. Yang, Z. (2010). The roots and scope of application of Panofsky's iconological method. *Xinmeishu, 6*, 38–41.
4. Wen, H. (2009). *Painting as heart's imprint—A study of Huang Gongwang's dwelling in the Fuchun mountains.* China Academy of Art.
5. Ni, Z. (2008). *Hydraulic history of the Fuyang region.* Hohai University Press.
6. Hou, X., & Guo, W. (2015). Polder landscape studies: Morphology, function, and impact. *Landscape Architecture, 2015*(6), 123–128.
7. Liang, S. (1998). *History of Chinese architecture.* Baihua Literature and Art Publishing House.
8. Wang, W., & Jiang, J. (2010). *Fuyang County Gazetteer of the Guangxu Era.* Shanghai Ancient Books Press.

Conclusion: Toward a Globally Relevant Terrapoetics

This book examines the development of human settlement environments from a regional perspective. Through concrete historical cases, it explains the composition, evolution, and constructive processes of regional landscape systems and the urban formations built upon them. This process is defined as Terrapoetic Urbanism, through which the ecological wisdom and cultural character embedded in traditional landscape-based urban construction are revealed, and a systematic research framework for terrapoetic urbanization is proposed.

1 The Regional Landscape System as the Medium of Urban Development

In this study, the regional landscape system is understood as the primary medium through which cities develop. Landscape construction is regarded as a key mechanism for mediating between nature and human intervention, supporting urban site selection and functional zoning. From a regional and dynamically adaptive perspective, the book analyzes how Shanshui environments promote the sustainable evolution of urban spatial form. Cities exist within nature, and natural resources constitute the material foundation of urban life; however, natural environments are not inherently suited to human habitation. Through sustained human intervention that works in accordance with natural processes, nature can be reorganized and reshaped to guide environmental conditions toward human settlement needs, thereby forming regional landscape systems. These systems include low-mountain structures extending from peri-urban areas into cities, as well as water networks of rivers and lakes that, through artificial regulation, connect urban interiors with surrounding landscapes. Together, they establish the ecological substrate of urban development. Urban growth relies on regional landscape systems in multiple dimensions, including site selection, functional zoning, and the cultivation of inner city gardens and extra-urban scenic landscapes. The regional landscape system thus operates not merely as scenery but as the ecological and spatial foundation of the city.

X. Zheng, *Terrapoetic Urbanism*, https://doi.org/10.1007/978-981-95-8676-9

2 Dynamic Adaptation: The Poetic Construction of Urbanization

Such regional landscape systems are conceptualized as shanshui, while the process by which Shanshui environments evolve to support urban development and foster integrated city–suburb construction is defined as urbanization. Urbanization here denotes a constructive process in which cities develop through artificial intervention grounded in natural conditions, continuously adjusting in response to emerging challenges and achieving a dynamic balance between built form and natural systems. Within terrapoetic urbanization, urban site selection and functional zoning including commercial markets, administrative precincts, residential areas, temples and pagodas, transport wharfs, scenic sites, and gardens are closely dependent on the regional landscape system. For example, canals and transport networks generate waterfront commercial districts; gardens are established along urban water systems; scenic landscapes are cultivated in low-mountain areas; and elevated structures such as temples and pagodas function as visual anchors that integrate urban and extra-urban landscape systems. Terrapoetic urbanization demonstrates how humans utilize and transform nature in ways that allow natural and artificial environments to mutually adapt and become structurally coupled. Through this process, an integrated "mountain–water–city" configuration emerges, characterized by poetic spatial qualities, resilient structures, and long-term vitality.

3 Resilient Systems: Responding to Disaster and Change

Many renowned Chinese cities such as Suzhou, Hangzhou, Beijing, and Nanjing have experienced periods of remarkable prosperity across different historical eras and continue to exhibit strong vitality and potential for sustainable development today. Throughout their histories, these cities have repeatedly confronted disasters and upheavals, yet have demonstrated a remarkable capacity for recovery. As political, social, economic, demographic, and spatial conditions changed, the integrated "mountain–water–city" model functioned as a resilient system capable of withstanding environmental and societal disruptions. Favorable surrounding Shanshui environments, shaped through ecological wisdom and human regulation, provided both natural and ecological support systems for urban emergence and long-term development, as well as a cultural and poetic foundation for habitation and identity.

4 Toward the Future: Constructing Shanshui Cities for a New Era

Culture, ecology, and livability have become central themes of contemporary urban development. Cultural continuity reflects cultural confidence; ecological systems provide resilient strategies for addressing global challenges and regional pressures; and livability serves as an indicator of urban health. The regional landscape system of a city's geographical unit should therefore be recognized as the fundamental substrate of urban development. Its functions, meanings, and values extend far beyond those of a scenic backdrop or a single-purpose green space. By coordinating natural and artificial urban resources, cities can respond to contemporary social and spatial challenges. In the context of ecological, cultural, and livability-oriented urbanism, urban development should proceed in accordance with Shanshui, drawing upon the ecological wisdom and cultural genes of traditional Chinese urban construction to build Shanshui cities for a new era.

Cities are not isolated entities detached from nature. Urban development should respect and follow existing Shanshui configurations, utilizing natural systems while seeking enduring principles within dynamic processes of balance and change to guide urban conservation and construction. From the perspective of regional landscape systems, cities must be integrally connected with their surrounding natural environments, elevating urban planning to a regional scale and allowing cities to be immersed within Shanshui landscapes. During urban expansion, the integrity of city–suburb regional landscape systems must be maintained to achieve long-term prosperity. At different stages of development, and under varying natural conditions, every Chinese city retains the opportunity to establish a complete internal–external regional landscape system. Each city holds the potential to become a place of harmony between humanity and nature, endowed with distinctive cultural value, humane scale, poetic atmosphere, and enduring vitality.

Viewing cities at the regional scale under the framework of terrapoetic urbanization enables the synthesis of traditional urban construction wisdom, the preservation of Shanshui relationships within urban environments, and the safeguarding of both everyday life and landscapes of natural and cultural significance. Through this approach, integrated and permeable regional landscape systems can be constructed. As the saying goes, to understand the present, one must study the past. Research on traditional urban systems provides valuable insight for future-oriented urban development, offering a regional landscape perspective on urban–regional relations and facilitating the transmission of the philosophy of tianren heyi. As Academician Meng Zhaozhen of Beijing Forestry University has emphasized, urbanization should "proceed in accordance with mountains and waters," and urban planning should be grounded in the governance of mountains and water systems. Urbanization and natural environments should not stand in opposition, but rather adapt to and reinforce one another. Urban development must elevate the construction and evolution of regional landscape systems to a strategic level, using landscape as a critical means of integrating complex, multilayered, and multifunctional relationships. Through this approach, spatial structures of natural resources can be maintained, ecological and built spaces can be coordinated, and the long-term, healthy development of urban form and socio-economic and cultural systems can be sustained.

The manufacturer's authorised representative in the EU is Springer Nature Customer Service Centre GmbH, Europaplatz 3, 69115 Heidelberg, Germany. If you have any concerns regarding our products, please contact ProductSafety@springernature.com

Printed and bound by CPI Group (UK) Ltd, Croydon, CR0 4YY
21/09/2026
02220013-0003